Advances in Soil Organic Matter Research:
The Impact on Agriculture and the Environment

Advances in Soil Organic Matter Research: The Impact on Agriculture and the Environment

Edited by
W.S. Wilson
Department of Biology, University of Essex

Advisory Editors
T.R.G. Gray
Department of Biology, University of Essex
D.J. Greenslade
Department of Chemistry and Biological Chemistry, University of Essex
R.M. Harrison
School of Biological Sciences, The University of Birmingham
M.H.B. Hayes
School of Chemistry, The University of Birmingham

The Proceedings of a Joint Symposium organised by The Agricultural and Environment Groups of The Royal Society of Chemistry, The British Society of Soil Science, and The International Humic Substances Society. University of Essex, 3rd-4th September 1990.

Special Publication No. 90

ISBN 0-85186-387-6

A catalogue record for this book is available from the British Library

© The Royal Society of Chemistry 1991

All Rights Reserved
No part of this book may be reproduced or transmitted in any form
of by any means — graphic, electronic, including photocopying,
recording, taping or information storage and retrieval systems —
without written permission from The Royal Society of Chemistry

Published by The Royal Society of Chemistry,
Thomas Graham House, Science Park, Cambridge
CB4 4WF

Printed by Redwood Press Ltd,
Melksham, Wiltshire

Preface

Soil organic matter research generally covers a wide range of phenomena involving numerous parameters. Some aspects of the research are fundamental and theoretical, others are applied and practical. The aim of this book is to provide a balanced account of some recently reported research with particular emphasis on new information, novel methods of analysis, and changes in approach to outstanding problems.

The contents comprise thirty-five papers differentiated into five themes, two of which are basically fundamental and the others applied. Research progress in the fundamental themes such as composition and structure of soil organic matter and turnover of soil organic matter tends to be slow. This is due to the complexity of the individual humic substances, humic acid, fulvic acid, and humin, in the former and to the multiplicity of variable influences such as climatic conditions, agricultural systems, plant associations, soil textures, and pH levels in the latter. The applied themes - on the effect of soil organic matter on soil structure, soil fertility, and water quality, respectively - achieve more rapid advances due to the pressing needs of agriculture and the management of the environment. These are illustrated by the effects of soil organic fertility to aid better land use, and increasing the adsorption of chemical pollutants to provide water of higher quality. New soil organic matter data can also be used as concomitant information for interpreting nitrogen research especially involving organic nitrogen in chemical and biological equilibria.

The structure of the book consists of five sections, corresponding to the aforementioned themes, with appropriate individual papers arranged in logical order. Each section is preceded by an 'Introductory Comments', guide explaining the inter-connection of the individual papers. Throughout the text there are numerous accounts of experimental procedures and instrumentation which include: several spectroscopic measurements such as solid state ^{13}C nuclear magnetic resonance and Fourier transform infrared, pyrolysis mass spectrometry, labelled ^{13}C, ^{14}C (also bomb) ^{14}C), and ^{15}N techniques, and diverse soil microbiology preparations. These developments have further substantiated the concept of humus quality especially in relation to enzymic activities which presage changes in humus levels under different soil management. Intermixed with classical field crop and afforestation experiments are reports on investigations into subjects of public interest such as characterisation of unstable soils of the Amazon Basin, the effect of global warming on soil organic matter levels, organic farming, incorporation of straw in soil, conversion of civic waste to humus, and the safe disposal in soil of meat processing by-products.

The book is based on the proceedings of a symposium of the same title held on 3rd-4th September, 1990, at the University of Essex. The event was sponsored by The Royal Society of Chemistry Agriculture and Environment Groups, The British Society of Soil Science, and The International Humic Substances Society. It was also financially supported by BASF plc. The published papers, which incorporate valuable discussion points by participants from seventeen countries, have been strictly refereed and almost all present new research.

I would like to thank R.L. Richards and P.B. Barraclough, of The Royal Society of Chemistry Agriculture Group, C.D. Vincent, of the British Society of Soil Science, and T.R.G. Gray, D.J. Greenslade, R.M. Harrison, and M.H.B. Hayes, the Advisory Editors, for help in preparing the manuscript. In addition, I am grateful to Catherine N. Lyall, of The Royal Society of Chemistry Publications, for helpful advice and guiding the material through the publication process.

Colchester,
March, 1991.

W.S. Wilson

Contents

Section 1   Advances in our Understanding of the Composition and Structures of Soil Organic Matter

Introductory Comments by M.H.B. Hayes	1
Concepts of the Origins, Composition, and Structures of Humic Substances *M.H.B. Hayes*	3
The Individuality of Humic Substances in Diverse Environments *R.L. Malcolm and P. MacCarthy*	23
Novel Methods of Soil Organic Matter Analysis: A Critique of Advanced Magnetic Resonance Methods *D.J. Greenslade*	35
Organic Matter as Seen by Solid State ^{13}C NMR and Pyrolysis Tandem Mass Spectrometry *J.A. Baldock, G.J. Currie, and J.M. Oades*	45
Fate of Plant Components during Biodegradation and Humification in Forest Soils: Evidence from Structural Characterization of Individual Biomacromolecules *I. Kögel-Knabner, W. Zech, P.G. Hatcher, and J.W. de Leeuw*	61
Spectroscopic Characterization of Organic Matter from Soil with Mor and Mull Humus Forms *D.W. Hopkins and R.S. Shiel*	71
Influence of Soil Management Practices on the Organic Matter Structure and the Biochemical Turnover of Plant Residues *K. Haider, F.-F. Gröblinghoff, T. Beck, H.-R. Schulten, R. Hempfling, and H.D. Lüdemann*	79

Section 2   Soil Organic Matter and Water Quality

Introductory Comments by H.A. Anderson	93
Organic Nitrogen in Soils and Associated Surface Waters *H.A. Anderson, M. Stewart, J.D. Miller, and A. Hepburn*	97
Organic Matter Turnover and Nitrate Leaching *M.J. Goss, B.L. Williams, and K.R. Howse*	107

The Influence of Soils on Organic Matter in
Water   115
P.N. Nelson, J.M. Oades, and E. Cotsaris

Dissolved Organic Matter as Carrier for
Exogenous Organic Chemicals in Soils   121
I. Kögel-Knabner, P. Knabner, and H. Deschauer

The Role of Soil Organic Matter in Pesticide
Movement via Run-off, Soil Erosion, and Leaching   129
T.R. Harrod, A.D. Carter, and J.M. Hollis

Section 3 Organic Matter and Soil Structure

Introductory Comments by D.A. Rose   139

Soil Oragnic Matter and Water Erosion Processes   141
D. Gabriels and P. Michiels

Effects of Humic Substances and Polysaccharides
on Soil Aggregation   153
R.S. Swift

Organic Matter in Water-stable Aggregates   163
A.G. Waters and J.M. Oades

Microalgal and Cyanobacterial Soil Inoculants
and Their Effect on Soil Aggegate Stability   175
S.L. Rogers, K.A. Cook, and R.G. Burns

Organic Matter and Chemcial Characteristics of
Aggregates from a Red-Yellow Latosol under
Natural Forest, Rubber Plant, and Grass in
Brazil   185
E. de S. Mendonça, W. Moura Filho, and L.M. Costa

The Effect of Long-continued Organic Manuring on
Some Physical Properties of Soils   197
D.A. Rose

The Influence of Soil Compaction on Decomposition
of Plant Residues and on Microbial Biomass   207
E.A. Kaiser, G. Walenzik, and O. Heinemeyer

The Cotton Strip Assay: Field Applications and
Global Comparisions   217
G. Howson

Section 4 Soil Organic Matter Turnover

Introductory Comments by T.R.G. Gray   229

Sources and Fate of Soil Organic Matter   231
J.M. Lynch

The Potential of Bomb-^{14}C Measurements for
Estimating Soil Organic Matter Turnover   239
D.D. Harkness, A.F. Harrison, and P.J. Bacon

The Influence of Soil Organic Carbon on
Microbial Growth and Survival   253
T.-H. Anderson, and T.R.G. Gray

Carbon Dynamics of a Soil Planted with Wheat
under an Elevated Atmospheric CO_2 Concentration   267
P.J. Kuikman, L.J.A. Lekkerkerk, and J.A. Van Veen

Solubilisation of Wheat Straw by Actinomycete
Enzymes   275
A.S. Ball and M. Allen

Dynamics of Organic Carbon and Nitrogen
Accumulation and Distribution in Soils Following
Farm Woodland Planting   285
G.M. Whiteley

Changes in Organic Matter in Fen Silt Soils   293
P.A. Johnson and J.M. Prince

Section 5   Fertility and Soil Organic Matter

Introductory Comments by A.E. Johnston   297

Soil Fertility and Soil Organic Matter   299
A.E. Johnston

Mineral Nitrogen Arising from Soil Matter and
Organic Manures Related to Winter Wheat
Production   315
L.V. Vaidyanathan, M.A. Shepherd, and
B.J. Chambers

The Effect of Long-term Applications of
Inorganic Nitrogen Fertilizer on Soil Organic
Nitrogen   329
M.J. Glendining and D.S. Powlson

Straw Incorporation into Soils Compared with
Burning during Successive Seasons - Impact on
Crop Husbandry and Soil Nitrogen Supply   339
J.S. Rule, D.B. Turley, and L.V. Vaidyanathan

Soil Organic Matter: Its Central Position in
Organic Farming   355
R.D. Hodges

Refuse-derived Humus: A Plant Growth Medium   365
A.A. Keeling, J.A.J. Mullett, I.K. Paton,
N. Bragg, B.J. Chambers, P.J. Harvey, and
R.S. Manasse

Land Disposal of Meat Processing Plant Effluent   375
M.R. Balks and R.F. Allbrook

The Influence of Organic Matter and Clay on
Absorption of Atrazine by Top Soils   381
L.V. Vaidyanathan and D.J. Eagle

Subject Index   393

Section 1

Advances in Our Understanding of the Composition and Structures of Soil Organic Matter

Introductory Comments

by Michael H.B. Hayes, The University of Birmingham

Awareness of the composition and structures of major naturally occurring polymers, such as proteins, polysaccharides, and nucleic acids, has been essential to our understanding of how these react and interact in their natural environments. The same should be true for humic substances, the most abundant of the naturally occurring organic macromolecules of nature, and the major components of the organic matter in well humified soils. However, the structures of humic substances are far more complex than those of biopolymers synthesized under genetic or biological control.

In the first paper in this Section, emphasis is placed on the formation of humic substances from a variety of precursors in a broad spectrum of biological and chemical processes, and on the likelihood, because of the heterogeneity of the substrates, of the soil biota, and of the soil environments where degradative and synthetic processes take place, that it is highly unlikely that two humic molecules in any batch are exactly the same. It would be impossible to fractionate such gross mixtures into homogeneous components in sufficient yields to allow detailed structural studies. Nevertheless, fractionation can decrease polydispersity significantly. Thus emphasis is given in the opening paper to classical and to modern procedures for the isolation and the fractionation of humic substances, and to studies of structures.

In the presentation by Malcolm and MacCarthy there is proof, from cross polarization magic angle spinning (CPMAS) ^{13}C nuclear magnetic resonance (NMR) spectroscopy, and from sugar and amino acid analyses of humic substances from different soils and waters, that each humic component in each environment possesses an individuality that distinguishes it from other components in the same environment, and from the same humic components in different environments.

Modern concepts of humification suggest that the predominating processes include selective preservation of refractory plant (and animal) components, the direct transformation of some of these to humic macromolecules, and synthesis by microorganisms proliferating on organic residues. Chemical synthesis of humic macromolecules from reactive precursors may also occur; the Maillard or Browning reaction is an example that has been widely favoured. It is clear enough that lignin residues can contribute to the composition of humic substances, but a number of the contributions in this Section would suggest that the involvement of lignin-type structures is characteristic of newly formed humic components. CPMAS ^{13}C NMR data in the contributions by Kögel-Knabner et al., Haider et al., and Hopkins and Shiel suggest that, as humification proceeds, the contribution of phenolic groups, such as those in lignin, to the humic structures decreases. This is described as a process of lignin mineralization, and Haider et al. show that it is accelerated in intensively managed soils, such as those subjected to bare fallow and single crop culture. The total C and N contents of such soils are less than for those in which crop rotations, and/or organic manures are added.

Considerations of soil organic matter should include soil polysaccharides which are also major components of humus, and have important roles in soil environments. Baldock et al. stress that it is unlikely that it will be possible to extract the entire organic fraction in an unaltered state from soil, but they show how soil fractions of different sizes can be obtained, using non-chemical treatments. Applications of CPMAS ^{13}C NMR spectroscopy and of pyrolysis tandem mass spectrometry to such fractions can give useful indications of the composition of organic matter in soils. Their results show conclusively that the compositions of the organic matter associated with the different size fractions of the same soils are different. Contents of phenols, with probable origins in lignins, increase in the smaller size soil particles. In contrast the contents of O-alkyl carbon, presumably attributable to carbohydrates, were most abundant in the largest fractions studied. The fact that aromatic structures were not detected when ^{13}C-labelled glucose was metabolized for 34 days by soil microorganisms would cast doubt on the role of soil microbial activity in contributing to the aromatic structures in soil organic matter.

Considerable excitement has been generated during the past decade by data obtained from NMR, and especially CPMAS ^{13}C NMR, by pyrolysis mass spectrometry, and by Fourier Transform Infrared (FTIR) spectroscopy. New developments in electron spin resonance (ESR), outlined by Greenslade, may yield further significant data. Instrumental techniques, and especially the NMR procedures, have provided useful quantitative measurements of different components in the macromolecular structures. For example, from examinations of the chemical shift assignments, and from comparisons between the CPMAS ^{13}C NMR spectra in the contributions by Malcolm and MacCarthy, and by Kögel-Knabner et al., it is clear that quantitative measurements can be made of, for example, aromatic, carboxyl, and carbohydrate components in a variety of humic substances from different environments.

It is not possible at this time, however, to provide from spectroscopy measurements details of the structural units which compose the macromolecular components of soil organic matter. In fact, there is no direct method available now which can provide such structural details. In the view of this author, identification of components in the digests of chemical degradation reactions provides the most reliable information about the 'building blocks' of humic substances. The majority of the compounds in such digests are derivatives, and often 'distant relations' of the molecules that compose the macromolecular structures. Thus it is important, through an awareness of the degradation mechanisms, to derive plausible structures from which the products in the digests could have formed.

Concepts of the Origins, Composition, and Structures of Humic Substances

Michael H.B. Hayes

SCHOOL OF CHEMISTRY, THE UNIVERSITY OF BIRMINGHAM, EDGBASTON, BIRMINGHAM B15 2TT, UK

1 INTRODUCTION

Knowledge of the structures of polymers and macromolecules helps the understanding of their interactions and reactivities in biological and environmental processes. For example, a knowledge of all aspects of the structures of proteins and enzymes and of nucleic acids allows valuable predictions to be made of their involvements in biological processes. However, because protein biosynthesis is controlled genetically, all proteins formed from the same 'template' will be exactly the same, and it is generally it is relatively easy to isolate these from other proteins and components of biological media.

Because peptide linkages between the amino acids in proteins are labile in hydrolysis conditions, the amino acid contents, and their proportions in the protein substrates can readily be determined, and there are procedures available which allow the sequencing of the amino acids to be determined, and thus their *primary structures* to be established. *Secondary structures* of proteins refer to the first hierarchical level of conformational shapes which the stretches of polypeptide molecules in the primary structures can adopt. Regular coiling of the polypeptide chain along one dimension, as in fibrous proteins, for example, where intramolecular hydrogen bonding between the H attached to N in the peptide bond on one amino acid residue and the carbonyl on the third amino acid residue beyond it gives helical conformations characteristic of α-keratin proteins. In the *tertiary structures* of proteins, the polypeptide chains fold, and form compact, three dimensional structures, such as those which are characteristic of globular proteins. The folding is specific, and may include stretches of coiled secondary structures. *Quaternary structures* refers to associations of proteins with other macromolecules, but without the formation of covalent linkages. Such associations are relevant to biological function.

1.1 Origins of Humic Substances

There is no evidence to suggest that there is genetic or biological control of the synthesis of humic substances. Hence they lack regularity in the sequencing of the molecules which compose the macromolecules, and therefore fail to meet the criteria for primary structure. Intramolecular and intermolecular associations can take place within and between humic (and other) macromolecules to give a semblance of secondary and tertiary structure. However, there are no rigid regularites between such associations.

Although associations can form between humic substances and other organic and inorganic materials, such associations are invariably random, and are not necessarily a part of any biological function which humic substances may have. Hence, humic substances do not meet any of the criteria for structure, as they apply for proteins.

Two broadly based processes are considered to give rise to the humic substances of nature. The first of these, the *degradative process*, considers that biological transformations of the refractory organic macromolecules of nature, such as lignins, paraffinic macromolecules from algae and plant cuticles, N-containing paraffinic substances, cutins, melanins, suberins and other substances give rise to humic substances.[1] Polysaccharides and proteins form readily available substrates for microorganisms, but these too give rise to humic substances, although the origins of such substances are more likely to be from the microorganisms which proliferate on the labile substrates than from the substrates themselves. There is the possibility also that amino acids and peptides released from proteins, and sugars and oligosaccharide released from polysaccharides can undergo 'browning' reactions to give rise to humic substances. Quinones from oxidized phenols are also known to give rise to humic-type substances. Such pathways are part of the second or *synthetic process* for the genesis of humic substances.

Maillard[2] introduced the 'browning' reaction or the 'melanoidin' theory when he observed that monomeric reducing sugars, such as glucose, could condense with amino acids, such as glycine, to form brown macromolecular substances. Later[3] he suggested that the colour in the water soluble extracts from straw in the early stages of decomposition resulted from the interactions of xylose from gums and glucose from cellulose with the amino acids from proteins. Several since Maillard have shown that some browning reaction products have certain properties similar to those of soil humic substances.[4]

Polycondensation reactions between glycine and 2-oxopropanal (methyl glyoxal) were considered by Enders and his colleagues[5] to provide plausible processes for the formation of humic substances. By regulating the ratios of the reactants, the macromolecular substances produced can have elemental contents and charge and other characteristics similar to soil humic acids.[6]

More recently there has been emphasis on the role of quinones from di- and polyhydroxybenzene structures, with -OH groups in the 1,2- and 1,4- ring positions, on the synthesis of humic substances.[7] Lignins can give rise to the appropriate phenols, and fungi are known also to synthesize phenols, many of which are components of melanins, the coloured secondary metabolites formed during fungal degradations of saccharides.[8]

Thus it is clear that soil humic substances can be formed from a variety of precursors in a broad spectrum of biological and chemical processes. The presence in the humic macromolecules of components attributable to origins in lignins need not imply that these macromolecules were formed directly by peripheral modifications to a parent lignin structure. Recombinations, in biological and/or chemical synthesis processes, of derivatives of component molecules released from liganins, and of lignin fragments, are equally probable. Biological oxidation of peripheral parts of such macromolecules can also be expected after the recombinations have taken place. Thus, because of the heterogeneity of the substrates, of the soil biota, and of the soil environments in which the degradation and synthesis processes take place, it is highly unlikely that two molecules in any batch are exactly the same. Hence, it must be realised that in any study of humic substances we are dealing with mixtures.

2 HUMIC SUBSTANCES: A PROBLEM OF MIXTURES

Aiken et al.[9] have considered *humic substances* to be a 'general category of naturally occurring, biogenic, heterogeneous organic substances that can generally be characterized as being yellow to black in colour, of high molecular weight, and refractory'. Hayes and Swift[10], based on proposals by Kononova[11], considered humic substances to be the amorphous, macromolecular, brown-coloured components of soil organic matter which bear no morphological resemblances to the plant or animal tissues from which they were derived, and which can be differentiated into broad general classes on the basis of solubility differences in aqueous acids and bases. Thus soil humic acids are the components of humic substances precipitated when extracts in dilute aqueous alkali are acidified to pH 1. *Fulvic acids* are the components soluble in aqueous acids and bases, and *humin* is the term applied to the components which are insoluble in aqueous acids and bases. However, these three terms refer to gross mixtures, and the elemental composition of these mixtures and their chemical and physicochemical properties can vary with their origins, and with the environments in which they were formed.

In order to study the composition and structure of any organic compound it is important to be able to isolate the compound free from contaminants. Such is impossible in the cases of the components of humic substances where the almost infinite variety of macromolecules within the humic and fulvic acid fractions have broadly similar physicochemcal properties. Hence, it will not be possible to establish accurate primary structures (Section 1) for soil humic and fulvic acids. Nevertheless, there is much that can be learned about the structures in mixtures from studies of fractions with broadly similar compositional and physicochemical properties. There follows an outline of procedures for the isolation and fractionation of humic substances, and of studies which have led to the evolution of concepts of structures which help to explain and to predict some of the interactions involving humic substances in the soil environment.

3 ISOLATION OF HUMIC SUBSTANCES FROM SOIL

Hayes[12] has reviewed the properties of solvents used for the isolation, of humic substances from soil, and the processes by which dissolution of the macromolecules take place in the different solvent systems. At the pH of most agricultural soils, humic substances are significantly ionized, and the negative charges of the conjugated bases in the acid groups are balanced mainly by divalent and polyvalent metal cations. These cations form bridges between charges on the same strand, and between those on adjacent strands, and between charges on the strands and on the inorganic soil colloids. Such effects cause the structures to be pulled together and water to be partially excluded from the macromolecular matrix. The macromolecules in this shrunken state are relatively insoluble in aqueous solvents. When the divalent and polyvalent cations are replaced by monovalent metal cations, the bridging effect is lost, and the resulting ionized expanded structures can solvate in water. This explains the mode of dissolution in diute solutions of sodium pyrophosphate ($Na_4P_2O_7$). The pyrophosphate complexes the divalent and polyvalent metals allowing sodium to neutralise the negative charges freed by the complexation process. Usually the pH of the pyrophosphate solution is adjusted to neutrality (using phosphoric acid) in order to avoid alkaline oxidation effects.

When the metal cations are replaced by hydrogen ions, the undissociated acidic groups allow the humic polyelectrolytes to behave essentially as neutral, polar, macromolecular species. Intra- and intermolecular hydrogen bonding, however, causes the macromolecules to remain shrunken, and relatively insoluble in water. As the pH of the medium is raised the acid groups ionize, with the strongest acids ionizing first and the weakest last at the higher (pH 10, and even above) pH values. Solvation of the anionic species takes place, and when the macromolecule-solvent interactions are appropriate the intra- and intermolecular forces holding the macromolecules together are broken and dissolution takes place.

Aqueous sodium hydroxide is a good solvent for humic and fulvic acids, provided the macromolecules are H^+-exchanged prior to addition of base. In such circumstances it is clearly a better solvent than sodium pyrophosphate (pH 7) because more of the acidic groups are ionized at higher pH. Neutral pyrophosphate solutions dissolve only the more highly oxidized (and acidic) and lower molecular weight humic substances.[13,14] Increased amounts of humic aicds are isolated when non-neutralized pyrophosphate is used[15], presumably because of increased ionization (and solvation) of weaker acid groups.

Alkaline and aerobic conditions give rise to the oxidation of humic substances, and hence to the formation of artefacts.[16] Thus, extractions in base are generally carried out in an atmosphere of dinitrogen gas. Such precautions do not eliminate artefact formation, however, and hydrolysis and other base catalyzed reactions can occur. Hence, it is safer to use non-alkaline media, or organic solvents which do not bond to the macromolecules.

Hayes[12] has reviewed the properties of organic molecules which are good solvents for humic substances. Effective solvents have high electrostatic factor (EF, the product of relative permittivity and dipole moment) and high base parameter, pK_{HB} (which measures the relative strength of the acceptor when a hydrogen-bonded complex is formed[17]), values relative to the less efficient solvents. Dipolar aprotic solvents, and especially dimethylsulphoxide (DMSO, EF = 209.2, pK_{HB} = 2.53) satisfy the criteria.

Anions are very sparingly soluble in DMSO, but cations are readily solvated in it.[18] DMSO was effective for the isolation of humic substances from soil when the solution contained acid which H^+-exchanged with the cations neutralizing the carboxylate charges. The acidic groups were thus largely in the carboxyl (-COOH) and hydroxyl (phenolic and enolic) forms, and dissolution of the H^+-exchanged humic macromolecules were similar to that for polar polymers. Hayes[12] has discussed how DMSO can associate with the phenolic and carboxyl groups in humic structures to break the inter- and intrastrand hydrogen bonds which render the macromolecules insoluble in water. Also, DMSO has a non-polar 'backbone' (from the $-CH_3$ groups) which can associate with non-polar groups on the humic macromolecules and thereby enhance the dissolution effect. Because of the strong association between DMSO and water, however, the presence of excess water in the medium inhibits the interactions between the $S = O$

groups and the acid groups in the macromolecules. Swelling of the dry macromolecular systems is a precursor to dissolution, and because DMSO is slow to penetrate into the interior of the dry humic macromolecular matrix, the swelling and dissolution processes are slow.

When acidified DMSO was applied to the clay fraction of a Waukegan soil which had been exhaustively extracted with 0.1M sodium pyrophosphate (first at pH 7, than at pH 10.7) and then with 0.1M NaOH (exhaustively), a substantial amount of organic matter was dissolved.[15] In the classical definitions this organic matter would be classified as humin. However, the properties of the isolates resembled those of fulvic acids. It seems likely that these so-called humin materials were associated with clays through cation-bridge mechanisms. In this way the polar acidic groups were orientated towards the clay surface, and the relativley non-polar moieties were orientated to the outside. Such non-polar moieties could swell in DMSO and eventually solvate in it. It is possible also that the macromolecules were protected from solvation in aqueous media by associations with non-polar waxy, and paraffinic non-humic substances.

4 'PURIFICATION' AND FRACTIONATION OF EXTRACTS

Extracts from soil in organic and in aqueous solvents are mixtures containing humic and non-humic substances. Water scientists have developed a resin treatment process to separate hydrophilic substances from the humic substances in waters. The poly(methylmethacrylate) resin, XAD-8, for example, binds the H^+-exchanged humic substances, but it allows salts and small molecules and macromolecular organic substances (such as polysaccharides) to pass through the resin column. The humic substances are recovered by raising the pH, causing the acidic groups to ionize and the macromolecules to desorb from the resins.

This procedure is applicable to the so-called *fulvic acid fraction* contained in the supernatants when aqueous alkaline extracts from soils are precipitated at pH 1. Thus, the substances retained when this fraction is applied under acidic conditions to XAD-8 resins, and subsequently eluted when the pH of the eluent is raised, are true fulvic acids.

Humic acid precipitates cannot, of course, be applied directly to resins. However, these acids can be dissolved in DMSO, and passed into XAD-8 resin columns. Elution with acidified water (pH 1-2) removes the DMSO and polar substances, and the humic acids are then eluted as the pH is raised. This procedure allows fractionation of the humic acids using a pH gradient system. It also allows humic substances to be recovered from the acidified DMSO extracts of soils. These extracts are applied to XAD-8 columns, the DMSO is removed, and the humic substances are recovered as described. Humic acids are precipitated when the pH of the eluate from the resin is adjusted to 1.

Swift[19] has reviewed procedures for fractionating soil humic substances by the use of pH or salt gradients to elute humic substances adsorbed on XAD-8 or similar resins, or held by anion-exchange resins. Such techniques fractionate on the basis of acid strength or charge density differences. Should all other relevant parameters be equal, the strongest acids, or the most ionizable substances would be eluted first from XAD-8 resins. Electrophoresis and isoelectric focusing also fractionate on the basis of charge density differences.

Fractional precipitation provides a range of fractionation procedures. One of these involves the isolation of precipitates formed as the pH is lowered. In that instance, the strongest acids are precipitated last. Addition of salt suppresses the electrical double layer. When that suppression allows the attractive forces to predominate as molecules of similar charge approach, precipitation occurs. Fractional precipitation results also when divalent and polyvalent cations are used.

Fractionations on the basis of molecular size differences are most frequently used. The procedures generally involve uses of gel permeation chromatography,[19,20] ultrafiltration, or centrifugation. For gel permeation and ultrafiltration procedures to be effective, it is important that the humic substances should not interact (either adsorb to or be rejected by the gel or membrane) with the media used.

Appelqvist[21] has fractionated sodium humate preparations using ultrafiltration (with Sartorius membranes of pore sizes with nominal molecular size exclusion values of 5,000, 20,000, and 100,000 daltons), and gel chromatography (using Sephacryl S-200 gel, a cross-linked dextran from Pharmacia). There were definite differences between the samples excluded by the gel (MW >150,000) and those retained by the membranes which nominally retained materials of 100,000 MW and above (>100,000), 20,000 and above (100,000 - 20,000 MW fraction), and 5,000 and above (20,000 - 5,000). Table 1 gives cation-exchange capacity (CEC) data for the different fractions.

Table 1 pH dependence of CEC of humic acids of different molecular sizes (MW)

	CEC value (mmol H^+Kg^{-1})			
pH	$>1.5 \times 10^5$ MW	$>10 \times 10^4$ MW	$10 \times 10^4 - 2 \times 10^4$ MW	$2 \times 10^4 - 5 \times 10^3$ MW
4	2851	2622	1945	1861
5	3465	3299	2791	2538
6	4338	3807	3384	3130
7	4754	4145	3807	3468
8	5012	4399	3976	3997
9	5321	4822	4043	4230
10	5648	5837	4653	5245

From these data it can be deduced that the higher molecular weight fractions contained more strong acid groups than the lower molecular weight materials. This would suggest, for example, that more aromatic carboxylic acids were activated by hydroxyl groups in the 2- and/or 4- ring positions. As the molecular sizes decreased, the carbon contents also decreased (52.5, 51.6, 49.2, 47.4% for the fractions in Table 1) and the E_4/E_6 ratios increased. Such data would support the concept of greater aromaticity for the higher molecular weight materials[22], and this is also in keeping with the concept of greater numbers of aromatic carboxylic acids in the higher molecular weight substances. Fourier transform infrared spectroscopy (FTIR) indicated increased aliphatic character for the lower molecular weight components. ^{1}H-NMR spectra for the high molecular weight materials showed clearly that phenols were present in the aromatic structures, and this too would substantiate the date from the potentiometric titrations.

There were differences too between the >150,000 MW (obtained by gel filtration) and >100,000 MW (obtained by ultrafiltration) fractions. Some of these differences can be deduced from examinations of the titration data. Others were evident in the amino acid and sugar analyses. For example, the total amino acid contents of the high molecular weight materials from ultrafiltration and from gel filtration were 43.24 and 38.25 nmol mg^{-1}, respectively, and the corresponding values for the total sugars were 20.6 and 25.7 µg mg^{-1}. Prior to all analyses, the humic acids (which were isolated in 0.1M NaOH from a sapric histosol) were dissolved in DMSO–HCl (1% v/v) and adsorbed on, and recovered (by back elution with 0.1M NaOH) from XAD–8 resin. The resin treatment had separated from the humic acid fraction sugar – and amino acid-containing residues which were not covalently bonded to the humic acid 'core' or 'backbone' structures (see Hayes et al.[23]).

5 STUDIES OF STRUCTURE: PHYSICOCHEMICAL INVESTIGATIONS

The most useful of the physicochemical procedures for studies of structures of humic substances focus on molecular size, shape, and charge characteristics. For such studies, it is desirable to work with molecules that are relatively homogeneous with respect to sizes and shapes. There has been only one exhaustive attempt made to fractionate humic acids on the basis of molecular size. Cameron et al.[12] subjected the humic acids from the neutral pyrophosphate-, NaOH-, and NaOH at 60 °C - extracts of a highly organic soil to exhaustive fractionation on the basis of molecular size using gel filtration and ultrafiltration procedures. Eleven fractions of low polydispersities were obtained in this way, amd these were subjected to analysis by use of equilibrium ultracentrifugation. The molecular weight values determined ranged from 2.6 x 10^3 to 1.36 x 10^6 daltons. Two fractions (of lowest molecular weight, 2.6 x 10^3 and 4.4 x 10^3) were isolated in pyrophosphate (pH 7), seven fractions of higher molecular weight (12.8 x 10^3 – 412 x 10^3) values were isolated in dilute NaOH (25 °C), and two fractions (40.8 x 10^3 and 1.36 x 10^6 MW) were isolated in the NaOH solution at 60 °C. When frictional ratio values (f/f$_{min}$, where f is the frictional coefficient of the molecule in question, and f$_{min}$ is the coefficient for a condensed sphere, containing no solvent, occupying the same volume), obtained from the equilibrium ultracentrifugation data, were plotted against the molecular weight values (Figure 1) a linear relationship was obtained for the data for the eight fractions having the lowest molecular weight values (the highest of which had a MW value of 199 x 10^3). Interpretations of these data have been discussed in terms of the conformations which humic molecules adopt in solution.

Comparisons of the frictional ratio data (from experimental measurements) with the theoretical relationships for various models of shapes indicated that only oblate ellipsoids and flexible random coil shapes were relevant humic acids in solution. The humic acids were unlikely to possess the degrees of order needed to maintain rigid conformations. Their modes of formation, heterogeneity, and their polyelectrolytic properties in solution would give structures most likely to lead to random coil conformations in solution. It is seen from Figure 1 that the plot of the frictional ratios versus molecular weight values followed the theoretical (solid) line for random coil structures in the MW range of 2.6 x 10^3 to 199 x 10^3. Values for the 408 x 10^3, 412 x 10^3, and 1.36 x 10^6 MW fractions deviated from the linear relationship, but this was considered to be attributable to the influences of branching rather than to any regualr inter-

and intramolecular associations that would give rise to more rigid structures.

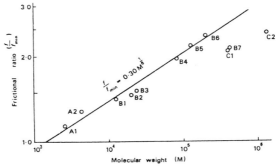

Figure 1   Plot of frictional ratio versus molecular weight values (circles) compared to the theoretical linear relationship (solid tine) for random coil structures under theta solvent conditions   (from Cameron et al.)[12]

6 STUDIES OF STRUCTURE: DEGRADATIVE PROCESSES

Because, as referred to in Section 1, the links holding the 'building' blocks in proteins and polysaccharides are hydrolyzable, it is relatively easy to assign primary structures to such biological polymers.   In humic substances, the amino acid and sugar residues released during hydrolysis in aqueous acids and bases are peripheral to the 'core' or backbone structures. These could arise from the linking of separately and biologically synthesized amino acid and peptide and/or protein, and of sugar, and of oligosaccharide and/or polysaccharide units to the 'backbone' structures.   Such linkages could, for example, involve amide structures formed between carboxyl groups on the 'backbone' and free amino groups on the amino acid and peptide structures, or the formation of $>C=N$ - structures between the amino groups and the carbon α - to the carbonyl of quinone structures.[9] Sugars, and oligosaccharides or polysaccharides could, for example, form phenolic glycosides between phenolic hydroxyls on the 'backbone' and reducing sugars, as well as sugars at the reducing ends of oligosaccharides and polysaccharides.   Because amino acid and sugar residues are decreased when humic acids dissolved in DMSO/HCl (1%, Section 2) are treated with XAD-8 resin, it can be concluded that all components containing sugar and amino acid residues are not held to the humic backbone by covalent linkages, but by secondary forces (such as hydrogen bonding and van der Waals forces), or that some of the residues become separated from the macromolecules during the course of treatments with the solvent and resin.

No single reagent or procedure can cleave all of the component molecules in humic structures.   In fact, the positions at which cleavages of the macromolecules take place depend on the energy inputs from the degradative reagents and processes.   Hence, the building blocks or units released by any one degradative procedure may well be different from those released by another.   Also, the influences of the reagents are unlikely to cease as soon as the structural units are cleaved from the macromolecules, and changes to these units are likely to take place during their residence time in the degradation digests.   This is especially relevant to degradations using oxidative processes.   Thus, the products identified in degradation digests are often different from the parent components in the macromolecules.   Therefore, it is often wrong to assign directly to structures in the macromolecules compounds identified in the degradation digests.

Awareness of the mechanisms of organic reactions allows predictions of the possible origins of compounds identified in digests of degradation reactions. However, when dealing with complex structures and highly reactive conditions, there are numerous precursors which could give rise to the compounds identified. Thus it is important to combine information from a variety of degradation reactions and reagents when trying to assign generalized 'primary-type' structures to humic macromolecules. There follows a summary of some useful information provided by hydrolysis, oxidative and reductive processes, and pyrolysis techniques.

6.1 Information from Hydrolysis Processes

Parsons[24] has reviewed the mechanisms of hydrolysis as they apply to soil humic substances, and he has listed the types of products identified in the hydrolysis digests. Reflux conditions, and the use of 6M HCl can give rise to the release of 40-50 per cent of the masses of the macromolecules as CO_2, and as soluble digest products. The soluble components would include amino acids, sugars, phenolic compounds, some aliphatic carboxylic acids, and small amounts of purine and pyrimidine bases. Condensation reactions between the sugars and amino acids in the digests can give secondary macromolecules of the melanoidin type (see Section 1). Clearly, the sugars and amino acids have origins in structures bound to or associated with the 'core' or backbone as described, and the aliphatic carboxylic acids could arise from cleavages of esters. Despite the losses of CO_2 during hydrolysis, the cation-exchange capacity values of the macromolecules are not altered appreciably. That suggests that new acid groups are released, and such would be consistent with the hydrolysis of esters. CO_2 release could arise from processes such as the decarboxylation of β-keto acids, and of activated (with -OH and $-OCH_3$ groups in the ring) benzenecarboxylic acids.

Hydrolysis in dilute alkali can release long chain monocarboxylic acids and nitrogen-containing substances from humic macromolecules. The long chain acids could be present as phenolic ester structures. It is, however, difficult to prevent oxidative processes during alkaline hydrolysis.

6.2 Information from Oxidative Degradation Processes

More than 100 compounds have been identified in the digests of oxidative degradations of humic substances, but that does not confirm that so many components are necessarily contained in the 'backbone' structures of the macromolecules. In any particular process several products could arise from the same precursors, should the time of the reaction, the temperature, the concentration of reagents, etc., be altered.

Examples of the types of compounds identified in the digests from degradations of humic substances in alkaline permanganate, and in alkaline cupric oxide media are shown in Figure 2. These summary structures are from data by Hayes and Swift[10] and Griffith and Schritzer,[25] who have discussed the mechanisms.

Benzenedi- and benzenepolycarboxylic acids predominate in the aromatic structures in the digests of permanganate and alkaline cupric oxide, and most of the structures are tricarboxylic, tetracarboxylic and pentacarboxylic. Aliphatic dicarboxylic acids (structure type **IV**, Figure 2) ranging from ethanedioic (where n = 0) to decanedioic (where n = 8) acids are the second most abundant structures in these digests. There is evidence

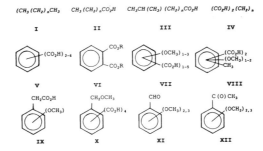

Figure 2 Representative structures in digests of degradations of humic substances in alkaline permanganate and alkaline cupric oxide media.

also for aliphatic tricarboxylic acids, for long chain monocarboxylic acids (structure type II, where n = 10 to 22 and higher), and for long-chain aliphatic hydrocarbons (structure type I, where n = 12 to 38), especially in degradations in alkaline cupric oxide media.

It is tempting to infer that the benzenepolycarboxylic acids in the alkaline permanganate digests might have their origins in fused aromatic structures. Such is unlikely, however, because evidence from other experimental procedures suggests that there are not significant amounts of fused aromatic structures in the composition of humic substances. In any case, the same benzenecarboxylic acids are found also in the digests of alkaline cupric oxide degradations of humic substances, and fused aromatic structures would not be degraded to benzenepolycarboxylic acids under such conditions. Hence, it is more likely that the carboxylic acids were formed from the oxidation of aliphatic side chains. However, the necessary experimentation has not yet been done to eliminate the possibility that the carboxyl substituents did not arise from carbonylation processes under the reaction conditions in the digest.

Carboxyl groups are generated during oxidative cleavages of alkenes, and the separations of the alkene structures would be indicated by the number (n) of CH_2 groups in structure type IV, Figure 2. The carboxyl groups could, of course, arise from other oxidizable functional groups, such as aldehyde and primary alcohol structures. There is evidence also for tri- and tetracarboxylic acids, and these could arise from oxidizable functional groups as branches from the major chain structures.

Structure types indicated in Figure 2 were identified as products of methylation. The methyl esters were formed from carboxylic acids, and thus the acid structures are shown. However, methoxy structures could be present as such, or as hydroxyl groups, and hence methoxy structures are shown. From one to three methoxy substituents have been detected in aromatic structures in the digests, and these variously were accompanied by one to five carboxyl groups (structure type VII, Figure 2).

Hayes and O'Callaghan[26] have discussed the mechanisms involved in reactions of sodium sulphide with organic substances at elevated temperatures. The mechanisms indicate that some oxidation of the digest products can take place. From the types of structures identified in the Na_2S digests of humic acids (Figure 3) it can be seen that aliphatic alcohols (when n = 1 to 4 in structure type XIII), acids (ethanoic, XIV, in considerable abundance, arising, possibly, from carbohydrates, as well as the acids represented by structure type II, where n = 1 to 4), and hydroxy and keto acids (types XV, XVI and XVII) were present in the digests. The aromatic structures were, however, significantly different from those in the digests of the classical oxidative degradation reactions. Most of the aromatic structures in the methylated Na_2S digests contained one or two methoxy substituents, one or two methyl and other aliphatic substituents, and with the exception of benzene-1,2-dicarboxylic acid (XXII), there was never more than one carboxyl group attached to the aromatic nuclei (Figure 3). This contrasts with the numbers of benzenedi- and of benzenepolycarboxylic acids (as indicated by structure type V) in the digests from oxidations with permanganate and with alkaline cupric oxide.

Hayes and O'Callaghan[26] have discussed the involvements of quinone methide intermediate structures in degradations with sodium sulphide at elevated temperatures. They have shown how a methyl substituent would form from a hydroxy or ether functional group on carbon alpha to the aromatic ring and *ortho* or *para* to a phenolic hydroxyl. (It is possible, however, to introduce methyl groups as artefacts in these ring positions during the course of methylation of phenols.) Ether, and other alkyl substituents were also detected in the same ring positions in compounds identified in the digests (in structures of the types represented by XIX, XX and XXI, Figure 3), and these would not be formed as artefacts during methylation. Ethyl, or other alkyl substituents could arise from appropriate secondary alcohol or from ether structures on the carbon alpha to the ring, and *ortho* or *para* to a phenolic hydroxyl. The evidence provided by Hayes and O'Callaghan would suggest that the quinone methide mechanism played an important role in the cleavage of aromatic moieties from the humic 'core' or 'backbone' structures. They reasoned that many of the compounds identified had origins in phenylpropane structures of the types associated with lignins, and in which two or three hydroxy or ether substituents were contained in propyl side chains.

6.3 Information from Reductive Processes

Stevenson[27] and Hayes and Swift[10] have described products obtained from the classical experiments where reductive procedures were used for the degradation of humic substances. The types of products identified in the digests of degradations in sodium amalgam are shown in Figure 4. Although aliphatic constituents in the digests were indicated by infrared spectroscopy, the techniques available at the time did not identify these substances. All of the compounds identified were phenols or derivatives of phenols. Model studies had indicated that the linkages most susceptible to cleavage are phenolic ether, and biphenyl structures bearing activating (hydroxyl, methoxy, methyl) substituents *ortho* or/and *para* to the carbons linking the biphenyl. Aliphatic to aromatic ether linkages could be expected to be removed as well.

Mention was made in Section 6.2 of the substitution of methyl groups in aromatic nuclei during the course of methylation of phenols. Structure

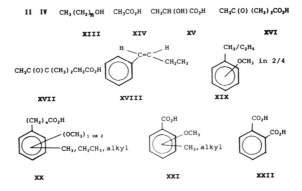

Figure 3 Representative structures in digests of degradation of humic acids in Na_2S (10% solution) at elevated temperatures.

on the nuclei, and these were unlikely to be artefacts because the substituents were found in phenolic structures in digests which had not been methylated, and were separated by liquid chromatography.

Although several of the aromatic structures identified in digests of degradations with sodium amalgam had carboxyl substituents (structure type XXIX) no compound was identified which had more than one such substituent. Because the degradation processes were reductive, it is clear that the carboxyls were not artefacts of the degradation processes. It is likely that some of the compounds represented by structure types XXIII, XXIV, and XXV were also derived from benzene carboxylic acids with two to three activating ($OH/OCH_3/CH_3$) substituents. Such substituents would promote decarboxylation under the reflux conditions used in degradation.

Some of the phenols and methoxybenzene compounds identified could be considered to be components of phenylpropane-type structures (compound types XXVIII, XXX, XXXI), and the positions in the rings of the hydroxy and/or methoxy substituents (when 3- and 4, and 3-, 4-, and 5-) would suggest that the substances might have had origins in altered lignins. Some of the compounds in the digests had hydroxyl groups in the 3- and 5-ring positions only, and that would suggest that the parent macromolecules had their origins in microbial synthesis processes.

Sodium in liquid ammonia provides another useful method for reductive cleavages of humic substances. However, only limited applications have been reported, and there has not been enough emphasis on studies of the mechanisms likely to be involved in the degradation processes. The procedure has provided some limited information which further suggests that ether linkages are important in humic structures.

Interpretations based on identifications of polycyclic aromatic structures in digests of zinc dust distillation and fusion reactions caused Haworth and his colleagues[28,29] to propose that humic acids have polycyclic aromatic 'cores' to which are attached saccharides and peptides, metals, and phenols.

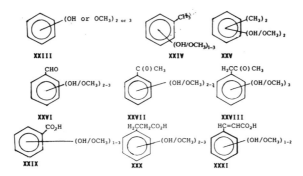

Figure 4 Representative structures in digests of degradations of humic acids with sodium amalgam preparations

However, the harsh conditions in such degradation processes inevitably gave rise to polycyclic aromatic artefacts because such compounds were formed when furfural and even cellulose were subjected to the reaction conditions.

6.4 *Information from Pyrolysis*

There is evidence in the pyrolyzates of alkali-extracted soil humic acids for components rich in substances of polypeptide, and of lignin or polyphenol origins, and to a lesser extent for compounds with origins in polysaccharides or pseudopolysaccharides (i.e. materials derived from polysaccharides which give rise on pyrolysis to furan, methylfuran, dimethylfuran, furfural, and methyl furfural structures). Evidence for peptides and polysaccharides is greatly diminished following hydrolysis in 6M HCl, but the contributions of phenol pyrolysis products (indicating that altered lignins and microbially synthesized polyphenols contributed to the 'backbone' or 'core' structures[30]) are increased. When the pyrolyzate products of the different fractions of soil humic substances are compared, strong similarities are seen between the compounds from humic acids and those from the residual humins after the humic acids have been extracted. Such observations support the concept (Section 2) which suggests that there are close similarities between the soluble humic and fulvic acids and the residual humin substances held by the soil mineral colloids. These observations could also support the view that humic acids have their origins in microbial transformations of humin parent materials. (Section 1.1).

Pyrolyzates of soil fulvic acids suggest that these macromolecules might have origins in substances with enrichments of polysaccharides and pseudopolysaccharides, and with lesser amounts of lignin-derived substances than are contained in the whole soil organic matter. However, when the fulvic acid fraction is treated with XAD-8 resin, and the polysaccharide-type contaminants are removed (Section 4) the contributions to the pyrolyzates of compounds with origins in lignins are enhanced.

It is clear that modern pyrolysis procedures, and especially soft ionization pyrolysis, can play an important role in determining the component structures in the humic macromolecules. The procedure allows the release of components of several hundred molecular weight, but it will be necessary to identify such structures before the capability can be applied with maximum effect. Thus a very considerable research effort is required, using model compounds relevant to structures in humic substances, before such pyrolyzate products can be related to structures in the humic macromolecules.

6.5 Information from Spectroscopy

MacCarthy and Rice[31] have discussed applications and limitations of spectroscopic procedures, other than nuclear magnetic resonance (NMR), in studies of functionality of humic substances. Their study took account of the problems of mixtures. Although all of the spectroscopic procedures have, at one time or another, been used on humic substances, the most useful data have been obtained from infrared (IR) and NMR procedures.

When applied to appropriately derivatized samples, IR spectroscopy can provide useful compositional data for carboxyl, ester, hydroxyl, hydrocarbon, and keto groups, but the procedure has limited value in determining the composition of the structural units (or 'building blocks') to which such functional groups are attached.

Electron spin resonance (ESR) spectroscopy has provided useful information about the influences of solvents and of chemical and physical treatment processes on the free radical contents of humic substances. It provides, for example, unambiguous evidence for the presence of unpaired electrons in only a small fraction of the components of the structures. The most likely source of free radicals are quinone–hydroquinone moieties in the macromolecules. However, quantitative analysis of the data indicate that the radicals are relatively few, and are probably far between.[32]

Applications to studies of humic substances of the theory and practice of liquid and of solid state NMR have been discussed by Wershaw[33], Malcolm[34], Wilson[35] and by Steelink et al.[36]. Progress in the states of the arts of liquid and solid state NMR analyses makes it possible now, when due care is taken, and when samples are irradiated with the proper pulse sequences, to obtain reasonably quantitative and well resolved spectra for humic substances. The availability of cross polarization magic angle spinning (CPMAS) ^{13}C NMR has done much to improve resolution, and applications of this procedure to studies of humic substances in the solid state are discussed by Malcolm in this series.

The presence in the sample of paramagnetic ions, and especially iron and copper, influences the mode of carbon relaxation and the cross-polarization efficiency within the sample. Thus the ash content, and especially the contents of paramagnetic species in the ash, can strongly influence the extents to which the carbon in samples can be observed in the CPMAS ^{13}C NMR spectra of solid humic substances. However, when the contents of ash and of paramagnetic species are low, it is possible to observe up to 97 per cent of the carbon.

The reader is referred to the subsequent contribution by Malcolm for a more comprehensive account of the information which ^{13}C NMR can provide. It will be seen there that the procedure provides an excellent method for quantitative comparisons between different humic fractions, and between the similar fractions from different environments. Compositional assignments are made in eight chemical shift regions, and these have been described in the references given above.

As the contribution from Malcolm will show, fulvic acids do not have detectable methoxy groups, are less aromatic than humic acids from the same environments, and have low phenolic contents. There can, however, be very significant differences between the same gross fractions from dissimilar environments. In general, the spectra for humic acids are more complex than those for fulvic acids. Nevertheless, there are similarities between some aspects of the ^{13}C NMR spectra of all humic substances. Invariably, there are intense peaks in the 0-50 ppm region representing unsubstituted aliphatic hydrocarbon structures, in the 60-110 ppm region, which has been attributed primarily to saccharides (the contribution from this region can be expected to be decreased when samples are subjected to treatment with XAD-8), and at the 160-170 ppm region, attributable to carboxyl and ester groups, and there is always a moderate to strong aromatic signal at 115-145 ppm, and a weak ketonic peak in the 190-215 ppm region.

With the exception of information concerning aromaticity, compositional data relevant to functionality, as revealed by ^{13}C NMR, as well as by IR have limited applications for considerations of structures of humic substances. In the case of aromaticity, however, the NMR data have shown that the aromaticity fraction (f_a, the fraction of the carbon content in the aromatic structures) is lower than might be concluded on the basis of compounds detected in digests of degradation reactions. For example, the values for aromaticity of soil humic acids are seldom greater than 40 per cent. In the view of this author there is not a firm basis for making comparisons between aromaticity, as inferred from degradation studies and that detected by ^{13}C NMR spectroscopy. The spectroscopic studies have invariably been made using samples that were not hydrolyzed, or subjected to chemical and heat treatments which would lead to decarboxylation and the losses of volatile and degradable aliphatic compounds. As indicated already, hydrolysis of humic acids in 6M HCl can result in losses in mass (in the forms of CO_2, amino acids, sugars and phenols) amounting to as much as 40-50 per cent of the mass of the starting material. The non-hydrolyzable residues would be expected to be enriched in aromatic components, and calculations based on the masses of such residues would give high values for aromaticity.

The NMR data suggest that from three to five of the ring positions in the aromatic nuclei of soil humic substances bear substituents. This would substantiate information from degradation studies and especially from oxidative degradation studies which indicate substantial substitution in the aromatic nuclei. Also, a level of aromaticity of 40 per cent or less places constraints on the possibilities that fused aromatic structures could contribute substantially to the structures of soil humic substances. Considerations of the aromatic constraints, along with the constraints placed by elemental and functional group analyses, would suggest that the contributions to the humic macromolecular structures of fused aromatic components would be very small, even in the cases of humic acids with high molecular weight values.

7 CONCEPTS OF STRUCTURE AND ASPECTS OF FUNCTION

Throughout this contribution due emphasis has been placed on the polydispersity in composition, and in the structures which compose humic substances. It has been emphasized that such polydispersity is inevitable because of the heterogeneity of the substrates and of the processes involved in their genesis. Hence, it would be pointless to attempt to interpret the data which are available in terms of accurate structural arrangements. However, it is not necessary to know the structures accurately in order to obtain an acceptable understanding of the important interactions and reactions involving humic substances. These include the binding of metals and of anthropogenic xenobiotic chemicals, adsorption to clays and hydr(oxide) minerals, and reactions with compounds used in water disinfection processes, such as chlorine, permanganate, or ozone. Nevertheless, it will not be possible to provide unambiguous interpretations of the interactions involving humic substances until there is a better awareness of the component molecules, of the linkages between these molecules, and of the juxtapositions and spacings of reactive functional groups on the macromolecular structures.

The information provided in Section 3 suggests that there is good awareness of the conformations or shapes of humic macromolecules. High molecular weight humic acids can be expected to take up random coil solution conformations, with negative charges distributed along the lengths of the strands, and with the strands coiled randomly with respect to time and space. The structures which result are enclosed by shapes that are roughly spherical, with Gaussian distributions of molecular masses, and with the mass densities greatest at the centre and decreasing to zero at the outer edges. The shapes of the Gaussian mass distributions will vary, depending on whether the molecules are tightly or loosely coiled, and this in turn will depend on the extent of solvent penetration, the charge densities, the degrees of dissociation, the nature of the counterions, and the extents of crosslinking. The evidence which is available suggests that the extents of crosslinking are small.[10,13]

A high degree of dissociation of the component acid groups is necessary for the humic macromolecules to dissolve. Considerable energy is needed to overcome the forces holding the macromolecules together in the solid state. As acid is added to the dissolved macromolecules H^+–exchange takes place, and this allows the formation of intermolecular and intramolecular hydrogen bonding processes causing the structures to shrink, and water to be excluded from the macromolecular matrix. Eventually, precipitation takes place. Similar effects result when divalent and polyvalent cations form bridges between the charges within and between the strands. As drying proceeds, it would appear that the polar groups orientate towards the interiors of the structures exposing to the exterior the more hydrophobic moieties. That would explain why dried humic acids are difficult to rewet.

Fulvic acids are smaller, more highly charged, and more polar than humic acids. Repulsion between the more closely arranged negative charges on the strands cause the macromolecules to have structures that are more linear than the random coil (spherical) arrangements of humic acids in solution. Fulvic acids do not precipitate when acidified and these should be susceptible to losses in drainage waters. The colour of drainage waters from upland acid peats is testimony to the loss of water soluble humic components. It would seem that hydrogen bonding and van der Waals forces between the soluble fulvic and the insoluble humic acids are not sufficient to hold both types of macromolecules together. Losses of fulvic acids from mineral soils are less severe. In these soils, humic substances are bound to mineral colloids, are attached to each other by divalent and polyvalent cation bridges, and resist dissolution in the soil solution.

For interactions to take place within the macromolecular matrix, between the macromolecules and anthropogenic chemicals and metals, it is necessary for the chemical or metal to diffuse into the matrix. The presence or absence of water in the macromolecular structures will determine how the charge characteristics are expressed, as well as the flexibilities of the components of the macromolecules, and their abilities to assume arrangements which would allow component structures and functional groups to interact with the guest species. The information which we have suggests that aromatic structures contribute significantly to the reactivities of the macromolecules. There is reason to believe, for example, that these bind appropriate aromatic guest molecules through charge transfer processes. It is clear too, from information in Section 6, that the aromatic structures provide strong acid groups through, for example, carboxyl groups *ortho* and/or *para* to hydroxyl or methoxy groups. These structures can also provide powerful ligands for metal complexation (e.g. hydroxyl *ortho* to the carboxyl group). There is evidence to suggest that 35-40 per cent of the structures of humic acids are aromatic, that these aromatic components are single-ring structures. The hydrogens in three to five of the ring positions may be replaced by substituent groups, and these groups (in addition to hydroxyl groups and methoxy structures) may often include linkage structures (between aromatic groups, and between aromatic/aliphatic groups) essential for extending the macromolecular chains. There is clear-cut evidence too for the presence of aldehyde and keto functionalities on the aromatic nuclei, and for phenylpropane units, or structures derived from phenylpropane units, and for hydroxyl groups and methoxy substituents on the 3-, 4-, and (sometimes) 5-ring positions. Such structures would suggest origins in lignins. Aromatic structures bearing aliphatic hydrocarbon substituents, and with hydroxyl and methoxy groups in 3- and 5-ring positions, suggest that some of the components found in digests have origins in the metabolic products of microorganisms. There is evidence to indicate that the aromatic units are linked through aromatic ether structures, and possibly by aromatic-aliphatic ether linkages, and by aliphatic hydrocarbons. Hayes and O'Callaghan[26] have discussed how degradation reactions in phenol can provide evidence for hydrocarbon linkages between aromatic units. It is not necessary, however, to consider that all of the hydrocarbon substituents are involved in inter-aromatic linking processes, and many could be present as isolated side chains. These hydrocarbons can contain olefinic structures, as well as carboxyl and other polar substituents.

Sugar and amino acid residues, which may be components of oligosaccharides and polysaccharides, and of peptide structures, are linked to the humic acid 'backbone' through covalent linkages. Amino acids invariably contribute more abundantly to the structures than do sugars, and the combined contributions of sugars and amino acids to the macromolecular structures are likely to be less than 20 per cent of the mass. The lesser contributions of saccharides to the structures might reflect the transposing of these to the so-called pseudopolysaccharides, or polar components which are not true saccharides but which give fragments characteristic of sugars when subjected to pyrolysis.

Fatty acids found in digests of degradation reactions of humic substances may well be released from esters of phenols or other hydroxyl groups in the 'backbone' structures. These may also be components of waxes associated with the macromolecules, but not a part of the humic structures. Long chain hydrocarbons have also been detected in some digests, but these are not necessarily present in all digests. Such compounds might be held by physical adsorption processes, although there is evidence to support the view that long-chain hydrocarbon structures are components of macromolecules from plant cuticles and algal metabolites. The fatty acids and hydrocarbons could contribute to the hydrophobic characteristics which humic substances display in some circumstances. At this time we do not have enough information to conclude whether or not the fatty acids and hydrocarbons are contained in all humic acid structures, or are components only of macromolecules characteristic of certain environments.

Titration data show that the acid groups in humic and fulvic acids provide a continuum of dissociable protons representing a range of acid strengths, from the strong to the very weak. The strongest acids are carboxylic, and the dissociations of the carboxyl groups are influenced by the proximity of neighbouring activating groups. Many of the aromatic structures have carboxyl substituents. Phenolic hydroxyls contribute to the acidity, and more so for humic than fulvic acids, and most, perhaps, in the cases of newly formed humic acids. Again, neighbouring substituents influence dissociations of the phenolic hydroxyls, and some are relatively strongly acidic. Enols, and other weakly dissociable groups are considered also to contribute to the charges on humic acids.

Although fulvic acids have some of the characteristics of humic acids, they have significant variations as well. Reference has been made to differences in size, shape, polarity, and charge characteristics. Fulvic acids are less aromatic than humic acids, and aromatic structures may account only for 25 per cent or so of their macromolecular masses. True fulvic acids have less carbohydrate and peptide structures than the so-called fulvic acid fraction, although true fulvic acids may contain significant amounts of the components tentatively classed as pseudopolysaccharides.

Fulvic acids are known to enhance the solubilities or complexations of sparingly soluble anthropogenic organic chemicals, and this enhancement is considered to aid the transport of such chemicals through soils and into waterways. The mechanisms of the transport processes are not fully understood, but it is possible that the flexible fulvic acids assume the shapes and properties of surface active agents when reacting with such chemicals.

REFERENCES

1. M. Nip, E.W. Tegelaar, J.W. de Leeuw, and P.A. Schenck, *Naturwissenschftenen*, 1986, **73**, 579.
2. L.C. Maillard, *C.R. Acad. Sci., Paris,* 1912, **154**, 66.
3. L.C. Maillard, *Ann. Chim.*, 1917, **7**, 113.
4. M.H.B. Hayes, J.E. Dawson, J.L. Mortensen, C.E. Clapp, and M.J. Hausler, In: M.H.B. Hayes and R.S. Swift (eds.) *Volunteered Papers 2nd. Intern. Conf. Intern. Humic Substances Soc.* (Birmingham, 1984), 1985, 157.
5. C. Enders, M. Tschapek, and R. Glane, Kolloid Z. 1948, **110**, 240.
6. A.C. Schuffelen and G.H. Bolt, *Landbouwk tidjschr.* 62, ste. Jaargang No. 4/5.
7. W. Flaig, In: F.H. Frimmel and R.F. Christman (eds.) *Humic*

substances and their role in the environment. Wiley, Chichester, *1988*, 59.

8. J.P. Martin, K. Haider, and G. Kassim, *Soil Sci. Soc. Am. J.* 1980, **44**, 1250.
9. G.R. Aiken, D.M. McKnight, R.L. Wershaw, and P. MacCarthy, In: G.R. Aiken et al. (eds.) *Humic substances in soil, sediment, and water*. Wiley, New York, 1985, 1.
10. M.H.B. Hayes and R.S. Swift, In: D.J. Greenland and M.H.B. Hayes (eds.) *The chemistry of soil constituents*. Wiley, Chichester, 1978, 179.
11. M.M. Kononova, *Soil organic matter*. Pergamon Press, Oxford, 1966.
12. M.H.B. Hayes, In: G.R. Aiken, D.M. McKnight, R.L. Wershaw, and P. MacCarthy (eds.) *Humic substances in soil, sediment, and water*. Wiley, New York, 1985, 329.
13. R.S. Cameron, B.K. Thornton, R.S. Swift, and A.M. Posner, *J. Soil Sci.* 1972, **23**, 394.
14. R.S. Swift and A.M. Posner, *J. Soil Sci. 1972, 23, 381.*
15. C.E. Clapp and M.H.B. Hayes, *Unpublished data*.
16. J.M. Bremner, *J. Soil Sci.* 1950, 1, 198.
17. R.W. Taft, D. Gurka, L. Joris, R. Von Schleyer, and J.W. Rakshys, *J. Am. Chem. Soc. 1969, 91, 4801.*
18. D. Martin and H.G. Hauthal, *Dimethyl sulphoxide*. Van Nostrand-Reinhold, New York.
19. R.S. Swift, In: M.H.B. Hayes, P. MacCarthy, R.L. Malcolm, and R.S. Swift (eds.) *Humic substances II. In search of structure*. Wiley, Chichester, 1989, 467.
20. M. DeNobili, E. Gjessing, and P. Sequi, In: M.H.B. Hayes, P. MacCarthy, R.L. Malcolm, and R.S. Swift (eds.) *Humic substances II. In search of structure*. Wiley, Chichester, 1989, 561.
21. I.A.M. Appelqvist, *The binding of Cu^{2+} and Al^{3+} ions by a high molecular weight humic acid fraction using continuous flow stirred cell-flow injection analysis*. Ph.D. Thesis, The University of Birmingham, *1990*.
22. Y. Chen, N. Senesi, and M. Schnitzer, *Soil Sci. Soc. Am. J.*, 1977, **41**, 352.
23. M.H.B. Hayes, P. MacCarthy, R.L. Malcolm, and R.S. Swift, In: M.H.B. Hayes, P. MacCarthy, R.L. Malcolm, and R.S. Swift (eds.) *Humic substances II. In search of structure*. Wiley, Chichester, 1989, 689.
24. J.W. Parsons, In: M.H.B. Hayes, P. MacCarthy, R.L. Malcolm, and R.S. Swift (eds.) *Humic substances II. In search of structure*. Wiley, Chichester, 1989, 99.
25. S.M. Griffith and M. Schnitzer, In: M.H.B. Hayes, P. MacCarthy, R.L. Malcolm, and R.S. Swift (eds.) *Humic substances II. In search of structure*. Wiley, Chichester, 1989, 69.
26. M.H.B. Hayes and M.R. O'Callaghan, In: M.H.B. Hayes P. MacCarthy, R.L. Malcolm, and R.S. Swift (eds.) *Humic substances II. In search of structure*. Wiley, Chichester, 1989, 143.
27. F.J. Stevenson, In: M.H.B. Hayes, P. MacCarthy, R.L. Malcolm, and R.S. Swift (eds.) *Humic substances II. In search of structure*. Wiley, Chichester, 1989, 121.
28. M.V. Cheshire, P.A. Cranwell, C.P. Falshaw, A.J. Floyd, and R.D. Haworth, *Tetrahedron* 1967, **23**, 1669.
29. M.V. Cheshire, P.A. Cranwell, and R.D. Haworth, *Tetrahedron* 1968, **24, 5155.**
30. J.M. Bracewell, K. Haider, S.R. Larter, and H.-R. Schulten, In: M.H.B. Hayes et al. (eds.) *Humic substances II. In search of*

structure. Wiley, Chichester, 1989, 181.
31. P. MacCarthy and J.A. Rice, In: G.R. Aiken, D.M. McKnight, R.L. Wershaw, and P. MacCarthy (eds.) *Humic substances II. In search of structure.* Wiley, New York, 1985, 527.
32. N. Senesi and C. Steelink, In: M.H.B. Hayes *et al.* (eds.) *Humic substances II. In search of structure.* Wiley, Chichester, 1989, 372.
33. R.L. Wershaw, In: G.R. Aiken, D.M. McKnight, R.L. Wershaw, and P. MacCarthy (eds.) *Humic substances in soil, sediment, and water.* Wiley, New York, 1985, 561.
34. R.L. Malcolm, In: M.H.B. Hayes *et al.* (eds.) *Humic substances II. In search of structure.* Wiley, Chichester, 1989, 339.
35. M.A. Wilson, In: M.H.B. Hayes *et al.* (eds.) *Humic substances II. In search of structure.* Wiley, Chichester, 1989, 309.
36. C. Steelink, R.L. Wershaw, K.A. Thorn, and M.A. Wilson, In: M.H.B. Hayes, P. MacCarthy, R.L. Malcolm, and R.S. Swift (eds.) *Humic substances II. In search of structure.* Wiley, Chichester, 1989, 281.

The Individuality of Humic Substances in Diverse Environments

Ronald L. Malcolm[1] and Patrick McCarthy[2]

[1] US GEOLOGICAL SURVEY, MS 408, DENVER, COLORADO 80225, USA
[2] COLORADO SCHOOL OF MINES, GOLDEN, COLORADA 80401, USA

1 ABSTRACT

Measured differences among different humic components (e.g. humic acid, fulvic acid, and humin) isolated from the same environment are often diminished to some degree by contamination of the humic samples with nonhumic materials. Characteristic differences among samples of a given humic component (e.g. humic acid) from environment to environment are frequently obscured for this same reason. Careful isolation of humic materials and separation from the nonhumic components by XAD-8 resin techniques eliminates these problems, and allows the individuality of various humic isolates to be recognized. Data are presented to show that each humic component in each environment possess an individuality that distinguishes it from other components in the same environment, and from the "same" component in different environments. The data presented in this chapter supporting these conclusions consist of amino acid and sugar contents, and ^{13}C nuclear magnetic resonance spectra. Additional data from ^1H nuclear magnetic resonance spectroscopy, carbon isotopic measurements ($\delta^{13}C$ and ^{14}C), and pyrolysis/mass spectrometry, in referenced literature support these conclusions. Environments investigated for this chapter are soil, stream water, and ocean water; humic components investigated are humic acid, fulvic acid, and humin. In the case of soil, fulvic acid extracted from the bulk or solid soil, as well as fulvic acid isolated from the interstitial soil water are included in the study. It is concluded that each humic component in each environment is unique and possesses an individuality that is characteristic of the particular component of the particular environment.

2 INTRODUCTION

Because of the complicated, heterogeneous nature of humic substances, one cannot meaningfully compare humic substances from different sources in terms of discrete structural differences.[1-3] Nevertheless, one can validly compare different humic substances on the basis of <u>compositional</u> differences. Compositional differences among humic acids from soils, sediments, and waters are reportedly large. For a given parameter, such as nitrogen content or cation exchange capacity, fifty percent or more differences are common. These large differences are evidently due to many factors, including variations in extractive methods, "purification" techniques, analytical methodology, and perhaps natural climatic factors. Because of these inherent problems and the wide range in compositions accepted for all humic substances:

(1) the differences in humic acid, fulvic acid, and humin within the same soil were often obscured;

(2) true differences among the humic substances from various soils due to pedogenic factors were difficult to determine; and

(3) the possible differences between a given humic component (humic acids, fulvic acids, or humin) from different environments (soil versus stream, soil solution versus bulk soil, stream versus groundwater) were obscured.

It has recently been shown that the elemental composition of humic substances from a wide diversity of sources displays a remarkably small degree of dispersion or scatter in terms of <u>standard deviation</u> even though the compositional <u>ranges</u> are large.[4] When the humic substances are segregated on the basis of humic acid, fulvic acid and humin fractions the degree of dispersion in the elemental compositions is diminished. The dispersions decrease even more when each of these fractions is further segregated on the basis of environment--soil, peat, freshwater, and marine sediments. These observations demonstrate statistically significant differences in the <u>elemental</u> <u>compositions</u> of humic substances from different environments.

This contribution focuses on <u>compositional</u> differences among humic substances from different environments based on the content of discrete molecules, such as amino acids and sugars, which can be hydrolytically cleaved from the humic materials. It also treats compositional differences in terms of functionality (e.g. carboxyl content) and other structural moieties (e.g. aromatic and aliphatic contents) that are part of the humic substances, as determined quantitatively by proton (^{1}H) and ^{13}C nuclear magnetic resonance (NMR) spectroscopy. Other compositional parameters discussed in this contribution are δ^{13}C (which relates to mechanisms of formation) and ^{14}C content (which relates to the age of the material).

Accordingly, our objectives are:

(1) to introduce some of the recent advances that have occurred in the field of humic substances chemistry;

(2) to reshape some of the paradigms that currently prevail in humic substances research; and

(3) to review recently-reported findings that humic substances from different natural environments (soils, streams, groundwaters, soil solution, and seawater) are truly different in composition.

Due to limitations of space in this contribution, the majority of the data presented or reviewed will be the those of the authors.

The emphasis in this contribution is on demonstrating the individuality of humic substances in diverse environments, and not on the details of the techniques themselves.

The many recent advances in humic substances chemistry have been due to several factors. The importance of humic substances in water chemistry, environmental sciences, and health sciences has been increasingly recognized over the past 15 to 20 years.[5] Because of these factors, significantly increased funding has become available in this area of research, and consequently a host of new well-trained researchers has entered the field of humic substances research. What in the past was almost exclusively the domain of the soil scientist, is now well represented by scientists from a wide diversity of disciplines. New methods of humic isolation and "purification", which provide more homogeneous isolates free of non-humic components have been developed. The use of XAD-8 resin for isolation and "purification" provides one example of these advances. New instrumental methods have been applied in humic substances research. In particular, ^{1}H-NMR spectroscopy, ^{13}C-NMR spectroscopy, ^{14}C and δ^{13}C analysis by nuclear accelerator techniques, and pyrolysis/mass spectrometric (Py/MS) analysis of humic substances are now commonplace.

3 RESULTS AND DISCUSSION

Isolation of Humic Substances Using XAD-8 Resin

All of the humic and fulvic acids from the various environments used for comparisons in this study were isolated by methods involving the same XAD-8 resin procedure.[6,7] This isolation procedure results in a <u>quantitative</u> concentration and recovery of humic and fulvic acids from waters and alkaline extracts of soils and sediments. Quantitative isolation enables the direct comparison of all samples, without losses or fractionation, which are common to most isolation methods. The procedure also yields materials of low ash contents which are relatively free of unhumified components. These aspects of the isolation procedure are critical to the characterization of relatively pure humic isolates rather than variable mixtures of humic substances with non-humic components. The importance of working with low-ash humic materials has previously been discussed.[8,9] The problems of working with humic substances that are contaminated with nonhumic materials often go unrecognized, or are simply ignored.

A rather clear-cut example as to how the XAD-8 procedure for the isolation and purification of humic substances enables the procurement of definitive data is shown in Table 1 and Figure 1. The soil humic components discussed in this contribution were extracted by 0.1 $\underline{M}$ NaOH from mallic epipedons of the Harps soil (Minnesota) and the Webster soil (Iowa). Alkaline extracts of soils contain variable amounts of sugars co-extracted with humic substances. Upon acidification both the sugars and the fulvic acids occur in the <u>fulvic acid fraction</u>. In the past, a distinction between fulvic acid <u>per se</u> and the fulvic acid fraction has generally not been made. The development of the XAD-8 resin isolation techniques allows this distinction to be readily made.[10, 11] The Webster soil fulvic acid fraction contains high concentrations of all six sugars listed in Table 1. Treatment of the Webster soil fulvic acid fraction with XAD-8 resin separates the free sugar molecules from the fulvic acid by allowing the sugars to pass directly through the resin while the fulvic acid is retained (at low pH). The fulvic acid is subsequently eluted with a basic solution. The sugar content of the fulvic acid is reduced by approximately 70 percent compared to that of the fulvic acid fraction. The sugars remaining in the fulvic acid following XAD-8 separation, and subsequently released by hydrolysis, were only those sugars covalently bonded in the Webster fulvic acid structure.

Sugar Contents of Fulvic Acids and Fulvic Acid Fractions

After removal of the unhumified free sugars which are mixed with the fulvic acids in the fulvic acid fractions, the Webster soil fulvic acid has a covalently-bonded sugar content similar to that of Harps soil fulvic acid (Table 1) and to those of the other six soil fulvic acids which were analyzed. All other soil fulvic acid fractions had lesser and variable amounts of free sugars, but the corresponding purified fulvic acids possessed very similar sugar contents. The removal of the free sugars from the fulvic acid fractions was corroborated by cross polarization magic angle spinning (CPMAS) ^{13}C-NMR spectroscopy (Figure 1), free-flow electrophoretic analysis, and dissolved organic carbon (DOC) data.

Table 1 Monosaccharide content of selected fulvic acids and fulvic acid fractions

Sample source	Concentration in $\mu g\ mg^{-1}$					
	Fucose	Galactose	Glucose	Mannose	Rhamnose	Xylose
Webster soil FA fraction	26	41	62	55	30	21
Webster soil FA	7.7	12	17	8.7	12	5.8
Harps soil FA	8.9	9.8	17	5.3	11	6.6
Harps soil interstitial FA	1.5	0.9	1.5	0.8	1.3	1.1
Missouri River FA (1/82)	0.4	0.2	0.3	0.4	0.4	0.2

Fulvic acids, free of the variable amounts of admixed sugars, can be directly compared for differences in composition of structurally-bound sugar moieties. These comparisons for six hydrolyzed sugars (Table 1) demonstrate that sugar analysis is a definitive indicator of the environment from which the fulvic acid was isolated. The individual sugar contents in soil fulvic acids are 3 to 5 times greater than those of sugar contents in the corresponding fulvic acids from river waters. Sugar content is shown only for one stream (Missouri River fulvic acid) which contains Harps soils in its watershed; however, the sugar content of the Missouri sample is similar to those of more than 20 other stream fulvic acids which were analyzed.

The sugar content of Harps soil interstitial water fulvic acid is very different from both that of Harps soil fulvic acid and of the Missouri River fulvic acid. The intermediate sugar composition of fulvic acid from the Harps soil interstitial water compared to those of the fulvic acids from soil and water is an indication, along with the ^{13}C-NMR spectroscopic data to be presented later in the contribution, that soil is not a <u>direct</u> source of stream fulvic acids. These data also indicate that soil interstitial fulvic acids are very different from both soil and stream fulvic acids.

Amino Acid Contents of Fulvic and Humic Acids

The hydrolyzable amino acid compositions of representative soil and stream humic substances are given in Table 2. The marine fulvic acid data in Table 2 are for a single seawater isolate from the mid-Pacific. Almost all of the individual amino acid contents are 5 to 10 times higher for soil fulvic acids than the corresponding values for stream fulvic acids. The total amino acid contents of soil fulvic acids are more than 10 times higher than those for stream fulvic acids. The individual and total amino acid contents are 3 to 5 times higher for soil humic acids than for stream humic acids. The amino acid contents of seawater fulvic acids are

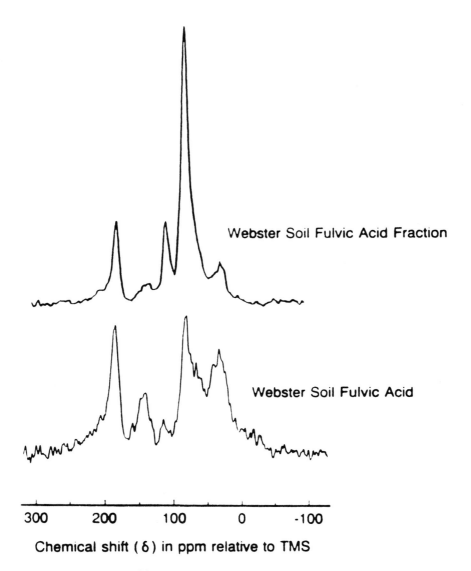

Figure 1   CPMAS ^{13}C-NMR spectra of Webster soil fulvic acid and Webster soil fulvic acid fraction

Table 2 Amino acid concentrations (nmol mg^{-1}) after hydrolysis of selected soil, stream, and marine humic substances

Amino acid	Sanhedrin soil		Ogeechee River		Seawater
	Fulvic acids	Humic acids	Fulvic acids	Humic acids	Fulvic acids
Alanine	66	66	4.8	22	5.7
Arginine	3.3	14	0.5	3.3	1.0
Aspartic acid	135	97	8.1	27	7.6
Cysteine	0.2	0.2	0.1	0.6	0.3
Glutamic acid	102	74	4.6	20	7.2
Glycine	88	83	13	37	14
Histidine	32	21	3.1	7.9	3.2
Isoleucine	13.5	23	2.0	7.6	2.3
Leucine	18	41	1.3	11	3.5
Lysine	12	23	2.2	6.2	1.8
Methionine	2.6	5.8	0.1	1.6	0.6
Phenylalanine	20	32	2.2	9.6	1.8
Proline	22	35	2.5	15	2.9
Serine	43	40	4.6	20	5.3
Threonine	37	40	4.2	20	3.9
Tyrosine	21	12	1.1	6.4	1.6
Valine	32	40	2.9	14	1.0
Total	647.6	597.0	57.3	231.2	63.7

very similar to those of stream fulvic acids. These data suggest that stream fulvic acids may be the major source of seawater fulvic acids.

Thus, amino acid analysis provides another independent method to definitively establish that soil fulvic acids and soil humic acids are different in composition from stream fulvic acids and stream humic acids, respectively. Amino acid analysis is not a new approach in the study of humic substances. Studies describing the amino acid analysis of humic substances have appeared in the literature for decades, but the variation in amino acid content was great due to the inclusion of variable amounts of non-humic components into the humic isolates.

CPMAS ^{13}C-NMR Spectroscopy

Cross polarization magic angle spinning ^{13}C nuclear magnetic resonance (CPMAS ^{13}C-NMR) spectroscopy is now the most widely used and definitive instrumental technique for the study of humic substances. After 10 years of development, CPMAS ^{13}C-NMR spectroscopy can produce spectra of humic and fulvic acids that are generally highly resolved as shown in Figure 2. A table of chemical shift assignments and interpretations for humic substances is given in several references.[2,12,13] Liquid-state ^{13}C-NMR spectroscopy has been used as frequently as solid-state CPMAS ^{13}C-NMR spectroscopy with equal success in elucidating the composition of humic substances.[14,15]

The most striking conclusion to result from the ^{13}C-NMR spectral studies of humic substances is that these materials are predominantly aliphatic in composition. Interpretations of other types of data (e.g. analyses of degradation products of humic substances) had previously led to the conclusion that most humic substances (especially humic acids and the organic fraction of humin) were predominantly aromatic in composition. Recognition of the predominantly aliphatic content of humic substances has significantly changed our general concept of the compositional nature of humic substances.

Prior to the development of ^{13}C-NMR spectroscopy, it was well established that humic acids were very different from fulvic acids in composition and reactivity. Recent ^{13}C-NMR data support this conclusion. From ^{13}C-NMR spectroscopic data, it can readily be shown that humic acids are very different in composition from fulvic acids in each environment (stream, soil, seawater, groundwater, etc.).[10,11,12,16]

Recent CPMAS ^{13}C-NMR spectra have also demonstrated that fulvic acids from all environments are not identical, contrary to some previously-held views. The CPMAS ^{13}C-NMR spectra shown in Figure 2 and published elsewhere[10,11,12,16] have established that fulvic acids from soils, streams, groundwaters, and seawaters are very different in chemical composition. CPMAS ^{13}C-NMR spectra have been presented in the references to support the same conclusion for humic acids. It can also be concluded that fulvic acid from each specific environment can be considered as an individual entity; that each fulvic acid is closely related to other fulvic acid individuals within a given environmental group; and that each fulvic acid can be distinguished as to its environment by CPMAS ^{13}C-NMR spectroscopy. These conclusions have been shown to be consistent within the limited sample size analyzed to date; due to the possible broad range of sources within each environment, it is anticipated that exceptions to these generalizations may occur.

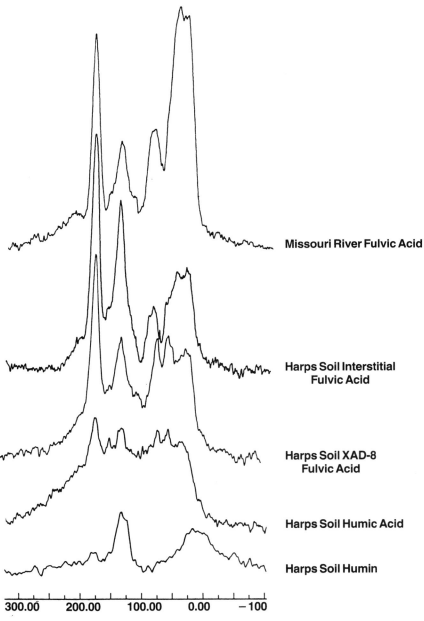

Figure 2. CPMAS ^{13}C-NMR spectra of Missouri River fulvic acid and Harps soil humic components

The Individuality of Humic Substances in Diverse Environments 31

The upper three CPMAS ^{13}C-NMR spectra illustrated in Figure 2 show marked differences between fulvic acid from three different environments (soil interstitial waters, soil, and adjacent stream waters); these differences are evident by casual observation of the spectra without an in-depth knowledge of ^{13}C-NMR spectroscopy. The major differences occur in the chemical shift ranges of 0-50 ppm (unsubstituted aliphatic carbons), 50-100 ppm (carbons singly substituted with O or N), and 108-162 ppm (aromatic carbons). All stream fulvic acids have very high unsubstituted aliphatic carbon contents, the Harps interstitial fulvic acid sample is exceptionally high in aromatic carbons compared to most fulvic acids, and soil fulvic acids are highly substituted with O and/or N.

The enormous differences in the CPMAS ^{13}C-NMR spectra between the soil interstitial water fulvic acid and soil fulvic acid, and between soil interstitial water and stream fulvic acids have many important pedogenic, hydrologic, and geochemical implications. This finding, pertaining to fulvic acids in interstitial soil waters, supports considerable other data, including the amino acid and sugar content data presented in this paper, that:
 (1) the source of stream fulvic acids is not direct leaching of humic substances from soils;
 (2) the soil fulvic acids or soil interstitial water fulvic acids leached or exuded from soils into streams are changed by a different humification process in streams; and
 (3) the reactivity of soil interstitial water fulvic acids may be very different from the reactivity of the stream water fulvic acid or bulk soil fulvic acids.
These differences in chemical composition and reactivity of soil interstitial fulvic and humic acids may help to explain many aspects of solute movement in soils; in particular, these differences in composition and reactivity may account for the apparently large effect which the chemical partitioning of certain nonionic organic solutes by humic substances in soil solution has on organic contaminant movement from soils to groundwaters.

Comparison of the lower four CPMAS ^{13}C-NMR spectra in Figure 2 shows the marked compositional differences among humic components from the same soil. By CPMAS ^{13}C-NMR spectral comparison, the four components can be said to have four distinct and different compositions equivalent to four separate individuals. The spectral differences occur in both chemical shift and intensity.

<u>Proton Nuclear Magnetic Resonance, Pyrolysis-Mass Spectrometry, and Carbon Isotopic Analysis of Humic Substances</u>

^{1}H-NMR spectroscopy, Py/MS, and carbon isotopic analyses (δ^{13}C and ^{14}C measurements) by nuclear accelerator techniques provide additional compositional information on the nature of humic substances. These independent techniques provide further evidence that humic acids, as well as fulvic acids, from various environments are of different composition.[12] ^{1}H-NMR spectra of humic and fulvic acids not only confirm, but further augment the compositional differences shown by ^{13}C-NMR spectroscopy; ^{1}H-NMR spectroscopy has superior resolution to ^{13}C-NMR in certain cases, and consequently yields more high-resolved spectra for certain functional groups in humic substances.

It is recommended that δ^{13}C and ^{14}C measurements be used more commonly for humic substances because:
 (1) a total of less than 2 mg of sample are required for both analyses;
 (2) the cost is not prohibitive using accelerator techniques;

(3) the combination of the two data parameters has been shown to be definitive for the environment from which humic substances were isolated; and
(4) each parameter alone has other geochemical implications.
For more extensive discussions of ^{1}H-NMR and carbon isotopic data as indicators of different environments for humic substances, refer to other recent references.[12,15]

Py/MS data on the fragmentation patterns of humic substances have considerable potential for research in soil and water sciences. Curie point Py/MS analysis has been shown to be definitive for identifying the environment from which the humic substance was isolated.[17] The new temperature programmable pyrolysis/soft ionization MS techniques developed by Shulten[18] have been used successfully to determine the sources of many types of leaf and wood fragments.[19] Py/MS patterns obtained by this technique have also been used in new approaches in soil classification.[20] These novel Py/MS techniques will be used on humic substances in the near future.

4 SUMMARY

The combination of improved methods of isolation and new instrumental techniques has produced rapid advances in humic acid chemistry during the past decade. The XAD-8 isolation procedure has made it possible to obtain "pure" isolates of humic substances from alkaline extracts of soils and sediments. This procedure has also enabled, for the first time, the quantitative concentration and separation of humic substances from natural waters. The humic isolates are relatively free of nonhumic constituents, the presence of which often obscured the contribution of humic substances or resulted in incorrect data interpretations. The application of new instrumental techniques (such as ^{1}H-NMR, ^{13}C-NMR, Py/MS, and nuclear accelerator technology for isotope measurements) and the resulting data obtained from their use, have led to significant changes and rapid advances in the science of humic substances.

^{13}C-NMR spectroscopy has been the most useful, the most frequently used, and the most accepted of the new instrumental techniques. ^{1}H-NMR spectroscopy and nuclear accelerator techniques for isotope determination are less frequently used. Py/MS techniques have enormous potential in humic acid research, but it has only been slightly used because of the inordinately high cost of the equipment and the sparsity of operating equipment.

The application of the new instrumental techniques to humic and fulvic acids from various environments (soils, soil interstitial waters, streams, groundwaters, seawater, etc.) which are relatively free of nonhumic components has produced data indicating that fulvic acids from each environment are different in chemical composition. In other words, fulvic acids from each environment have a unique individuality. The composition of fulvic acids from various sampling sites within each given environment will naturally have a narrow range of variability in composition, but the individuals collectively comprising the group from a given environment will be recognized as belonging to that environmental group by definitive data attributes. Correspondingly, the same definitive compositional differences can also be used to recognize humic acids from different environments.

Well-recognized data patterns which are characteristic of humic isolates from different environments were first recognized from ^{13}C-NMR spectra. The differences in composition for each environmental type are supported by data from five other independent instrumental technique and chemical methods (^{1}H-NMR spectroscopy, Py/MS, sugar analysis, amino acid analysis, and carbon isotopic measurements).

In conclusion, it must be emphasized that humic acids are not just humic acids from anywhere, but there are compositional differences among humic acids from different environments. These compositional differences may result in sufficient changes in the chemical, hydrological, or geochemical reactivity of the humic substances that each component should be designated as to the environment from which it was isolated.

5 REFERENCES

1. P. MacCarthy and J.A. Rice, 'Humic Substances in Soil, Sediment, and Water--Geochemistry, Isolation, and Characterization', G.R. Aiken, D.M. McKnight, R.L. Wershaw, and P. MacCarthy (eds.), Wiley and sons, New York, 1985, Chapter 21, p. 527.
2. M.H.B. Hayes, P. MacCarthy, R.L. Malcolm, R.S. Swift (eds.), 'Humic Substances II--In Search of Structure', Wiley and Sons, Chichester, 1989.
3. P. MacCarthy, R.L. Malcolm, C.E. Clapp, and P.R. Bloom, 'Humic Substances in Soil and Crop Sciences--Selected Readings', P. MacCarthy, C.E. Clapp, R.L. Malcolm, and P.R. Bloom (eds.), American Society of Agronomy and Soil Science Society of America, Madison, 1990, Chapter 1, p. 1.
4. J.A. Rice and P. MacCarthy, Organic Geochemistry, in Press.
5. P. MacCarthy and I.H. Suffet, 'Aquatic Humic Substances--Influence on Fate and Treatment of Pollutants', I.H. Suffet and P. MacCarthy (eds.), Advances in Chemistry Series 219, American Chemical Society, Washington, D.C., 1989, Introduction, p. xvii.
6. E.M. Thurman and R.L. Malcolm, Environ. Sci. Technol., 1981, 15, 463.
7. R.L. Malcolm, 'Proceedings of the August 1989 International Symposium on Humic Substances in the Aquatic and Terrestrial Environment', B. Allard, H. Boren, J. Ephraim, and A. Grimwall (eds.), University of Linkoping, Sweden, 1990, p. 390.
8. R.L. Malcolm, U.S. Geological Survey J. of Res., 1976, 4, 37.
9. R.L. Malcolm and P. MacCarthy, Environ. Sci. Technol., 1986, 20, 904.
10. R.L. Malcolm, 'Humic Substances in Soil and Crop Sciences--Selected Readings', P. MacCarthy, C.E. Clapp, R.L. Malcolm, and P.R. Bloom (eds.), American Society of Agronomy and Soil Science Society of America, Madison, 1990, Chapter 2, p. 13.
11. R.L. Malcolm, 'Proceedings of the Fifth International Soil Correlation Meeting (ISCOM), Characterization, Classification, and Utilization of Spodosols in Maine, Massachusetts, New Hampshire, New York, Vermont, and New Brunswick, October 1-4, 1988, Part A (papers), J.M. Kimble and R.D. Yeck (eds.), USDA-SCS, Lincoln, Nebraska, 1990, p. 200.
12. R.L. Malcolm, Anal. Chim. Acta, 1990, 232, 19.
13. M.A. Wilson, 'Humic Substances in Soil and Crop Sciences--Selected Readings', P. MacCarthy, C.E. Clapp, R.L. Malcolm, and P.R. Bloom (eds.), American Society of Agronomy and Soil Science Society of America, Madison, 1990, Chapter 10, p. 221.
14. R.L. Wershaw, 'Humic Substances in Soil, Sediment, and Water--Geochemistry, Isolation, and Characterization', G.R. Aiken, D.M.

McKnight, R.L. Wershaw, and P. MacCarthy (eds.), Wiley and sons, New York, 1985, chapter 22, p. 561.
15. R.C. Averett, J.A. Leenheer, D.M. McKnight, and K.A. Thorn (eds.), 'Humic Substances in the Suwannee River, Georgia--Interactions, Properties, and Proposed Structures', U.S. Geological Survey Open-File Report 87-557, 1989.
16. R.L. Malcolm, 'Humic Substances in Soil, Sediment, and Water--Geochemistry, Isolation, and Characterization', G.R. Aiken, D.M. McKnight, R.L. Wershaw, and P. MacCarthy (eds.), Wiley and Sons, New York, 1985, Chapter 7, p. 181.
17. P. MacCarthy, S.J. DeLuca, K.J. Voorhees, R.L. Malcolm, and E.M. Thurman, Geochim. et Cosmochim. Acta,. 1985, 49, 2091.
18. H.R. Shulten, J. of Anal. and App. Pyrol., 1987, 12, 149.
19. N. Simmleit and H.R. Shulten, Environ. Sci. Technol, 1989, 23, 1000.
20. H.R. Shulten, R. Hempfling, and W. Zech, Geoderma, 1988, 41, 211.

Novel Methods of Soil Organic Matter Analysis: A Critique of Advanced Magnetic Resonance Methods

David J. Greenslade

DEPARTMENT OF CHEMISTRY AND BIOLOGICAL CHEMISTRY,
UNIVERSITY OF ESSEX, CO4 3SQ, UK

Introduction

Soil organic matter contains free radicals and can also act as a complexing agent for paramagnetic transition ions. As a result it exhibits electron paramagnetic (or spin) resonance (EPR or ESR) signals; the study of humic substance EPR was recently reviewed by Senesi and Steelink[1]. Unfortunately, the resolution of the organic radical signals is poor and although there is some evidence that these may arise from semiquinone radicals, no further information may be derived about their structure. The main method of identifying unknown free radicals is by the hyperfine splitting patterns of their EPR spectra[2], which are due to magnetic nuclei in the vicinity of the unpaired electron of the free radical. Such nuclei produce intramolecular magnetic fields which may be isotropic (contact) or anisotropic (dipolar) with respect to the orientation of the molecular unit in the externally applied magnetic field, B_o. The dipolar field is distance and orientation dependent, averaging to zero in a sufficiently rapidly tumbling molecule, such as an aromatic radical ion in solution. The contact field comes from spin (unpaired electron) density at the nucleus[3], which seems to imply that the unpaired electron is in an s-orbital on the nucleus, however, as was found both for transition ions[4] and organic radicals[5], core polarisation by spins in p- or d-orbitals gives rise to complications. In the field, B_o, a magnetic nucleus will align and so give an additional field, ΔB, at the unpaired electron in a p-orbital, say, centred on a nucleus at a distance r such that the internuclear vector is at an angle θ to B_o. This picture, shown in figure 1, indicates the origin of the anisotropic coupling, although a full technical description uses a spin Hamiltonian, the meaning of which is clearly explained by Whiffen[2b].

Structural information may also be obtained from time dependent studies, that is relaxation from non-

$$\Delta B = \frac{\mu_n(3\cos^2\theta - 1)}{r^3}$$

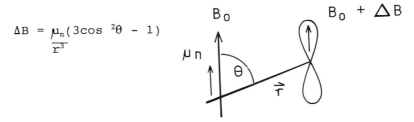

Figure 1. Classical picture of anisotropic hyperfine interaction of electron in p-orbital with a nuclear magnetic moment μ_n.

equilibrium spin states[6].

In order for the sensitive technique of EPR to be of further use in soil organic matter analysis, it is necessary to use methods giving higher resolution such as electron nuclear double resonance (ENDOR) or electron spin echo modulation spectroscopy. The latter has been applied to a wide range of systems in recent years, including complex biological systems, see Hoff[7], and in radiation damage studies[8]. ENDOR has been improved in recent years by the use of echo detection, for which Mehring[9] claims great advantages over the continuous microwave method, in which varying radio-frequency (r.f.) waves irradiate the nuclei in the sample whilst a partially saturated EPR signal is observed. In such an experiment high power r.f. is needed and, for a signal to be observed, competition between relaxation rates has to be correct, which is achieved by adjustment of the temperature of the sample. In echo detected ENDOR echo height is plotted against r.f., which is pulsed, so that photo-excited triplet states may be studied[10].

It is appropriate to mention here the effects due to the unpaired spin on the nuclear magnetic resonance of the humic substances. Although it has been estimated that there are an enormous number of atoms per unpaired electron, many of these may be in magnetic contact with such spins, especially if some are delocalised in an extensive aromatic system. The CPMAS method relies upon suitable relaxation times, and its quantitative use depends on all protons in the material having the same $T_{1\rho}$ with the carbon nuclei in good contact. In the case of coal studies, there has been much debate in the past few years about the reliability of the technique. Wind and collaborators have shown that the use of cross polarisation from the electrons to the nuclei, is a powerful way of overcoming this difficulty.[11]

Electron Spin Echoes

In the spin echo experiment two or more short pulses of radiation are used. The first is of sufficient power-width product to tip the static, equilibrium magnetic moment of the unpaired electron spin system through an angle of π/2 radians measured in the frame of reference rotating with the mean spin precession, that is at the so-called Larmor frequency about the direction of the static field. This frame has the convenient property

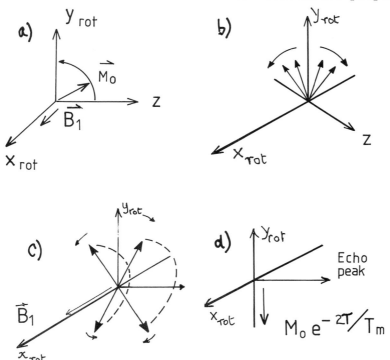

Figure 2. The formation of echoes in a rotating frame of reference, for the two pulse case, showing the refocussing of spins.

that the spins appear to precess little if at all, which implies the static field must be zero in that frame, and the microwave pulse appears as a magnetic field switched on for the time of the pulse and perpendicular to the axis of rotation, as in figure 2a. Following such a pulse, the spins dephase since their Larmor frequencies are proportional to their individual local fields, and have a range of values corresponding to the inhomogeneous linewidth of the EPR signal. The signal detected thus decays, this being the free induction decay (FID) well known in NMR. In EPR it is usually not of much value. In NMR the width of the spectrum is quite narrow compared to the frequency spectrum of the r.f. pulse and the sample circuit bandwidth, whereas in the EPR case this is not often true so that the Fourier transform of the FID may say more about the electronics than the sample! Fortunately, if at a time τ after the FID, we apply a second microwave pulse with twice the power-width product of the first, a micro- wave signal is emitted by the sample. This spin echo, demonstrated in figure 3 for a 2-pulse case, can be explained by returning to our rotating frame picture of the spins: the second pulse rotates all the spins by π radians and so reverses their relative precession, and the spins which

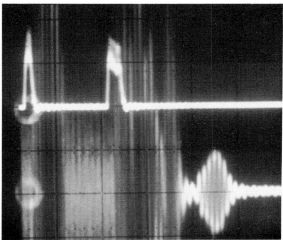

Figure 3. Oscilloscope presentation of spin echo at intermediate frequency (lower trace), with detected microwave pulses from transmitter. Horizontal 100ns/div.

have fanned apart (fig.2b) now come together (fig.2c): a rotating magnetic moment (fig.2d) is in the sample cavity! Some processes exist which lead to loss of phase coherence and so the echo envelope, a plot of echo height against τ, is an exponential decay with time constant T_m, the phase memory time. For some samples this may be very short, but sometimes may be artificially "extended" by splitting the second pulse into two equal halves: the first stores the magnetisation along the external field direction, then it relaxes with longitudinal, spin-lattice, time constant T_1, which for low enough temperatures can be of the order of several microseconds. A complicated pattern of echoes results but the required stimulated echo can be separated. Echo decays to 10 μs may be needed since Mims found that the envelope may be modulated with a period of order up to a microsecond. It is this modulation which makes the echo technique so interesting as regards molecular structure; it arises because the high power microwave pulses induce

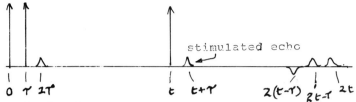

Figure 4. 3-pulse sequence and echoes.

forbidden as well as allowed transitions in the spin system (figure 5) and so give rise to a beat pattern. Figure 6 gives the decay curve for the iminoxyl radical, which shows hyperfine modulation from both nitrogens. The extensive work by Mims and co-workers and many other groups especially in the U.S.A., U.S.S.R. and Holland has shown that structural analysis based on the dipolar hyperfine splittings which may be deduced from echo modulation is capable of giving electron nuclear distances precisely and in the range up to five or more Angstroms (see ref. 5, chapters 1-3).

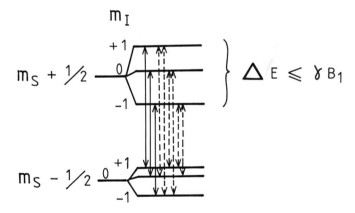

Figure 5. Energy levels of an I = 1 nucleus coupled to an electron. Full lines are allowed transitions, dotted forbidden m_s and m_I are the magnetic quantum numbers, γ the electronic magnetogyric ratio, and the nucleus is assumed to have a quadrupole.

Instrumental Technique

Although the electron spin echo technique has been known for some time, it is only slowly being applied and developed, since the technical difficulty of constructing an instrument is great and only recently

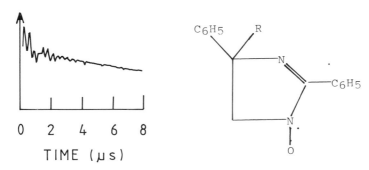

Figure 6. Echo height plotted against time for free radical structure shown.

has a commercial yet very expensive instrument become available. Sequences of microwave pulses of power 100 to 1000 watts and of duration 10 to 50 nanoseconds are required with positions in time requiring automatic programming to a precision of a nanosecond or better and in steps of no more than 10 ns. We have developed here a low cost instrument based on a phase primed, pedestal pulsed magnetron rather than the more expensive travelling wave tube used in most current instruments. Pulse position is timed by a PC-type computer coupled to a CAMAC crate containing a LeCroy 4222 timing module, in increments of 1 ns (24 bit word). Figure 7 is the block diagram of the construction of the instrument. It should be noted that a magnet of high field homogeneity is not required as the echo technique refocuses field inhomogeneities!

Complex High Molecular Weight Materials

The only examples which seem to have been studied by this technique are coals. The technique clearly should give more information on humic substances than past EPR studies using field scanning. It is also clear that there is great potential for combined chemical and echo studies, using probe radicals or spin labels. Appropriate reactive groups attached to these should enable the structure around, for example, phenolic groups to be probed. The sites at which transition metals are complexed may be established: the value of EPR for this has been established by previous workers[12], but the higher resolution of the echo method will lead to more confident conclusions. Figure 8 gives a preliminary result, recently obtained on our instrument: we cannot be sure as yet of the apparent modulation shown by the echo decay, but clearly the echo method is sensitive enough to be used.

Another use of the echo technique is the measurement of relaxation times. These can give information on the separation of radicals[13] and on delocalisation of the spins. This may be useful merely

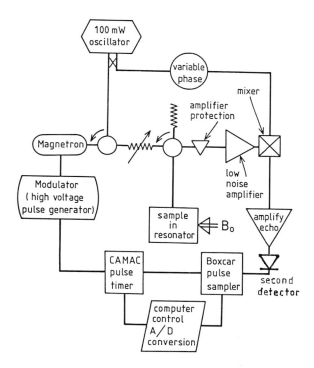

Figure 7. Outline construction of Essex electron spin echo spectrometer.

as a signature. Figure 9 gives the echo decay curves (envelopes) of two coals which have more or less identical EPR signals, but which have different phase memory times. Other workers have found spin lattice relaxation times can also be useful in the same way.

Conclusion

Recent developments in EPR and related techniques give good reason for the rejection of the view of MacCarthy and Rice[14] that "ESR spectra of humic substances contain ... little data ... concerning the nature of these materials" providing echo methods are included within the term "ESR spectra".

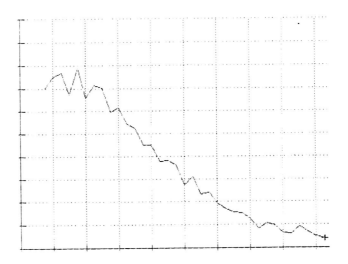

Figure 8. Echo decay envelope for a peat sample at room temperature, horizontal axis divisions 40 ns.

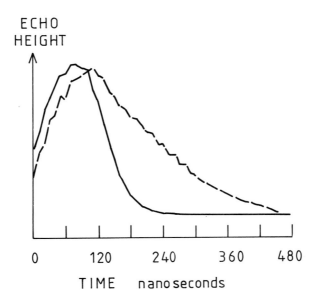

Figure 9. Two pulse echo decay curves for Merthyr (solid line) and Bagworth coals. The latter is of lower carbon content.

References

1. N. Senesi and C. Steelink, "Humic Substances II", M.H.B. Hayes, P. MacCarthy, R.L. Malcolm, and R.S. Swift, eds., J. Wiley, New York, 1989, Chapter 13, p.373.
2. a) J.E. Wertz and J.R. Bolton, "Electron Spin Resonance", McGraw-Hill, New York, 1979, p.165.
   b) D.H. Whiffen, "Problems in Molecular Structure~, G.J. Bullen and D.J. Greenslade, eds., Pion, London, 1982, p.212.
3. C.P. Slichter, "Principles of Magnetic Resonance", Springer-Verlag, Berlin, 1978, p.100.
4. A. Abragam, J. Horowitz and M.H.L. Pryce, Proc. Roy. Soc.]951, A230, 169.
5. H.M. McConnell, J. Chem. Phys., 1956, 24, 632.
6. V.V. Kurshev, A.M. Raitsimring and Y.D. Tsvetkov, J. Magn. Reson. 1989, 81, 441.
7. A.J. Hoff, ed., "Advanced EPR", Elsevier, Amsterdam, 1989, Chapters 1 to 4, especially.
8. M. Iwasaki, K. Toriyama, K. Nunome, J. Chem. Phys, 1987, 86, 5971.
9. M. Mehring, P. Hofer, A. Grupp, and H. Seidel, Phys. Lett. 1984, A106, 146, and also P. Hofer, A. Grupp and M. Mehring, Phys. Rev., 1986, A33, 3519.
10. D.J. Sloop, T-S. Lin, J. Magn. Reson., 1989, 82, 156.
11. R.A. Wind, M.J. Duijvestijn, C. van der Lugt, J. Smidt, H. Vriend, Fuel, 1987, 66, 876.
12. B.A. Goodman and M.V. Cheshire, J. Soil Sci., 1976, 27, 337.
13. L. Kevan and R.N. Schwartz, eds., "Time Domain Electron Spin Resonance", J.Wiley, New York, 1979.
14. P. MacCarthy and J.A. Rice in "Humic Substances in Soil Sediment, and Water", G.R. Aiken, D.M. McKnight, and R.L. Wershaw (eds.), J. Wiley, New York, 1985, p.558.

Organic Matter as Seen by Solid State ^{13}C NMR and Pyrolysis Tandem Mass Spectrometry

J.A. Baldock, G.J. Currie, and J.M. Oades

DEPARTMENT OF SOIL SCIENCE, WAITE AGRICULTURAL RESEARCH INSTITUTE, THE UNIVERSITY OF ADELAIDE, GLEN OSMOND, SOUTH AUSTRALIA 5064

1 INTRODUCTION

In its broadest sense the term soil organic matter encompasses all organic materials contained in soil and can be viewed as a complex heterogeneous mixture of living, dead and decomposing plant, animal and microbial tissues and detritus. The complexity of soil organic matter when considered in conjunction with the fact that a significant portion of it is bound to the soil mineral fraction has made its chemical characterization difficult. In the classical method of soil organic matter analysis a prerequisite to the selective chemical characterisation of soil organic matter has been its extraction and separation from the soil mineral fraction. The most common extractants have been alkaline solutions, such as sodium hydroxide and/or sodium pyrophosphate. The extracted organic materials which generally account for <50% of the total soil organic carbon are often fractionated into humic and fulvic acids on the basis of pH. The portion of the organic carbon not extracted by alkaline solutions has been termed humin. The use of alkaline reagents to extract soil organic carbon is inefficient and may create artefacts. In addition extraction schemes are entirely non-selective with respect to the various biological entities in soils.

Methods are required which can either extract the entire organic fraction from the soil in an unaltered state or methods which allow us to characterize the various organic fractions obtained without reactive chemicals. Based on the wide range of chemical structures contained in soil organic fractions it appears doubtful that an extractant which meets the above requirements will ever be produced. The objective of this study was to demonstrate how the need to extract organic matter from mineral soils can be eliminated by using pyrolysis tandem mass spectrometry and solid-state ^{13}C NMR spectroscopy to characterize the chemical composition of the soil organic matter contained in two soils. The ability of solid-state ^{13}C NMR to monitor the chemical changes associated following the incorporation of glucose ^{13}C-labelled into a soil organic fraction will also be described.

2 MATERIALS AND METHODS

Soils

The results presented in this paper have been collected from experiments using three soils: Millicent clay, Urrbrae fine sandy loam and Meadows fine sandy loam. A description of the three soils and some relevant chemical and physical properties are presented in Table 1. For each soil a composite sample of the 0-10 cm portion of the A horizon was collected, air-dried and sieved ≤2.0 mm.

Table 1: Description of the soils used in this study

Soil	Climate	Vegetation	Organic Carbon (%)	pH	Sand (%)	Silt (%)	Clay (%)
Millicent	temperate	Cultivated (wheat/pasture)	3.7	8.5	16.4	29.6	54.0
Urrbrae	mediterranean	Old pasture (>40 years)	2.3	5.6	50.3	33.6	16.1
Meadows	mediterranean	Mixed forest with grass	2.7	6.4	64.5	26.2	9.3

Solid-State ^{13}C NMR Spectrometry

In solid-state ^{13}C NMR analyses of mineral soils two limitations exist. Firstly, the amount of signal detected and the spectral quality (signal:noise ratio) are dependent upon the amount of ^{13}C contained in the sample. By definition mineral soils contain <17 % organic carbon by mass of which only 1.1% exists as ^{13}C. The low carbon content coupled with the low natural abundance of ^{13}C nuclei therefore makes the acquisition of spectra with an adequate signal resolution difficult unless long scan periods are used. The detection problems are compounded by the wide range of carbon structures present in soils which means that the concentration of each type of ^{13}C nucleus is low. Various particle size and density fractionation schemes have been used successfully to concentrate soil organic carbon, and thus ^{13}C nuclei, into specific soil fractions in an attempt to improve spectral characteristics[1,2,3,4,5]. The Millicent and Urrbrae soils used in this study were quantitatively fractionated on the basis of particle size and density according to Oades et al.[3].

An alternative method of increasing the concentration of ^{13}C nuclei in soil is to incubate soil with a ^{13}C-labelled substrate. Incubation of a ^{13}C-labelled substrate in soil appears attractive because, as well as increasing the concentration of ^{13}C nuclei in the soil which would make the acquisition of solid-state ^{13}C NMR spectra much easier, changes in the chemical structure of the substrate carbon as it is decomposed by soil microbes and incorporated into the soil organic fraction can be monitored. An example of the results which can be obtained by incubating ^{13}C-labelled glucose in the Meadows soil is described. Detailed accounts of this work have been published[5,6,7].

The second limitation during solid-state ^{13}C NMR analysis of mineral soils is the presence of paramagnetic species containing unpaired electrons, which reduce the effficency of signal acquisition such that ^{13}C nuclei in proximity to paramagnetics may be rendered NMR "invisible". The major paramagnetic component in mineral soils is Fe^{3+} but other transition metal cations containing unpaired electrons may also cause problems if present in high enough quantities. Oades et al.[3] showed that chemical reduction of paramagnetic Fe^{3+} to nonparamagnetic Fe^{2+} by treatment with dithionite markedly improved signal resolution and allowed recognition of carbohydrate (62 ppm), aromatic (127 ppm) and carboxyl (171 ppm) resonances which were absent in the untreated sample. Similar results were obtained by Arshad et al.[9] using sodium dithionite or stannous chloride to reduce Fe^{3+} to Fe^{2+}. Therefore, the total iron content of the particle size and density fractions was determined by X-ray fluorescence[10]. Fractions containing >3 per cent Fe_2O_3 by weight were treated with a dithionite solution[11] to reduce free Fe^{3+} to Fe^{2+} and decrease the influence of paramagnetic materials on the solid-state ^{13}C NMR analyses.

All solid-state ^{13}C NMR spectra were acquired using a Bruker CXP 100 instrument as described by Oades et al.[3]. For the particle size and density fractions, spectra were only collected for the samples listed in Table 2. The remaining fractions were not analysed because their contribution to the total carbon content of the soils was small and/or their low carbon contents precluded the acquisition of adequate spectra.

Table 2 Fractions for which pyrolysis tandem mass spectrometry and solid-state ^{13}C NMR analyses were performed.

Soil	Particle Size (μm)	Particle Density (Mg m^{-3})	Solid-State ^{13}C NMR Spectra Acquired	Pyrolysis Tandem Mass Spectrometry Analysis
Millicent	250 - 2000	≤2.0	yes	yes
	53 - 250	≤2.0	yes	yes
	20 - 53	≤2.0	yes	yes
	2 - 20	≤2.0	yes	yes
	0.2 - 2	whole fraction	yes	no
	<0.2	whole fraction	yes	yes
Urrbrae	250 - 2000	≤2.0	yes	yes
	53 - 250	≤2.0	yes	yes
	20 - 53	≤2.0	yes	yes
	2 - 20	≤2.0	yes	yes
	0.2 - 2	whole fraction	yes	yes
	<0.2	whole fraction	yes	yes

The total signal intensity and the proportion contributed by each type of carbon were determined by integration of the spectral areas delineated in Table 3. The chemical shift limits of each region are by no means definitive as there would undoubtedly be some overlap of adjacent regions. It is also important to note that the labels placed on each spectral region are only indicative of the dominant type of carbon present.

Table 3 Chemical shift limits and assignments of the spectral regions into which the the solid-state ^{13}C NMR spectra were divided.

Label	Chemical Shift Limits (ppm)	Assignment
Alkyl	10-45	Alkyl
N-Alkyl	45-60	Alkyl-amino, methoxyl
O-Alkyl	60-90	Oxygenated alkyl
Acetal	90-110	Acetal, ketal
Aromatic	110-140	Protonated and carbon substituted aromatics
Phenolic	140-160	Oxygenated aromatics
Carboxyl	160-200	Carboxylic, esters, amides

Pyrolysis Tandem Mass Spectrometry

Pyrolysis followed by chemical ionization tandem mass spectrometry (CIMS/MS) may be used for the characterization of organic compounds in whole soil samples. Organic material in soils may be divided into volatile and non-volatile, with the latter not always being considered amenable to mass spectrometry. However with the aid of pyrolysis it is possible to identify these large non-volatile components in a mass spectrometer.

During pyrolysis thermal degradation of large non-volatile compounds often leads to the production of volatile products which are characteristic of the original compound. These pyrolysis products may then be detected using a mass spectrometer. Pyrolysis mass spectra of soils are extremely complex mainly due to the number of pyrolysis products formed and hence the number of ions generated. In addition many ions of the same nominal m/z ratio may have different ion structures and originate from different compounds. This problem can be alleviated by the use of a soft ionization technique such as chemical ionization (CI). In this study isobutane was used as the reagent ion for chemical ionization. Isobutane has the advantage of producing predominately [M+H$^+$] ions for most compounds with very few if any fragmentations, hence spectra are greatly simplified. Even after soft ionization of the pyrolysis products the mass spectra are still very complicated. As a result much use has been made of multivariate analysis programs to identify clusters of ions, which can be assigned to particular types of compounds. Recently DeLuca et al.[12] showed that multivariate analysis with predefined clusters may be used with more accuracy for analysis of complex biomaterials than the general unrestricted multivariate analysis.

A variation of this approach has recently been developed in this laboratory. It involves the use of tandem mass spectrometry to detect and identify a particular ion or group of ions formed that are characteristic pyrolysis products of selected biomaterial (e.g. saccharides, proteins). The methodology is akin to multireaction monitoring commonly used in gas chromatographic mass spectrometry of complex samples, and involves selection of an ion which is believed to be a particular pyrolysis product by the first mass spectrometer, this ion is then collisionally activated and the second mass spectrometer is used to detect a fragmentation that would confirm the structure of the ion and hence the presence of the component being monitored. In this paper pyrolysis CI tandem mass spectrometry has been used to detect and monitor the following groups of compounds, saccharides (i.e hexoses and hexuronic acids), proteins, C_{10}-C_{28} saturated and monounsaturated fatty acids, and lignins/phenols in the different particle sizes of two soils.

All pyrolysis CI mass spectra were obtained from a Finnigan TSQ 70 triple quadrupole mass spectrometer, using a standard direct insertion probe and isobutane as the chemical ionization reagent gas. The conditions have been described in detail[13]. Typically one milligram of soil was placed in a crucible and this was placed on the probe directly within the ion source. The probe temperature was ramped from 100° C to 450°C in 10 minutes and the mass spectrometer was programmed to scan for numerous predefined ions and one or more of their collisionally activated fragmentations. An internal standard of 1 µg of benzophenone per mg of soil was added, spectra were normalized to this internal standard, and then each group of ions were expressed as a fraction of the total ions observed.

The particle size and density fractions of the two soils analysed by pyrolysis tandem mass spectrometry are indicated in Table 2.

3 RESULTS AND DISCUSSION

Fractionation of Millicent and Urrbrae Soils

The mass balance and carbon balance data obtained for the fractionation of the Millicent and Urrbrae soils are presented in Tables 4 and 5, respectively. Separating the particle size fractions >20 µm on the basis of particle density concentrated organic carbon successfully with 3.6 - 16.3 fold increases in organic carbon content being obtained. The carbon balance data indicated that most of the organic carbon contained in the particle size fractions >20 µm accumulated in the light fractions (87% for the Millicent soil, and 89% for the Urrbrae soil). Significant amounts of carbon were observed in the heavy fraction of the 2-20 µm material in both soils. Oades et al.[3] demonstrated that the chemical composition of organic carbon contained in 2-20 µm and 0.2-2 µm particles having densities ≤2.0 and 2.0-2.2 Mg m^{-3} was similar; however they also observed that particles having a density >2.2 Mg m^{-3} differed from the lighter particles. It therefore appears incorrect to assume that the chemical structure of the carbon contained in the light and heavy fractions of the 2-20 µm size fraction of the soils studied was similar. However, the low carbon content of the heavy fractions when considered in conjunction with the available instrumentation and technology precluded the acquisition of CP/MAS ^{13}C NMR spectra for these materials.

A decrease in the C:N ratio of both the heavy and light fractions, away from that of plant materials (30-80) and towards that of the soil microbial biomass (8-12), was observed with decreasing particle size. The observed changes in C:N ratio with decreasing particle size suggest that the organic materials contained in the finer particle size fractions of the soils examined were more decomposed ("humified") than those in the larger fractions. This suggestion is supported by results collected from peats. Based on numerous studies, Preston et al.[14] indicated that with increasing decomposition the proportions of coarse fibrous material and fine material <150 µm decrease and increase, respectively and that the finer particle size fractions are often more decomposed than the coarser fractions.

Table 4 Distribution of soil particles and organic carbon in the particle size and density fractions isolated from the Millicent soil.

Particle Size (µm)	Particle Density (Mg m^{-3})	Mass Balance (%)	Organic Carbon Content (%)	Carbon Balance (%)	C:N Ratio
250-2000	≤2.0	0.2	30.0	2.0	16.8
	>2.0	2.6	1.0	0.7	39.6
53-250	≤2.0	1.2	21.1	7.0	10.9
	>2.0	4.6	0.5	0.6	20.4
20-53	≤2.0	0.1	13.9	0.6	13.5
	>2.0	4.8	0.4	0.5	17.1
2-20	≤2.0	2.7	24.4	17.7	8.8
	>2.0	21.9	3.2	19.0	11.5
0.2-2	N.A.	18.9*	5.5	28.1	8.1
<0.2	N.A.	26.1*	2.6	18.5	7.5
Recovery		83.2		94.8	

* Some portion of these samples was lost during freeze drying which accounts for the low recovery.

Table 5 Distribution of soil particles and organic carbon in the particle size and density fractions isolated from the Urrbrae soil.

Particle Size (μm)	Particle Density (Mg m^{-3})	Mass Balance (%)	Organic Carbon Content (%)	Carbon Balance (%)	C:N Ratio
250-2000	≤2.0	0.33	37.5	5.6	16.7
	>2.0	2.49	0.2	0.2	19.6
53-250	≤2.0	0.71	32.8	10.6	13.9
	>2.0	20.42	0.1	0.9	10.0
20-53	≤2.0	0.59	8.3	2.2	18.4
	>2.0	23.50	0.1	1.1	5.0
2-20	≤2.0	2.80	18.4	23.5	17.0
	>2.0	29.24	0.9	12.4	21.1
0.2-2	N.A.	11.76	5.0	26.8	10.2
<0.2	N.A.	3.63	4.8	8.1	9.1
Recovery		95.46		91.4	

Solid-State ^{13}C NMR of Millicent and Urrbrae Soils

The composition of the organic carbon contained in the particle size and density fractions isolated from the Millicent and Urrbrae soils is shown in Fig. 1. For both soils, signals from all seven of the spectral ranges defined in Table 3 were observed. The coarser particle size fractions were dominated by signals originating from O-alkyl carbon presumably due to carbohydrate structures contained in plant fragements. When viewed under the light microscope or by scanning electron microscopy particles of organic matter >250 μm often contain recognisable cellular structures[3]. With decreasing particle size a decrease in the content of more readily decomposible O-alkyl carbon and an increase in the content of alkyl carbon was observed. The increase in alkyl carbon content was much more pronounced for the Urrbrae soil where the alkyl carbon in the fine clay fraction accounted for approximately 60% of the total carbon contained in that fraction. The high contents of alkyl carbon in the finer particle size fractions are thought to arise from the gradual accumulation of recalcitrant plant waxes strongly associated with soil clays[3]; however, as will be demonstrated later, a significant quantity of alkyl carbon is produced as soil microorganisms utilize a carbohydrate substrate. An increase in the content of alkyl carbon and decrease in the content of O-alkyl carbon with decreasing particle size has also been observed in studies focusing on the decomposition of peats[14]. In the Millicent soil an increase in the content of carboxyl carbon was also noted with decreasing particle size.

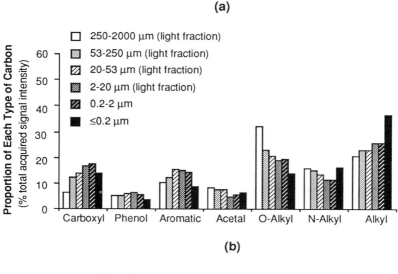
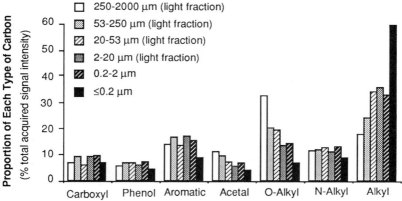

Fig.1 Composition of the organic carbon contained in the particle size and density fractions of the (a) Millicent soil and (b) Urrbrae soil as determined by solid-state ^{13}C NMR.

A concentration of aromatic carbon in the coarse clay and silt fractions was observed for both soils. Although the increase in aromatic carbon content was not large it is important since a large proportion of the total soil organic carbon in both soils was contained in these fractions.

It should be pointed out that the portion of the total organic carbon analysed by solid-state ^{13}C NMR using this technique was 74% and 77% for the Millicent and Urrbrae soils respectively. These values are substantially higher than that which would have been obtained if only the carbon extracted by the classical method was analysed. For the Millicent and Urrbrae soils the amount of carbon extracted using 0.5 M NaOH was 32% and 53%, respectively.

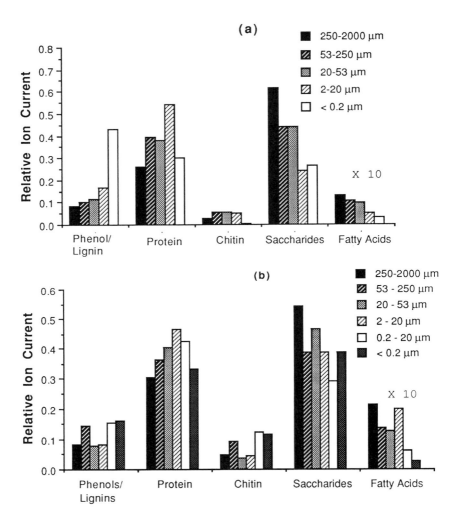

Fig. 2 Composition of the organic carbon contained in the several particle size and density fractions of the (a) Millicent soil and (b) Urrbrae soil as determined by pyrolysis tandem mass spectrometry.

Pyrolysis CI Tandem Mass Spectrometry of Millicent and Urrbrae Soils

The variation in the composition of the components monitored between the two soils and their different particle size fractions are shown in Fig. 2. A trend is apparent for the saccharides and fatty acids components in the different particle size fractions of both soils. Both classes of compounds decrease in their relative intensity in smaller size fractions. This result suggests that the majority of saccharides and fatty acids originate in plant cell structures, i.e. as the particle sizes increase there is a higher proportion of whole plant material and hence more saccharides and fatty acids. The opposite trend seems to hold true for the lignins/phenols, which increase in proportion in the smaller size fractions. The proportions of protein are greatest in the 2-20 μm size fraction. Since proteins produce phenols from pyrolysis of tyrosine it may be expected to see a correlation between phenols and proteins but this does not hold true for the smaller particle size fractions (e.g < 2.0 μm). The proportion of chitin (a fungal building block) does not appear to have any correlation with particle size fractions.

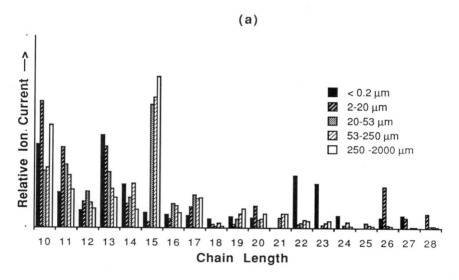

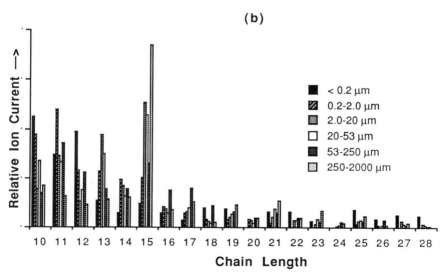

Fig. 3 Distribution of C_{10}-C_{28} saturated fatty acids in the particle size and density fractions of (a) Millicent soil and (b) Urrbrae soil as determined by pyrolysis tandem mass spectrometry.

Plots of the C_{10}-C_{28} saturated fatty acids distribution for the two soils and their different particle size fractions are shown in Fig. 3. The larger size fractions are dominated by the odd number carbons and in particular the C_{15} carbon, while the lower size particle fractions are dominated by the C_{10}-C_{13} range. It is suggested the fatty acids in the larger size fractions originate directly from plant material, whereas the fine particle fractions are dominated by fatty acids from bacteria, fungi and degraded fatty acids.

Comparison of Solid State ^{13}C NMR and Pyrolysis Tandem Mass Spectrometry Results

The results from pyrolysis CIMS/MS and NMR both show the same trends in the two different soils. The exception to this is the trend in alkyl of the NMR and the fatty acids trend in the pyrolysis MS. The reason for this is the alkyl region of NMR spectra is dominated by the recalcitrant waxes of the soil clays, and therefore the fatty acid trend is obscured. Similarly the pyrolysis CIMS/MS was used only to monitor for some fatty acids and therefore neglects waxes and other aliphatic moieties. This generalization/ selectiveness of the two techniques is likely to be the cause of other minor differences observed in the results from the two techniques e.g. while trends in the saccharides are apparent from both techniques, the plots of saccharides from pyrolysis MS of the Urrbrae particle size fractions are not as linear as those of O-alkyl from NMR.

Solid-State ^{13}C NMR Analysis of the Meadows Soil Incubated with ^{13}C-labelled Glucose

The use of ^{13}C-labelled substrates in conjunction with solid-state ^{13}C NMR poses a problem not encountered when non-enriched substrates are used. When ^{13}C nuclei are located next to each other in organic structures, as is the case in uniformly labelled ^{13}C substrates, they interact causing solid-state NMR signals to broaden, and spinning side bands to appear. In natural organic materials this problem does not exist because the chance of having two ^{13}C nuclei adjacent to one another is approximately one in 10^4. The broadening of peaks and production of spinning sidebands caused by adjacent ^{13}C nuclei on the solid-state ^{13}C NMR spectra of glucose can be seen in Fig. 4.

The solid-state ^{13}C NMR spectra acquired for the Meadows soil incubated with the ^{13}C-labelled glucose for 0, 4, 10 and 34 days are shown in Fig. 5. Other than a decrease in resolution, the spectra obtained after 0 days incubation strongly resembled that of the ^{13}C-labelled glucose indicating that any signals arising from the native organic ^{13}C in the soil were reduced to background noise.

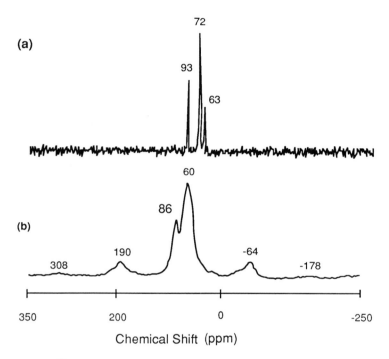

Fig. 4 Solid-state ^{13}C NMR spectra of (a) natural glucose and (b) uniformly labelled ^{13}C-glucose. Both spectra were collected using a contact time of 1 ms and a recycle delay time of 80 s. The number of scans collected for the natural and labelled glucose were 152 and 659, respectively. Line broadening was 2 Hz.

The spectra in Fig. 5 are therefore only indicative of changes in the chemical structure of the substrate carbon. As the duration of the incubation period increased resonances centred around 33, 102 and 177 ppm developed. These resonances correspond to the production of alkyl, acetal and carboxyl carbon. The strong O-alkyl resonance at 72 ppm in the day 0 spectrum originated from glucose carbon; however, since glucose decomposition in soils is generally complete within 4-7 days[6,15] the O-alkyl signal detected subsequent to day 4 must have arisen from structures synthesized by soil microbes. No strong signals were observed in the aromatic region of the spectra.

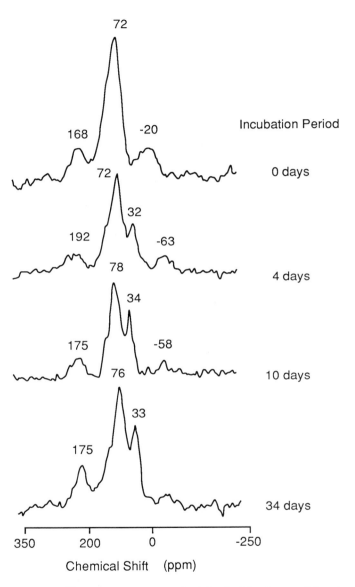

Fig. 5 Solid-state ^{13}C NMR spectra acquired for the Meadows soil incubated with uniformly labelled ^{13}C-glucose for 0, 4, 10 and 34 days using a 1 ms contact time and a 0.5 s recycle delay time. For each spectrum 10^5 scans were collected.

The proportion of substrate carbon found in alkyl, O-alkyl, aromatic and carboxyl structural groups after each incubation period and that found in the native soil organic matter are presented in Fig. 6.

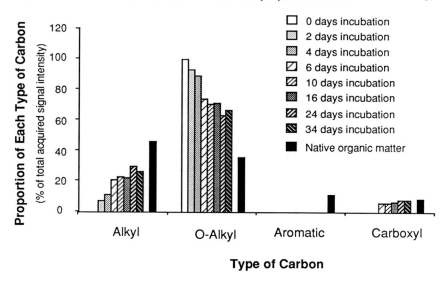

Fig.6 Distribution of substrate ^{13}C in alkyl, O-alkyl, aromatic and carboxyl fractions during incubation and that of the unincubated native soil organic carbon.

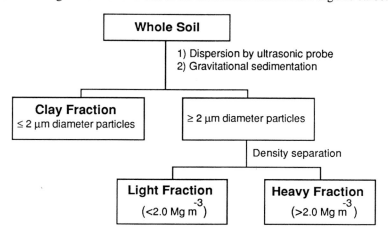

Fig. 7 Procedure used to fractionate the Meadows soil incubated for 34 days with uniformly labelled ^{13}C-glucose on the basis of particle size and density.

With increasing time of incubation the composition of the residual glucose ^{13}C tended towards that of the native soil organic matter with an increase in the content of alkyl and carboxyl carbon and a decrease in that of O-alkyl carbon. However, even after 34 days of incubation significant differences still existed between the recently synthesized residual glucose carbon and the older native soil organic carbon. The residual glucose carbon contained more O-alkyl and less alkyl carbon suggesting that it was not as humified as the native soil organic carbon. The absence of any aromatic carbon in the residual glucose carbon indicated that little if any aromatic structures were synthesized by the soil microbes utilizing the glucose carbon. The aromatic signal detected in the spectra acquired for the native soil carbon was therefore considered to arise from lignin contained in plant residues; however, a slow synthesis and selective preservation of aromatic carbon by soil microbes cannot be dismissed.

The soil incubated for 34 days with uniformly labelled glucose was fractionated on the basis of particle size and density to determine the distribution and and chemical structure of the residual substrate ^{13}C in the isolated fractions The procedure used to fractionate the soil is illustrated in Fig.7. The majority of the recovered residual substrate ^{13}C was found in the clay fraction (72%) with lesser amounts being found in the light and heavy fractions (19% and 9%, respectively).

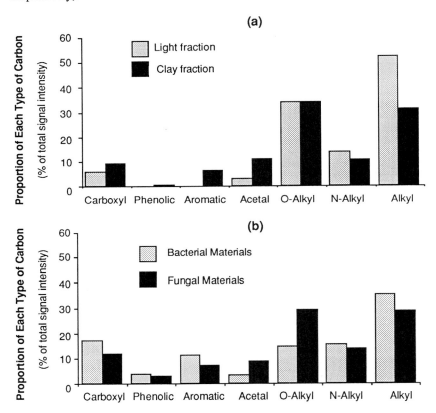

Fig. 8: Composition of (a) the residual glucose ^{13}C contained in the clay and light fractions isolated from the Meadows soil incubated for 34 days with uniformly labelled ^{13}C-glucose and (b) bacterial and fungal cultures isolated from the Meadows soil incubated with glucose.

The composition of the residual glucose ^{13}C found in the clay and light fractions was similar, the only major difference being the higher alkyl carbon content of the the light fraction (Fig. 8a).

Solid state ^{13}C NMR spectra were also collected for a bacterial and fungal culture isolated from the Meadows soil incubated with glucose. The composition of the organic materials synthesized by the two microorganisms differed significantly (Fig. 8b). The fungi contained more O-alkyl and less alkyl and carboxyl carbon than the bacteria. The larger proportion of O-alkyl carbon in the fungal materials suggested that the fungi synthesized more polysaccharide materials than the bacteria.

The composition of the residual glucose ^{13}C carbon in the soil was suspected to reflect the chemical structure of the organic materials synthesized by the soil microorganisms utilizing the added ^{13}C-glucose. When compared to the data collected for the fungal and bacterial cultures, the composition of the residual glucose ^{13}C in the clay fraction more closely resembled that of the fungal than bacterial carbon. This observation when considered in conjunction with the fact that 72% of the recovered residual glucose ^{13}C was contained in the clay fraction, indicated that the glucose added to the Meadows soil was utilized predominantly by soil fungi.

4 CONCLUSIONS

Characterization of organics in whole soil samples may now be accomplished without lengthy extraction/derivatization techniques In addition the possibility of altering the nature of the organics before characterization is greatly reduced. The combination of the two techniques solid state NMR and pyrolysis CIMS/MS has much greater potential than one technique alone. While both can give similar information, a combination of the two reduces the possibility of missing important groups of organics or of those groups being masked and increase the detail obtainable about any one specific class of compound. We believe that application of combinations of NMR, MS and IR to organic matter in soils and water will lead to significant advances in our understanding of natural organic materials, especially when done in conjunction with studies of the natural biological entities in soils.

5 REFERENCES

1. Barron, P.F., Wilson, M.A., Stephens, J.F., Cornell, B.A. and Tate, K.R. 1980. ^{13}C NMR spectroscopy of whole soils. Nature (London) 286, 585-586.
2. Skjemstad, J.O., Dalal, R.C. and Barron, P.F. 1986. Spectroscopic investigations of cultivation effects on organic matter of vertisols. Soil Sci. Soc. Am. J. 50, 354-359.
3. Oades, J.M., Vassallo, A.M., Waters, A.G. and Wilson, M.A. 1987. Characterization of organic matter in particle size and density fractions from a Red-brown earth by solid-state ^{13}C N.M.R. Aust. J. Soil Res. 25, 71-82.
4. Oades, J.M., Waters, A.G., Vassallo, A.M., Wilson, M.A. and Jones, G.P. 1988. Influence of management on the composition of organic matter in a Red-brown earth as shown by ^{13}C nuclear magnetic resonance. Aust. J. Soil Res. 26, 289-299
5. Baldock, J.A., Oades, J.M., Vassallo, A.M. and Wilson, M.A. 1990. Solid-State CP/MAS ^{13}C NMR analysis of bacterial and fungal cultures isolated from a soil incubated with glucose. Aust. J. Soil Res. 28, 213-225.
6. Baldock, J.A., Oades, J.M., Vassallo, A.M. and Wilson, M.A. 1989. Incorporation of uniformly labelled ^{13}C-glucose into the organic fraction of

a soil. Carbon balance and CP/MAS 13C NMR measurements. Aust. J. Soil Res. 27, 725-746.
7. Baldock, J.A., Oades, J.M., Vassallo, A.M. and Wilson, M.A. 1990. Solid-state CP/MAS ^{13}C NMR analysis of particle size and density fractions of a soil incubated with uniformly labelled ^{13}C glucose. Aust. J. Soil Res. 28, 193-212.
8. Baldock, J.A., Oades, J.M., Vassallo, A.M. and Wilson, M.A. 1990. Significance of microbial activity in soils as demonstrated by solid-state ^{13}C NMR. Environ. Sci. Technol. 24, 527-530.
9. Arshad, M.A., Ripmeester, J.A. and Schnitzer, M. 1988. Attempts to improve solid state ^{13}C NMR spectra of whole mineral soils. Can. J. Soil Sci. 68, 593-602.
10. Norrish, K. and Hutton, J.T. 1969. An accurate X-ray specrtrographic method for the analysis of a wide range of geological samples. Geochim. Cosmochim. Acta 33, 431-453.
11. Mitchell, B.D. and McKenzie, R.C. 1954. Removal of free iron oxide from clays. Soil Sci. 77, 73-184.
12. DeLuca, S., Sarver, E.W., Harrington, P. and Voorhees, K.J. 1990. Direct analysis of bacterial fatty acids by curie point pyrolysis tandem mass spectrometry. Anal. Chem. 62, 1465-1472.
13. Currie, G.J. and Oades, J.M. 1990. Direct characterization of organic material in soil and water samples using pyrolysis chemical ionization tandem mass spectrometry. Anal. Chem. submitted for publication.
14. Preston, C.M., Shipitalo, S.E., Dudley, R.L., Fyfe, C.A., Mathur, S.P. and Levesque, M. 1987. Comparison of ^{13}C CPMAS NMR and chemical techniques for measuring the degree of decomposition in virgin and cultivated peat profiles. Can. J. Soil Sci. 67, 187-198.
15. Coody, P.N., Sommers, L.E. and Nelson, D.W. 1986. Kinetics of glucose uptake by soil microorganisms. Soil Biol. Biochem. 18, 283-289.

Fate of Plant Components During Biodegradation and Humification in Forest Soils: Evidence from Structural Characterization of Individual Biomacromolecules

Ingrid Kögel-Knabner[1], Wolfgang Zech[1], Patrick G. Hatcher[2], and Jan W. de Leeuw[3]

[1] LEHRSTUHL FÜR BODENKUNDE UND BODENGEOGRAPHIE, UNIVERSITÄT BAYREUTH, POSTFACH 10 12 51, D-8580 BAYREUTH, GERMANY
[2] FUEL SCIENCE PROGRAM, THE PENNSYLVANIA STATE UNIVERSITY, 209 ACADEMIC PROJECTS BUILDING, UNIVERSITY PARK, PA 16802, USA
[3] ORGANIC GEOCHEMISTRY UNIT, FACULTY OF CHEMICAL ENGINEERING, DELFT UNIVERSITY OF TECHNOLOGY, DE VRIES VAN HEYSTPLANTSOEN 2, 2628 RZ DELFT, THE NETHERLANDS

SUMMARY

Solid-state ^{13}C NMR (nuclear magnetic resonance) spectroscopy, analytical pyrolysis, and chemical degradative methods are combined to elucidate the structural changes in different plant-derived biomacromolecules during biodegradation and humification in forest soils. The humification processes operating with different intensity on the plant compounds are selective preservation, direct transformation, and resynthesis by microorganisms. Polysaccharides of plant litter are decomposed intensively and substituted by microbial polysaccharides. Lignin is partly mineralized and the remnant molecule is transformed directly by ring cleavage and side-chain oxidation. Further modifications of the aromatic carbon in forest soils lead to an increase of C-substituted aromatic carbon and a concomitant loss of phenolic structures in the humic acid fraction during humification. The results indicate that the refractory alkyl carbon moieties in humified forest soil organic matter do not result from a selective preservation of plant-derived biomacromolecules. Cutin and suberin are easily mineralized or transformed and do not accumulate with depth. The recently described "aliphatic biopolymers" cutan and suberan were also not found to accumulate in the forest soils investigated. The process occurring most likely in the soils investigated is a transformation and increase in cross-linking of lipid- and/or cutin-type and suberin-type materials.

1 INTRODUCTION

Undisturbed forest soils show a characteristic distribution of undecomposed and decomposed litter and humus on and within their surface. This distribution is classified in the different forest humus types mull, moder, and mor. With increasing depth in the forest floor the organic matter is gradually mineralized and humified resulting in decreasing proportions of morphologically identifiable plant remains. The top layer is fresh plant litter (L horizon) and the bottom layer is fully humified material (Oh), with transitional intervening horizons (Of). These forest floor horizons overly the humified mineral soil horizon (Ah, Aeh). The degree of humification is directly related to the morphological changes observed[1]. The amount, chemical composition and distribution of the litter input is different for different ecosystems[2]. Fresh litter input in temperate forest ecosystems is composed mainly of leaves/needles and varying proportions (24 - 39 %) of woody debris[3]. Above-ground leaf litter contains about 50 - 60 % polysaccharides, 15 - 20 % lignin and other aromatic compounds, and about 15 - 20 % aliphatic compounds, such as lipids, cutin, and other aliphatic biopolymers[4]. The input of root litter is confined mainly to the Oh and mineral soil horizons and makes up 20 to 50 % of the total C input to temperate forest soils[5].

2 HUMIFICATION PROCESSES

Carbon from a variety of different plant and microbial sources contributes to the formation of humic matter. The litter input to soils is mineralized in a two-stage process. A rapid initial phase of plant litter (primary resources) decomposition and transformation is followed by a second much slower phase. In the second phase the microbial biomass and its metabolic products (secondary resources), which were built up during the initial phase, are decomposed[6, 7]. The preferential decomposition of easily mineralizable materials leads to the selective preservation of refractory plant or microbial components[8]. Consequently, the formation of forest humus can be viewed as the result of a combination of different processes:
- selective preservation,
- direct transformation, and
- resynthesis by microorganisms.

In the following we will discuss the susceptibility of different aliphatic and aromatic compound classes of forest litter to these processes. For this purpose we combine data from different investigations on the structural characterization of forest soil organic matter. Our approach to delineate the formation of humic substances is to follow the structural changes observed in plant-derived biomacromolecules at different stages of decomposition.

3 FATE OF INDIVIDUAL PLANT CONSTITUENTS

Polysaccharides

Polysaccharides are the major component of forest litter. The plant litter input of polysaccharides into forest soils is comprised of about 20 - 25 % cellulose and 20 - 30 % hemicelluloses. The major structural difference is that cellulose is a crystalline polymer of glucose, whereas hemicelluloses are composed of various pentoses and hexoses. The amounts and types of monomers are different for different tree species[9]. In soils, these plant-derived polysaccharides are decomposed preferentially compared to lignin[10]. This is reflected by the complete loss of cellulose with depth. The cellulose content in forest soils decreases from 20 - 25 % in the litter layer to less than 3 % in the A horizon[11, 12]. Similar results are obtained in laboratory experiments[10, 13, 14]. The concurrent increase of carbohydrates from microbial sources points to **microbial resynthesis** as the major humification process[15-17]. The polysaccharides comprise a high proportion of the O-alkyl carbon in the humified horizons (Oh, Ah). There is still a lack of data providing evidence for the processes which lead to a stabilization of these polysaccharides.

Aromatic compounds

CPMAS ^{13}C NMR spectroscopy shows that forest soil organic matter contains about 20-30 % aromatic carbon[11, 12]. The major aromatic components of plant litter are lignin and tannins. Lignin is a complex three-dimensional polymer produced by dehydrogenative polymerization of three phenylpropane monomers, commonly referred to as guaiacyl/vanillyl, syringyl, and p-hydroxyphenyl units. The monomers are linked together by several different C-C and ether linkages, most of which are not readily hydrolyzable. The relative proportions of each monomeric unit in the lignin of a particular plant species depend on its phylogenetic origin. The lignin of hardwoods such as beech consists of about equal proportions of guaiacyl and syringyl monomers; softwood lignin is composed mainly of guaiacyl units.

Lignin biodegradation is attributed mainly to white-rot fungi, but also bacteria, fungi imperfecti, and actinomycetes have been shown to be able to degrade lignin[10, 13, 18, 19, 20]. Studies of lignin breakdown in vitro have shown that the key reactions are oxidative cleavage of phenylpropanoid side-chains, demethylation of methoxyl groups, hydroxylation of aromatic rings, and cleavage of aromatic rings while they are still in the polymer. Lignin does not provide a source of carbon or energy for the soil microbial biomass. Lignin-derived carbon therefore remains in the non-hydrolyzable carbon fraction of soil organic matter[13]. Direct evidence for the pathway of lignin decomposition in forest soils is still lacking. Indirect evidence of the fate of lignin during decomposition in soils is obtained from chemical degradation and determination of the monomers released (e.g. CuO oxidation)[21-23] or from analytical pyrolysis[6, 11, 24]. Lignin is attacked preferentially at the ether-linkages. In most cases woody angiosperm lignin is decomposed at a higher rate compared to coniferous lignin. Fig. 1 shows the decrease of CuO lignin with increasing depth in forest soils. Biodegradation of lignin in forest soils leads to a modification of the remnant lignin polymer, which has a lower content of intact lignin moieties due to ring cleavage and a higher degree of side-chain oxidation[12, 14, 25]. This is reflected by changes in the acid/aldehyde ratio of the CuO oxidation products with depth in forest soils (Fig. 2), which can be used as an indication for the degree of side-chain oxidation of the remnant lignin[12, 21-23, 25]. Pyrolysis - field ionization - mass spectrometry shows[24, 26] that the more recalcitrant C-C linked moieties such as pinoresinol, phenylcoumaran, and biphenyl moieties are selectively preserved compared to the ether-linked moieties.

The changes in the aromatic C content and structures of forest soil humic acids at different stages of humification were studied using solid-state ^{13}C NMR and dipolar dephasing, a technique which differentiates between protonated and non-protonated C, in combination with molecular level characterization of lignin-derived phenols by CuO oxidation. The humic acid fraction isolated from fresh litter of European beech (*Fagus silvatica* L.) and Norway spruce (*Picea abies* Karst) shows mainly peaks attributable to aromatic C derived from lignin and tannin structures, which are both partly extractable by alkaline solvents. The most prominent feature of the NMR spectra is the decrease of phenolic C (150 ppm) and methoxyl C (56 ppm) with increasing degree of humification. Simultaneously the signal intensity at 130 ppm in the ^{13}C NMR spectra increases (Fig. 3). Detailed structural assignments for this signal can be obtained by measuring the percentage of signal intensity from protonated and non-protonated C by dipolar dephasing ^{13}C NMR spectroscopy. Then we can calculate the percentage of phenolic C (aryl-O) and C-susbtituted aromatic C (aryl-C) from the NMR data. Table 1 shows that the percentage of non-protonated aromatic C remains constant and the C-substituted aromatic C fraction increases when humification proceeds. We also find decreasing yields of lignin-derived CuO oxidation products and an increasing degree of oxidative decomposition (side-chain oxidation) in the lignin-derived structures as determined by acid/aldehyde ratios[27]. The higher number of carboxyl groups due to ring cleavage and side-chain oxidation results in an increasing percentage of aromatic C extractable with alkaline solvents[12].

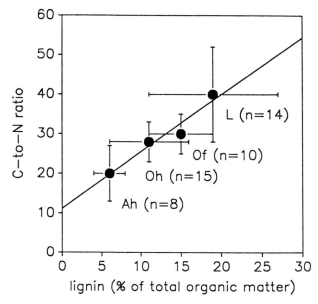

Figure 1: Decreasing contents of lignin (determined by CuO oxidation[12]) with increasing depth and decomposition, i.e. decreasing C/N ratio, in different forest soils (from 40).

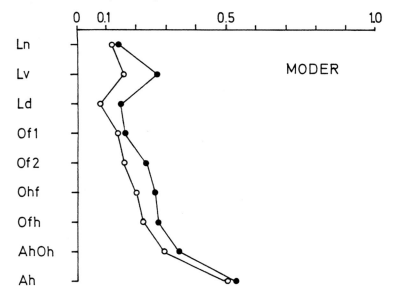

Figure 2: Acid/aldehyde ratio of the vanillyl (o) and syringyl (o) CuO oxidation products with increasing depth in a moder humus profile under beech.

Assuming that lignin is the primary precursor of the aromatic C components of humic acids in forest soils, then the lignin structure is altered considerably during humification, resulting in lignin-derived aromatic structures with a high degree of C-substitution and carboxyl functionality.

Table 1 Fraction of protonated ($f_a^{a,H}$) and nonprotonated ($f_a^{a,N}$) aromatic C, aryl-O C (ar-O/ar) and aryl-C C (ar-C/ar) of the total aromatic C signal intensity (calculated from 27)

		$f_a^{a,H}$	$f_a^{a,N}$	ar-O/ar	ar-C/ar
Eremitage	L	0.31	0.69	0.38	0.31
(beech)	Of	--	--	0.37	--
	Oh	--	--	0.31	--
	Ah	0.31	0.69	0.29	0.33
Farrenleite	Oh	0.27	0.73	0.34	0.39
(beech)	OhAeh	0.25	0.75	0.32	0.43
	Bsh	0.27	0.73	nd	0.73
Schneeberg	L	0.39	0.61	0.34	0.27
(spruce)	Of	--	--	0.37	--
	Oh	0.37	0.63	0.29	0.34
	Aeh	0.33	0.67	0.24	0.43

nd = ar-O not detected

Lignin, the major precursor of aromatic carbon in forest soils, undergoes several humification processes. There is evidence for a **direct transformation** of lignin, resulting in a decrease of the relative percentage of aryl-O C and a relative increase of C-substituted aromatic structures with depth. Concomitantly there is ring cleavage and side-chain oxidation. The more refractory lignin components are **selectively preserved**. It is clear that there is a degradation of lignin structural units in humic acids such that the aromatic units lose oxygen-functional groups as depth in the soil profile increases. Our results therefore do not corroborate models on humification pathways in soils, which would lead to an increase in aryl-O content when humification of lignin proceeds. The detailed nature of the aromatic C structures and the C-substituents remains to be investigated further. It would be too presumptive at this time, with so little evidence or data, to suggest possible reaction pathways which would result in loss of aryl-O C with concomitant retention or increase of aromatic C.

The leaves and barks of several tree species are high in tannins[28, 29]. Tannins have also been identified in the humic acid fractions extracted from forest soils[27]. Tannins are heterogeneous compounds and provide problems for analysis by wet chemical methods. Therefore the structural changes they undergo during humification have not been investigated in detail.

Cutin, suberin and other aliphatic biopolymers

CPMAS ^{13}C NMR spectroscopy has shown that forest soil organic matter contains significant amounts (15 - 30 %) of alkyl carbon[11, 12, 26, 30]. The major alkyl-C components of forest soil organic matter are extractable and bound lipids[31], the plant polyesters cutin and suberin, and other non-saponifiable aliphatic components[4, 7, 32]. The non-lipid components constitute a major fraction of soil organic matter and associated humic substances. Cutin is decomposed or transformed in forest soils and does not accumulate with depth[4]. This has been confirmed in litter-bag experiments in the field as well as in laboratory decomposition experiments[32, 33].

Analysis by solid-state CPMAS ^{13}C NMR and Curie-point pyrolysis-gas chromatography-mass spectrometry indicated that the alkyl carbon moieties are being altered significantly with increasing depth and decomposition in the soil profile. The NMR data show that the aliphatic structures in forest soil organic matter can be assigned to rigid carbon moieties and mobile carbon moieties[32]. Mobile carbons are compounds which are near their melting point or in gel-like structures. They have also been found in microbial residues isolated from soils[16, 34]. At depth in the soil profiles investigated the mobile components are lost but the rigid aliphatic components appear to be selectively preserved. We hypothesized that the mobile and rigid carbon types are possibly associated with different types of macromolecules. The polyesters cutin and suberin from leaves, barks, and roots show a high proportion of mobile carbon structures, whereas the resistant non-saponifiable aliphatic biomacromolecule which has been identified recently in the cuticles of several plants[35, 36-39] is composed almost exclusively of rigid carbon structures. This new non-saponifiable aliphatic biomacromolecule produces a homologous series of alkane/alkene pairs upon flash pyrolysis. Fig. 4 shows the pyrolysis-GC trace obtained from the residue of a mor Oh horizon under spruce after extraction and saponification. The alkane/alkene signals are present, but in very low amounts. This indicates that the aliphatic materials in the humified soil horizons bear no resemblance to the resistant aliphatic non-saponifiable biomacromolecules in fresh leaf cuticles. This lack of resemblance is

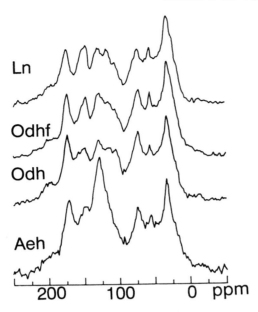

Figure 3: CPMAS ^{13}C NMR spectra of humic acid fractions extracted from selected horizons of a mor humus profile under spruce (from 20).

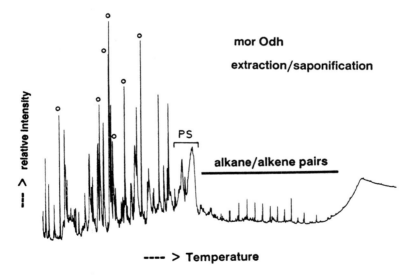

Figure 4: Pyrolysis-GC trace of the residue after extraction and saponification of a mor Oh horizon; PS = pyrolysis products from polysaccharides, o = lignin pyrolysis products; (from 21).

probably due to the fact that selective preservation of resistant, non-saponifiable plant macromolecules is not the dominant process leading to the accumulation of alkyl carbon moieties in forest soil organic matter. Alternatively, it seems possible that the structural differences observed between the alkyl carbon moieties in forest litter and humified soil horizons are due to a **direct transformation** of the material during humification resulting in increased cross-linking and therefore reduced mobility.

ACKNOWLEDGEMENTS

We thank Ludwig Haumaier, Erik Tegelaar, and Frank Ziegler for help and critical discussions. This work was supported by the Deutsche Forschungsgemeinschaft (SFB 137, C1) and NATO Grant 0799/89.

REFERENCES

1. C. Angehrn-Bettinazzi, P. Lüscher, and J. Hertz, Z. Pflanzenernähr. Bodenk. 1988, 151, 177.
2. J.M. Oades, Biogeochem., 1988, 5, 35.
3. M.E. Harmon, J.F. Franklin, F.J. Swanson, P. Sollins, S.V. Gregory, J.D. Lattin, N.H. Anderson, S.P. Cline, N.G. Aumen, J.R. Sedell, G.W. Lienkaemper, K. Cromack, and K.W. Cummins, Adv. Ecol. Res., 1986, 15, 133.
4. I. Kögel-Knabner, F. Ziegler, M. Riederer, and W. Zech, Z. Pflanzenernähr. Bodenk. 1989, 152, 409.
5. K.A. Vogt, C.C. Grier, and D.J. Vogt, Adv. Ecol. Res., 1986, 15, 303.
6. D.S. Jenkinson, In: A. Wild (ed.) Russel's soil conditions and plant growth. Longman, Essex, 1988, 564.
7. M.J. Swift, O.W. Heal, and J.M. Anderson, 'Decomposition in terrestrial ecosystems', Blackwell, Oxford, 1979.
8. P.G. Hatcher and E.C. Spiker, In: F.H. Frimmel and R.F. Christman (eds.) Humic substances and their role in the environment. Wiley, Chichester, 1988, 59.
9. D. Fengel and G. Wegener, 'Wood: chemistry, ultrastructure, reactions', de Gruyter, Berlin, 1984.
10. K. Haider, Trans. 13th Int. Congr. Soil Sci., Hamburg 1987, Vol. 6, 644.
11. R. Hempfling, F. Ziegler, W. Zech, and H.-R. Schulten, Z. Pflanzenernähr. Bodenk., 1987, 150, 179.
12. I. Kögel-Knabner, W. Zech, and P.G. Hatcher, Z. Pflanzenernähr. Bodenk., 1988, 151, 331.
13. K. Haider, In J.-M. Bollag and G. Stotzky (eds.) Soil Biochemistry Vol. 7, 1991, in press.
14. F. Ziegler, Bayreuther Bodenkundl. Ber., Vol. 13, Bayreuth 1990.

15. J.A. Baldock, J.M. Oades, A.M. Vassallo, and M.A. Wilson, Environ. Sci. Technol., 1990, 24, 527.
16. J.A. Baldock, J.M. Oades, A.M. Vassallo, and M.A. Wilson, Aust. J. Soil Res., 1990, 28, 213.
17. I. Kögel-Knabner, P.G. Hatcher, and W. Zech, Trans. 14th Int. Congr. Soil Sci., Vol. V, Kyoto, 1990, 217.
18. A.D.M. Rayner and L. Boddy, 'Fungal decomposition of wood', Wiley and Sons, Chichester, 1988.
19. H.E. Schoemaker and M.S.A. Leisola, J. Biotechnol., 1990, 13, 101.
20. W. Zimmermann, J. Biotechnol., 1990, 13, 119.
21. J.R. Ertel and J.I. Hedges, Geochim. Cosmochim. Acta, 1984, 48, 2065.
22. J.R. Ertel and J.I. Hedges, Geochim. Cosmochim. Acta, 1985, 49, 2097.
23. J.I. Hedges, R.A. Blanchette, K. Weliky, and A.H. Devol, Geochim. Cosmochim. Acta, 1988, 52, 2717.
24. R. Hempfling and H.-R. Schulten, Org. Geochem., 1990, 15, 131.
25. F. Ziegler, I. Kögel, and W. Zech, Z. Pflanzenernähr. Bodenk., 1986, 149, 323.
26. R. Hempfling and H.-R. Schulten, Sci. Total Environ., 1989, 81/82, 31.
27. I. Kögel-Knabner, P.G. Hatcher, and W. Zech, Soil Sci. Soc. Am. J. 1991, 54, in press.
28. R. Benner, P.G. Hatcher, and J.I. Hedges, Geochim. Cosmochim. Acta 1990, 54, 2003.
29. M.A. Wilson and P.G. Hatcher, Org. Geochem., 1988, 12, 539.
30. C.M. Preston, P. Sollins, and B.G. Sayer, Can. J. For. Res., 1990, 20, in press.
31. F. Ziegler, Soil Biol. Biochem., 1989, 21, 237.
32. I. Kögel-Knabner, P.G. Hatcher, E.W. Tegelaar, and J.W. de Leeuw, Sci. Total Environ., 1991, submitted.
33. F. Ziegler and W. Zech, Z. Pflanzenernähr. Bodenk., 1990, 153, 373.
34. M.A. Wilson, B.D. Batts, and P.G. Hatcher, Energy & Fuels, 1988, 2, 668.
35. M. Nip, E.W. Tegelaar, J.W. de Leeuw, P.A. Schenck, and P.J. Holloway, Naturwissenschaften, 1986, 73, 579.
36. E.W. Tegelaar, J.W. de Leeuw, and C. Saiz-Jimenez, Sci. Total Environ., 1989, 81/82, 1.
37. E.W. Tegelaar, J. W. de Leeuw, and P.J. Holloway, J. Anal. Appl. Pyrol., 1989, 15, 289.
38. E.W. Tegelaar, G. Hollman, P. van der Vegt, J.W. de Leeuw, and P.J. Holloway, Org. Geochem., 1990, submitted.
39. E.W. Tegelaar, J.H.F. Kerp, H. Visscher, P.A. Schenck, and J.W. de Leeuw, Palaeobiology, 1990, submitted.
40. F. Ziegler, I. Kögel-Knabner, and W. Zech, in preparation.

Spectroscopic Characterization of Organic Matter from Soil with Mor and Mull Humus Forms

D.W. Hopkins[1] and R.S. Shiel[2]

[1] DEPARTMENT OF BIOLOGICAL SCIENCES, THE UNIVERSITY, DUNDEE DD1 4HN, UK
[2] DEPARTMENT OF AGRICULTURAL AND ENVIRONMENTAL SCIENCE, THE UNIVERSITY, NEWCASTLE UPON TYNE NE1 7RU, UK

1 INTRODUCTION

Soil with a mor humus form is usually distinguished from that with a mull humus form by the presence of an Oa horizon and an abrupt transition into the underlying A horizon.[1] Although the term mor humus is generally applied to forest soils[1] where the plant litter entering the soil contains relatively large amounts of lignin,[2] both mor and mull humus forms have been identified in permanent grassland soils.[3] In acidic grassland soils a mat type of mor humus can develop in which an Oe horizon is also present.[3,4] Such mor humus soils have high cellulose, hemicellulose and lignin contents in the mat horizon, despite the relatively low lignin content of the grass litter.[4] The relative amounts of cellulose and hemicellulose decrease and that of lignin increases with depth.[4] These observations indicated that decomposition of selected humus components occurred[4] and this conclusion is supported by the measurement of reduced C:N ratio in the mat horizon compared with that of the plant litter and a reduction in C:N with depth.[5] Although acidic and near-neutral soils support different grass species, mor humus development appears to be independent of grassland vegetation because the litters from different grass species decompose at approximately the same rate under favourable soil conditions.[6] By comparison with a grassland soil with a mull humus form, the overall rate of organic matter accumulation is increased,[7] the rate of microbial respiration is reduced [7] and there is redistribution of organic matter towards the soil surface as coarse (>212 μm) particulate material[5] in mor humus soils.

In this study we have used two spectroscopic techniques to chemically characterize the organic matter associated with different particle-size fractions of soil from the surface horizons of two adjacent grassland profiles with either mor or mull humus forms. The techniques used were cross polarization magic-angle spinning ^{13}C nuclear magnetic resonance (CP MAS ^{13}C NMR) spectroscopy and diffuse reflectance infrared Fourier transform (DRIFT) spectroscopy.

2 METHODOLOGY

The soils were sampled from experimental grassland plots in Northumberland, U.K. Full descriptions of the site and soil properties have been given elsewhere.[5] The soil with mor humus has, as result of long-term (since 1897) annual ferilizer ($[NH_4]_2SO_4$) applications, a pH of 4.0 and the soil with mull humus, which has received annual basic slag until 1976, and triple superphosphate since then, and KCl dressing over the whole period, has a pH of 5.8.[5] The distribution of organic matter between particle size fractions from these soils has been reported elsewhere[5] and the samples used in this study were those from the 0-6 cm depths (0-3 cm and 3-6 cm samples added together in the appropriate proportions) reported in that paper. The soil was dispersed as completely as possible without the use of oxidizing agents, which would affect the organic matter chemistry. Mechanical dispersion methods were used and the the fractions obtained were inspected microscopically to ensure that there were no intact aggregates present. This is important when different soil samples are to be compared because aggregate stability may differ between samples and thus affect the observed particle size distribution.
CP MAS ^{13}C NMR spectra were obtained using a Bruker 300 MHz spectrometer and DRIFT spectra of the ground samples were recorded in air.

3 RESULTS AND DISCUSSION

Both sets of CP MAS ^{13}C NMR spectra (Figures 1 and 2) show similar features. The strong signals at chemical shifts of about 30 ppm, 74 ppm, 105 ppm and 173 ppm are characteristic of C in alkyl groups, C in carbohydrates and other O-alkyl groups, C_1 carbons of cellulose or hemicellulose and carbonyl C,[8,9,10,11,12] respectively. The presence of these groups is also indicated by the DRIFT spectra (Figures 3 and 4). The well-defined absorption bands at 2920 cm^{-1} and 2860 cm^{-1} indicate the presence of alkyl groups (C-H stretching), the band at 1159 cm^{-1} could be due to polysaccharides (C-O stretching) or SiO and the band at 1720 cm^{-1}[13,14] is indicative of carboxyl groups (C=O stretching).

To enable comparison of the CP MAS ^{13}C NMR spectra, they were divided into six regions. These regions corresponded to chemical shift ranges of 0-40 ppm (C in branched, straight chain and cyclic alkanes); 40-60 ppm (C in branched aliphatics, amino acids and methoxyl groups); 60-105 ppm (C in carbohydrates and aliphatic compounds containing either C bonded to OH, ether oxygens or occurring in five membered rings bonded to O); 105-150

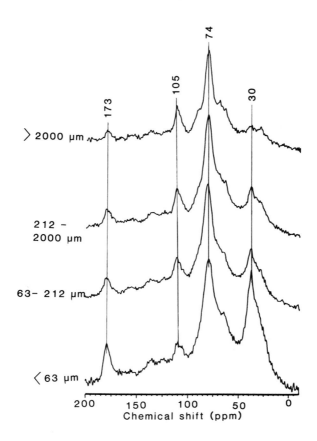

Figure 1. CP MAS ^{13}C NMR spectra for particle-size fractions from soil with mor humus

ppm (aromatic C); 150-170 ppm (phenolic C) and 170-190 ppm (carbonyl C).[11] With knowledge of the total C contents of each fraction, the sample bulk density and the particle size distributions of the soils,[5] it was possible to calculate the approximate of C in each of the chemical forms associated with each of the particle size fractions on a volumetric basis. Expression of the data in this way allowed comparisons to be made between the soils as they existed in the field.[5,7] However, accurate comparison between soil samples is only possible for ^{13}C with the same chemical shift values. The relaxation times of ^{13}C atoms with different chemical shifts will probably differ and this means that, despite the tendency of cross-polarisation to reduce this source of error and the choice of a relatively long relaxation delay, direct comparison of ^{13}C with different chemical shifts may not be completely accurate. However, the change in the relative proportions of one form of C in relation to another can still be judged.

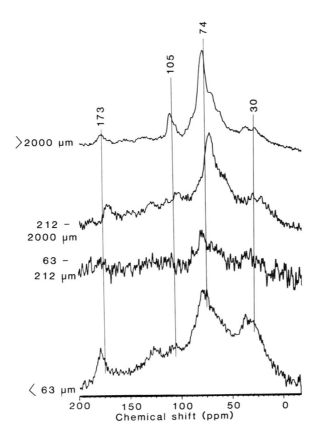

Figure 2. CP MAS ^{13}C NMR spectra for particle size fractions from soil with mull humus.

Distribution of organic matter between particle size fractions

In both soils the relative amounts of alkyl C (0-40 ppm) and carbonyl C (170-190 ppm) increased with decreasing particle size (Table 1). This trend was more readily seen in the mor humus soil (Figure 1) than in the mull humus soil (Figure 2). It is likely that the poorer definition of the CP MAS ^{13}C NMR spectra for the samples from the mull humus soil was a result of their lower C contents (Table 1). The increase in alkyl and carbonyl C with decreasing particle size is consistent with the reported aliphatic nature of organic matter associated with the clay fraction of soil.[10]

There was a marked decrease in the relative amounts of cellulose and hemicellulose C (approximately 105 ppm chemical shift) with decreasing particle size in both soils (Figures 1 and 2). This suggests that there was

Spectroscopic Characterization of Organic Matter from Soil 75

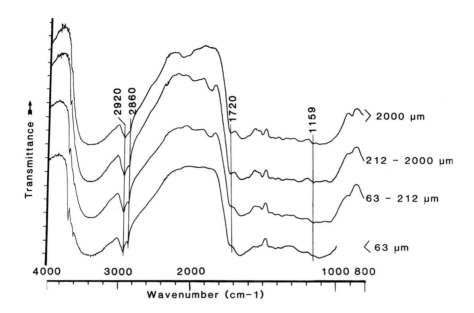

Figure 3. DRIFT spectra for particle-size fractions from soil with mor humus

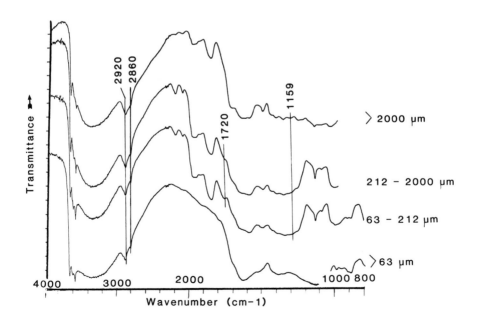

Figure 4. DRIFT spectra for particle-size fractions from soil with mull humus

Table 1. Approximate carbon contents by particle size and chemical shift

Soil fraction (μm)	Chemical shift range (ppm)						
	0-40	40-60	60-105	105-150	150-170	170-190	0-190
	(mg C ml^{-1} soil)						
Mor humus							
>2000	27	17	58	33	9.3	9.3	150
212-2000	13	8.3	26	13	4.5	6.0	71
63-212	10	5.0	18	10	3.2	3.2	49
<63	22	13	29	17	7.3	7.3	95
All fractions	72	43	130	73	24	26	370
Mull humus							
>2000	0.55	0.44	1.3	0.55	0.18	0.18	3.3
212-2000	2.1	1.4	4.3	2.4	0.68	0.85	12
63-212	7.8	5.1	14	9.7	3.2	3.9	44
<63	40	22	58	33	11	11	180
All fractions	51	29	78	46	15	16	240

Table 2. Approximate percentage of different forms of C in size fractions

Soil fraction (μm)	Chemical shift range (ppm)						
	0-40	40-60	60-105	105-150	150-170	170-190	0-190
	(%)						
Mor humus							
>2000	38	40	45	45	39	36	41
212-2000	18	19	20	18	19	23	19
63-212	14	12	14	14	13	12	13
<63	31	30	22	23	30	28	26
All fractions	100	100	100	100	100	100	100
Mull humus							
>2000	1.1	1.5	1.7	1.2	1.2	1.1	1.4
212-2000	4.1	4.8	5.5	5.2	4.5	5.3	5.0
63-212	15	18	18	21	21	24	18
<63	78	76	74	72	73	69	75
All fractions	100	100	100	100	100	100	100

a detectable amount of polysaccharide at the surface of both soils in the large size fractions, which is consistent with the presence of relatively unaltered plant litter.

Comparison of organic matter in soils with mor and mull humus

The approximate percentage of each form of C in the different particle size fractions is shown in Table 2. These data are presented so that relative accumulation or depletion of different chemical forms of C in particular size fractions can be identified by comparison of the values for the different forms of C with that for all forms of C (0-190 ppm).

These data indicate that all forms of C were relatively more abundant in the larger-sized fractions of the mor humus soil and also in the smaller-sized fractions of the mull humus soil. The low pH of the mor humus soil[15] limits the activity of the soil meso- and macro-fauna and the consequent reduction in comminution of the litter is a contributory factor in the accumulation of coarse particulate organic matter.[5] There was enhanced accumulation (relative to all forms of C) of aromatic C (105-150 ppm) in the >2000 µm particle size fraction and a corresponding depletion in the <63 µm particle size fraction in the mor humus soil. By contrast, in the mull humus soil there was no such preferential accumulation of aromatic C in the largest particle size fraction. The aromatic C (105-150 ppm) includes lignin and the data suggest that such compounds in the >2000 µm size fraction were selectively excluded from decomposition processes in the mor humus soil. Although the chemical distribution of C in the grass litter entering the different soils is not yet available, the process suggested above is consistent with the observed accumualtion of lignin in the Oe horizons of acidic grassland soils.[4]

Concluding remarks

There were strong trends of increasing aliphatic C and carbonyl C, and decreasing O-alkyl (carbohydrate) C contents of soil fractions with decreasing particle size in the mor humus soil. The same trends were present in the mull humus soil, but were not as strong. In the mor humus soil there was accumulation of all forms of organic matter compared with the mull humus soil. Furthermore, in the mor humus soil, all chemical forms of C were concentrated in the larger-sized fractions, whereas in the mull humus soil, all C forms were concentrated in the smaller-sized fractions. These findings are consistent with the low overall levels of micro- and macro-biological activity at the low pH of the mor humus soil. There was relatively greater accumulation of aromatic C, consistent with lignaceous material, in the largest size fraction of the mull humus soil, which is believed to be because this material is more recalcitrant and neither decomposes readily at the soil surface nor is comminuted and mixed in the intensely acidic mor humus soil.

ACKNOWLEDGEMENTS

The authors are grateful to Dr. J.A. Chudek and Mr. J. Anderson of the Department of Chemistry, Unversity of Dundee for considerable assistance with the spectroscopy.

REFERENCES

1. Soil Science Society of America, 'Glossary of Soil Science Terms', SSSA, Madison, Wisconsin, 1987.
2. R.L. Tate III, 'Soil Organic Matter, Biological and Ecological Effects', Wiley & Sons, New York.
3. B. Barratt, J.Soil Sci., 1964, 15, 342-356.
4. K. Shaw, PhD Thesis, University of London, 1958.
5. R.S. Shiel, J.Soil Sci., 1986, 37, 249-257.
6. D.W. Hopkins, R.S. Shiel and A.G. O'Donnell, J.Soil Sci., 1988, 39, 385-392.
7. R.S. Shiel and D.L. Rimmer, Plant and Soil, 1984, 76, 341-356.
8. M. Schnitzer and C.M. Preston, Soil Sci.Soc.Am.J., 1986, 50, 326-331.
9. M.A. Wilson, 'NMR Techniques and Applications in Geochemistry and Soil Chemistry', Pergamon Press, Oxford, 1987.
10. M. Schnitzer, J.A. Ripmeester and H. Kodama, Soil Sci., 1988, 145, 448-454.
11. M.A. Arshad, M. Schnitzer, D.A. Angers and J.A. Ripmeester, Soil Biol.Biochem., 1990, 22, 595-599.
12. R. Hempfling, F. Ziegler, W. Zech and H-R. Schulten, Z.Pflanzenernaehr.Bodenk., 1987, 150, 179-186.
13. F.J. Stevenson, 'Humus Chemistry', Wiley & Sons, New York, 1982.
14. A.U. Baes and P.R. Bloom, Soil Sci.Soc.Am.J., 1989, 53, 695-700.
15. D.W. Hopkins, D.M. Ibrahim, A.G. O'Donnell and R.S. Shiel, Plant Soil, 1990, 124, 79-85.

Influence of Soil Management Practices on the Organic Matter Structure and the Biochemical Turnover of Plant Residues

K. Haider[1], F.-F. Gröblinghoff[1], T. Beck[2], H.-R. Schulten[3], R. Hempfling[3], and H.D. Lüdermann[4]

[1] INSTITUTE OF PLANT NUTRITION AND SOIL SCIENCE, FEDERAL AGRICULTURAL RESEARCH CENTRE (FAL), D-3300 BRAUNSCHWEIG, GERMANY
[2] BAYERISCHE LANDESANSTALT FÜR BODENKULTUR UND PFLANZENBAU, D-8000 MÜNCHEN, GERMANY
[3] FACHHOCHSCHULE FRESENIUS, ABTEILUNG SPURENANALYTIK, DAMBACHTAL 20, D-6200 WIESBADEN, GERMANY
[4] INSTITUT FÜR BIOPHYSIK UND PHSYIKALISCHE BIOCHEMIE, UNIVERSITÄT REGENSBURG, GERMANY

1 INTRODUCTION

Soil humus is in a steady state equilibrium of formation and degradation. Humus contents can be shifted towards increase or decrease by climatic alterations, but also by soil cultivation or managment conditions. Soil organic matter as a critical component in soil productivity and fertility of croplands under defined climatic conditions can therefore best be sustained by improving management. There are, however, still severe deficiencies in our knowledge about the impacts of different kinds of management.

The influence of different soil management practices can best be investigated in croplands under continuous long term management conditions. Here, carbon and nitrogen contents, as well as microbial biomass and activities, turnover and humification rates of plant and organic residues are likely to be approaching equilibrium values[1]. For the present studies we have selected two sites which are kept by the Bavarian Provincial Agricultural Institute under long term continuous management. One is located at Puch near Fürstenfeldbruck (Southern Bavaria) and was established by Springer in 1949[2]. It is now under uniform continuous management since more than 35 years[3]. Several plots are under extreme management, including bare fallow, continuous monocultures of potatoes or wheat with and without additions of manure, and continuous grassland; other plots are under crop rotations with diverse organic and inorganic fertilization.

While the Puch site is more experimentally oriented, the second site at Neuhof near Donauwörth (Franconian Alb) is more related to agricultural practice. It was started in 1977 and is characterized by graded intensities of managements. These range from plots oriented at the requirements of 'organic' farming, a mixed organic/inorganic fertilization, an extensive integrated inorganic fertilization, to a very intense inorganic fertilized system comparable to modern industrial agriculture without livestock and preventive pesticide applications. The purpose in establishing this site was to characterize influences of different agricultural managements upon expenditures and returns as well as to recognize long term impacts of modern agriculture upon soil fertility and productivity[4].

2   IMPACTS OF SOIL MANAGEMENT UPON BIOCHEMICAL TURNOVER RATES OF ORGANIC RESIDUES

From 1987 - 1989 from each plot 400 soil cores (20 cm depth) were taken in spring before fertilization and in fall after harvest. They were carefully mixed and analyzed at field moist conditions. Due to the extreme differences in management at the Puch site, soil C- and N-contents as well as microbial biomass differed considerably (Table 1). C- and N-contents decreased

TAB. 1: MANAGEMENT TYPES OF THE PLOTS FROM PUCH

(Luvisol Derived From Loess Loam)

Abbreviation	Management	C_{org} (%)	N_t (%)	Biomass (mgC/100g)	
				Spring 87	Spring 88
1 BF	bare fallow	0.77	0.09	11	7
2 PM(−)	potatoe monoc. N-P-K	0.96	0.12	22	20
3 PM(+)	potatoe monoc. N-P-K + stable manure	1.04	0.13	22	22
	crop rotations[1]				
4 CR_1	barley, N-P-K	1.19	0.15	30	29
5 CR_2	winter wheat N-P-K, green manure	1.25	0.16	32	31
6 CR_3	potatoes N-P-K + stable manure	1.33	0.15	35	34
7 GL	contin. grassland	1.76	0.20	97	113
	C_t 1956	1.35 (see Diez and Bachthaler[3])			

[1] actual crops for 1988; more exact data about crop rotation and fertilisation see: Diez and Bachthaler[3]

by 50 % from the 1956 values in the bare fallow or the monoculture plots, and increased by about 10 % in the continuous grassland. Very similar biomass determinations in 1987 and 1988 by Beck[5] indicated that biological conditions in these plots approached equilibrium values.

TAB. 2: MANAGEMENT TYPES OF THE PLOTS FROM NEUHOF

(Gleysol Derived From Loess Loam)

Abbreviation	Management	C_{org} (%)	N_t (%)	Biomass (mgC/100g) Spring 87	Spring 88
8 CR/N_3 intense	crop rotation[2]) 182 kg/ha N,P-K straw + crop resid.	1.12	0.15	33	33
9 CR/N_2 integr.	crop rotation[2]) 124 kg/ha N,P-K straw + green man.	1.19	0.15	37	33
10 CR/N_1 org.min.	crop rotation[1]) 120 dt/ha manure + 37 kg/ha N, P-K	1.28	0.14	60	42
11 CR/N_0 org.	crop rotation[1]) 180 dt/ha manure + P-K	1.29	0.15	68	51
	C_t 1979	1.25	(see Kraus[4])		

[1]) crop rotation: summer barley, potatoes, winter wheat, oat, clover

[2]) crop rotation: summer barley, potatoes, winter wheat, oat, sugar beet

Soil C- and N-contents at the Neuhof site are more equal (Tab. 2) with a decreasing tendency in intensities of management. Biomass contents, however, decreased strongly, but the nearly equal biomasses in both 1987 and 1988 show again that the long term management resulted equilibrium in biological conditions. According to Beck[6] these lower biomass contents also are expressed by less enzymatic activities paralleled by an increase of the 'metabolic quotient', which means that more CO_2 was released per unit of biomass during incubation. A higher metabolic quotient expresses a faster degradation of the indigenous biomass and a faster decrease in the nutrient sources available to this biomass.

These alterations in the quality of biomass are also indicated in Table 3 and 4 where soil samples from Puch and Neuhof were incubated in the laboratory with uniformly ^{14}C-labeled wheat straw. Whereas the CO_2 release from soil samples of Puch decreased significantly with less soil C[7], the $^{14}CO_2$ release from the straw was rather equal. This is shown for selected soil samples from Puch in Table 3. The incorporation of the residual straw C into biomass after a 52 day incubation was determined by the $CHCl_3$-fumigation method[8] and showed a strong decrease with decreasing C- and biomass-contents. A decrease also was observed if the ^{14}C in biomass was related to the total metabolized ^{14}C. While 23 - 15 % of the metabolized ^{14}C was assimilated

TAB. 3: MINERALIZATION AND ASSIMILATION OF ^{14}C FROM UNIFORMLY ^{14}C-LABELED WHEAT STRAW, PUCH, MEANS FROM SPRING 1988 AND 1989

Soil samples with 0.1% straw (37.9% C) incubated 52 days at 25°C and 55% WHC

Management	%$^{14}CO_2$-C released	%^{14}C-assim in Biomass[1]	%$^{14}C_{ass}$ of C_{met} [2]	µg^{14}C-rel./ 100mg Biom. C
grassland	44.5	8.3	15.6	24.2
crop rotation winter wheat	43.1	6.3	12.9	70.9
potato monoc. + N-P-K	44.5	3.7	7.6	120.6
bare fallow	43.0	3.1	6.5	289.9
LSD 5%	2.3	1.5	2.1	

[1] Biomass values from table 1, spring 1988
[2] $C_{metabol} = C_{assim} + CO_2 - C_{released}$

by the biomasses of the grassland or crop rotation plots, it amounted to only 7 - 8 % for the monocultures or the bare fallow. Reasons for these observations can be either a faster degradation of the newly synthesized biomass after straw addition to the C-poor soils or their less efficient utilization of the substrate carbon. The latter reason appears probable by the values in the

TAB. 4: MINERALIZATION AND ASSIMILATION OF 14C FROM UNIFORMLY 14C-LABELED WHEAT STRAW, NEUHOF, MEANS FROM SPRING 1988 AND 89

Soil samples from plot 5 with 0.1% straw (37.9% C) incubated 52 days at 25°C and 55% WHC

Management	%$^{14}CO_2$-C released	%^{14}C-assim in Biomass[1]	%$^{14}C_{assim}$ of C_{metab} [2]	µg^{14}C-Releas./ 100mg Biom. C
1 org.	46.4	9.6	17.5	349
2 org.min.	45.8	8.4	15.5	414
3 integr.	48.5	8.4	14.8	550
4 intense	49.6	6.8	12.1	573
LSD 5%	1.5	1.1	1.9	

[1] Biomass values from table 2, spring 1988
[2] $C_{metabol} = C_{assim} + CO_2 - C_{released}$

last column of Table 3. They indicate a higher $^{14}CO_2$-release per unit of biomass in the C-poor soils. Domsch and Anderson[9] reported that glucose carbon was more efficiently incorporated into the biomass of various arable soils under crop rotations than under monocultures. That growth yields from a substrate applied to soil depend significantly upon climate and management factors was also reported by Coody et al.[10]. Holland and Coleman[11] suppose a different efficiency for C-assimilation by fungal or by bacterial populations.

Although the soil samples from the Neuhof site differ only slightly in C- and N-contents, they also show a similar tendency during incubation with ^{14}C-labeled straw (Table 4). The $^{14}CO_2$-releases during the 52 day incubation show a slight and insignificant decrease, but incorporation of ^{14}C into biomass decreases strongly with increasing intensities of management. A similar decrease could also be registered if assimilated ^{14}C was related to total metabolized carbon. Again ^{14}C-mineralization related to biomass carbon increased with increasing intensity of management. As with the Puch soils, a less efficient carbon utilization by the smaller biomass contents in the intensely managed plots can be suspected.

In the samples from both Puch and Neuhof lignin mineralization rates showed a significant dependency on management treatment. This is shown in Fig. 1 where

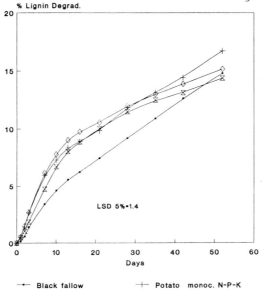

Figure 1 % $^{14}CO_2$-C release from corn stalk material ^{14}C-ring-labeled in the lignin portion, Puch, spring 1988 Incubation 52 d, 25 °C, 55 % WHC

Puch samples were incubated with corn straw material, specifically ^{14}C-labeled in the ring moieties of its lignin portion[12]. Samples from the bare fallow or potato monoculture plots released $^{14}CO_2$ from lignin more steadily than those from plots richer in soil C. Prolonged incubation, therefore, should lead to a greater mineralization. This tendency of a higher mineralization of lignin was also observed by the intense managed Neuhof samples during incubation[13]. Since lignin is an important source for humus formation[14,15] it is probable that intense management conditions leading to higher lignin mineralization result in less incorporation of modified lignin into humus. Powlson et al.[16] as well as Allison and Kilham[17] also reported that soils with a regular supply of straw contained a greater portion of ligninolytic fungi within their microflora thus leading to an enhanced straw mineralization.

At the Neuhof site, straw from the manure fertilized organically managed plots is regularly removed and only stubbles and leaves from tuber crops remain in the field. In the mineral fertilized and intensively cropped plots, crop residues are added to soil together with green manure from occasional intercroppings. Therefore, annual carbon inputs into the differently managed field plots always averaged around 3 Mg C per year (Table 5). From these annual inputs we estimated the

TAB. 5: MANAGEMENT SYSTEM NEUHOF: AVERAGE YEARLY C-INPUTS

($Mg\ C\ ha^{-1}\ a^{-1}$)

Management	Green Manure Crop Resid.	Straw Stubbles Roots	Stable Manure	Σ
1 org.	0.3	1.2	1.8	3.3
2 org. min.	0.3	1.4	1.2	2.9
3 integr.	0.5	2.7	-	3.2
4 intense	0.7	2.8	-	3.5

amounts of 'young' humus C formed during the last 10 years of continuous management[13]. The mineralisation rates of the various carbon amendments to soil were estimated by data from Kolenbrander[18] or Janssen[19]. In one year, additions of stable manure, straw or green manure were decomposed to 40, 60 and 80 %, respectively. Using those figures we estimated the 'young' humus portion remaining after 10 years of continuous management[19] (Fig. 2). They amounted to 8 Mg C · ha^{-1} in the organic fertilized plots and about 5 Mg C · C ha^{-1}

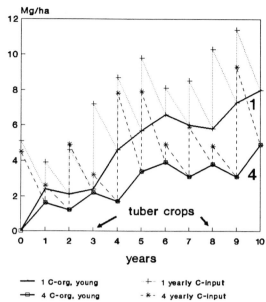

Figure 2   Accumulation of the 'young' humus portion
Neuhof, organic (1) and intense (4) managements

in the mineral fertilized plots. Considering an annual mineralization rate of the parent organic matter of about 1.5 % · y^{-1} 20, decreases or increases in humus carbon contents were calculated and compared with those determined by the actual C-contents (Table 6). This shows that the organic managed plots increased to about 1 - 2 Mg C · ha^{-1}, which is in a reasonable agreement with the humus determined by the actual soil carbon. Soil organic C in the integrated or intense managed plots should either remain constant or decrease by

TAB. 6: INCREASES OR DECREASES IN HUMUS CONTENTS AT NEUHOF DURING THE LAST TEN YEARS OF CONTINOUS MANAGEMENT

Calculated from annual C-inputs, mineralization rates and the actual humus contents observed in 1989; C_t 1979, 1.25%

Management	Calculated	Observed
	Mg C·ha^{-1}	
1 org.	+2	+2
2 org. min.	+1	+1
3 integr.	0	-2
4 intense	-2	-5

2 Mg C · ha^{-1}. The actual humus contents in the intense managed plots, however, decreased more strongly than this has been estimated by the carbon inputs. These higher C-losses probably are due to an accelerated mineralization of straw and crop residues by the specific microflora of the intense managed and inorganic fertilized plots. This agrees with observations by Franken[21] from the continuous management system at the Dikopshof near Bonn which show that application of crop residues together with intense inorganic fertilization could not maintain a steady humus equilibrium in the soil.

3 IMPACTS OF SOIL MANAGEMENT UPON HUMUS STRUCTURE

Structure of soil humus from the various plots was investigated by ^{13}C NMR-spectrometry or by field-ionization-mass-spectroscopy (FI-MS)[22]. By these methods either soil samples or the humus fractions extracted with NaOH were analyzed. ^{13}C NMR-spectra from both soil or humus samples from CP-MAS spectra or from solution spectra in NaOD were largely similar for each of the different managed systems. Generally, however, an integration of the distinct area revealed apparent differences in the aromatic and phenolic regions. This is shown in Table 7 from which a relative increase of the aromatic (110 - 140 ppm) and the phenolic regions (140 - 160 ppm) with increasing carbon contents can be recognized. The aliphatic as well as the C-O/C-N region showed a reverse tendency but not as straight as this from the aromatic and phenolic regions. A report by

TAB. 7: INTEGRATION OF THE CP-MAS ^{13}C-NMR RESONANCES OF HUMIC SUBSTANCES BETWEEN 110-140 AND 140-160 PPM

samples 1-7 from Puch 1988, 8-11 from Neuhof 1988

(assignments see tables 1 and 2)

Sample	Soil C_{org} %C	110-140 ppm Arom. Region	140-160 ppm Phenol. Region
		% of total area	
1 BF	0.70	10.4	4.0
2 PM(-)	0.92	11.1	4.6
3 PM(+)	0.97	12.9	5.4
4 CR_1	1.19	11.8	5.0
5 CR_2	1.25	14.2	6.3
6 CR_3	1.33	13.5	6.6
7 GL	1.76	14.5	6.6
8 CR/N_3	1.12	13.6	5.8
9 CR/N_2	1.18	12.4	5.2
10 CR/N_1	1.18	15.4	6.5
11 CR/N_0	1.27	15.4	6.7

Skjemstad et al.[23] from investigations of cultivation effects upon several soils of Australia by means of ^{13}C NMR spectroscopy indicated also a relative decrease of the phenolic region in intensively managed soils. These results indicate that the polyphenolic structure of humic compounds is less stable than was formerly suspected[24].

Pyrolysis mass spectrometry additionally showed that soil samples from differently managed plots could be distinguished from each other. The pyrolysis spectra generally revealed a similar pattern of signals in the mass range between m/z 50 - 500, with less numerous signals from the soil samples of the carbon poor plots (bare fallow monocultures) of the Puch site. By computer analysis, discriminating signals differing in signal heights could be registered. These signals were mostly coordinated to fragments from polysaccharides, lignin, proteins or other N-containing compounds. They are generally positively correlated to increases in C-contents and to a less intense management[22]. Fig. 3 shows an example, where signal intensities from polysaccharide-fragments (m/z 72, 74, 84, 85, 94, 110, 112, 114, 126, 132, 144, 162) are well correlated with increases in C_t-contents of the Puch and Neuhof plots. The results also allow the conclusion that polysaccharides were depleted faster than the total humus due to more intense management. Besides the decrease in the polysaccharide contents, a decrease of lipids in the intense managed soil samples seems to be a reason for the observed diminished aggregate stability in soil samples of intense managed fields[25].

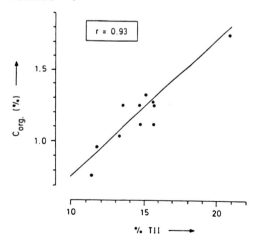

Figure 3   Correlation between soil organic carbon and the summed ion intensities (TII) of field ionization signals related to carbohydrates for the soil samples from Puch and Neuhof 87

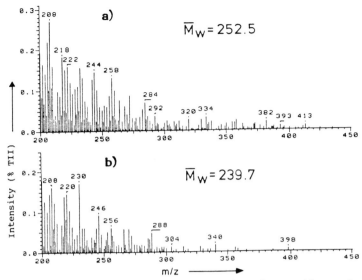

Figure 4   Integrated PY-FI mass spectra from soil samples Neuhof a) organic/mineral fertilization, b) integrated fertilization, mass signals m/z 200 - 400

An influence of management conditions could even be observed by pyrolysis mass spectra from samples of the Neuhof site where total C- and N-contents were nearly equal. This is shown in Fig. 4 for the plots with a mixed organic/inorganic fertilization (spectrum a) and a low dose of mineral fertilizer, respectively (spectrum b). The spectrum a) shows more and higher signals in the mass range from m/z 200 - 400 than spectrum b). In consequence, the mean molecular weight of the fragments from soil a) was higher ($\bar{M}_w$ = 253) than that from soil b) ($\bar{M}_w$ = 240). As indicated above, soil management influenced the degradation rates of plant residue materials including their lignin portion. Beside differences in quality of the added residues, alterations in degradation of plant residues also might be a reason for the observed qualitative differences in the soil humus developed under specific management conditions[26].

Pyrolysis-mass-spectroscopy also revealed differences in N-containing fragments under different managements. Fragments from amino acids or proteins were generally positively correlated to increasing N-contents of the soil samples. In contrast, however, fragments from heterocyclic N-compounds were negatively correlated. Fig. 5 shows that the relative contributions from pyridine or pyrrole increased with decreasing N-contents of the soil samples. Here the relative contributions of these fragments to the total ion current were plotted against the N contents in soils and resulted in a negative correlation of more than 90 %.

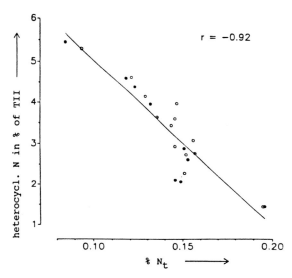

Figure 5 Correlation between soil N_t and relative contributions of heterocyclic nitrogen compounds (pyrrole, pyridine) pyrolysis-MS data for the sites Puch and Neuhof
o = 1987, ● = 1988

Since heterocyclic N-components in soil humus are probably not readily mineralized, this indicated that the easily mineralizable N-portion becomes smaller with decreasing N-contents. In samples from Puch, with significant differences in total N-contents, the potentially mineralizable N[27] decreased with decreasing soil N and microbial biomass[7]. As indicated by Table 8, the samples from Neuhof with similar N-contents, however, also showed decreases in the potentially mineralizable N-portion with increased management. In a similar manner also the N-mobilisation-immobilisation-turnover (N-MIT) decreased. For this purpose, a small amount of highly ^{15}N-enriched ammonia was added to each of the soil samples at zero time and the incorporation into biomass as well as the ^{15}N-dilution in ammonia and NO_3^--N was determined after several time intervals[28,29].

Both the potentially mineralizable N and the N-MIT rates are indicators for the capacity of a soil to supply nitrogen to crops and for the N-buffering capacities of such soils. Janssen[19] and Doran and Smith[30] reported from plots with only mineral or with additional organic fertilization and intermediate ley farming that after 25 years of continuous management total N-contents were similar. Potentially mineralizable N, however, was more tightly connected to the contents of 'young and active' humus than to total N. Also recently published investigations by Follett and Schimel[31] about the influence of soil cultivation upon microbial biomass and N-cyclation

TAB. 8: POTENTIALLY MINERALIZABLE N_{MIN} AND N-MOBILISATION-IMMOBILISATION TURNOVER (N-MIT) IN SOIL SAMPLES NEUHOF, SPRING 1987

Management	N_t mg/g	N_{min}[1] µg/g Soil	% N_{min} of N_t	%N-MIT of N_t[2]
1 org.	1.5	95.3	6.4	5.5
2 org. min.	1.4	89.6	6.4	5.4
3 integr.	1.5	73.8	4.9	5.0
4 intense	1.5	58.3	3.9	4.3

[1] Potentially mineralizable N_{min} determined according to Stanford and Smith[27]
[2] % N-MIT see: Gröblinghoff et al.[7]

showed that increased tillage intensities diminished both the microbial biomass and the capacity of the soil to immobilize and conserve mineral N.

4 CONCLUSIONS

In conclusion the experiments conducted by us or other authors about the impacts of management reveal that management conditions should not be considered only by a diminished and better timed addition of fertilizer, but also by their effects to improve the biological conditions of a soil. A better understanding of these effects with respect to their influences upon soil life, as well as on internal nutrient and carbon flows are, therefore, important requisites for an improvement of food, feed and fiber production and to reduce expenditures and negative effects of the agricultural production systems in future.

REFERENCES

1. T.-H. Anderson and K. Domsch, Soil Biol. Biochem., 1989, 21, 471.
2. H. Schaeffler and U. Springer, Bayer. Landw. Jahrb., 1957, 34, 131.
3. T. Diez and G. Bachthaler, Bayer. Landw. Jahrb., 1978, 55, 368.
4. A. Kraus, Bayer. Landw. Jahrb., 1984, 61, 6.
5. T. Beck, VDLUFA-Schriftenreihe, 1989, 28, 879.
6. T. Beck, Kali-Briefe (Büntehof), 1990, 20, 17.
7. F.-F. Gröblinghoff, K. Haider and T. Beck, VDLUFA-Schriftenreihe, 1989, 28, 893.
8. D.S. Jenkinson and D.S. Powlson, Soil Biol. Biochem., 1976, 8, 167.
9. K. Domsch and T.-H. Anderson, Abschlußbericht des DFG-Forschungsvorhabens Do 41/34-1, 1987.

10. P.N. Coody, L.E. Sommers and D.W. Nelson, Soil Biol. Biochem., 1986, 18, 283.
11. E.A. Holland and D.C. Coleman, Ecology, 1987, 68, 425.
12. K. Haider and J. Trojanowski, In: T.K. Kirk et al. (eds.): Lignin Biodegradation Vol. I, CRC-Press, Boca Raton Fl. USA, 1980, p. 111.
13. F.-F. Gröblinhoff, K. Haider and T. Beck, Mitteilgn. Dtsch. Bodenkundl. Gesellsch., 1989, 59/I, 563.
14. K. Haider and F. Azam, Z. Pflanzenernähr. Bodenk., 1983, 146, 151.
15. W.J. Parton, D.S. Schimel, C.V. Cole and D.S. Ojima, Soil Sci. Soc. Amer. J., 1987, 51, 1173.
16. D.S. Powlson, P.C. Brookes and B.T. Christensen, Soil Biol. Biochem., 1987, 19, 159.
17. N.E. Allison and K. Kilham, J. Soil Sci., 1988, 39, 237.
18. G.J. Kolenbrander, Trans. Xth Int. Congr. Soil Sci., Vol. 2, 1974, 129.
19. B.H. Janssen, Plant Soil, 1984, 76, 297.
20. D.S. Jenkinson and J.H.Rayner, Soil Sci., 1977, 123, 298.
21. H. Franken, Landw. Forschung, Kongreßband, 1985, 38, 19.
22. H.-R. Schulten, R. Hempfling, K. Haider, F.-F. Gröblinghoff, H.-D.Lüdemann and R. Fründ, Z. Pflanzenernähr. Bodenk., 1990, 153, 97.
23. J.O. Skjemstad, R.C. Dalal and P.F. Barron, Soil Sci. Soc. Am. J., 1986, 50, 354.
24. R.D. Haworth, Soil Sci., 1971, 111, 71.
25. P. Capriel, T. Beck, H. Borchert and R. Härter, Soil Sci. Soc. Am. J., 1990, 54, 415.
26. R. Hempfling, H.-R. Schulten and K. Haider, Mitteilgn. Dtsch. Bodenkundl. Gesellsch., 1989, 59/I, 575.
27. G. Stanford and S.M. Smith, Soil Sci., 1976, 126, 210.
28. D.S. Schimel, Biogeochem., 1986, 2, 345.
29. D.D. Myrold and J.M. Tiedje, Soil Biol. Biochem., 1986, 18, 559.
30. J.W. Doran and M.S. Smith, In: J.J. Mortvedt and D.R. Buxton (eds.): Soil Fertility and Organic Matter as Critical Components of Production Systems, Soil Sci. Soc. Amer., Spec. Publ. 19, Madison, Wisconsin, 1987, p. 53.
31. R.F. Follett and D.S. Schimel, Soil Sci. Soc. Amer. J., 1989, 53, 1091.

Section 2

Soil Organic Matter and Water Quality

Introductory Comments

by H.A. Anderson, Macaulay Land Use Research Institute

Soil organic matter (SOM) is capable of retaining solutes by a variety of sorption mechanisms, and in soil waters dissolved organic matter (DOM) is also capable of retaining extrinsic chemicals added during various land use practices. These co-solute interactions can lead to the transport into surface and ground waters of many pollutant chemicals of low water solubility, which would not be expected to be mobile in soils.

Anderson, Stewart, Miller and Hepburn contrast plant uptake of mineral N with its immobilization in soil organic matter, and the release of both mineral and organic N into water. Fertilizer N applied to a potato crop is partially immobilized in humic substances, mineralization then being dependent on drying and wetting cycles in the soil. In forest soils, N losses into surface waters are predominantly in non-humified organic forms. After felling, it is well-known that N is lost as nitrate; this appears to be a series of point-source losses, rather than a general catchment-wide feature. Relationships between the disturbance of organic soils and these point sources of nitrate is unclear, and is complicated by the relative unimportance of nitrate in drainage outputs from recently-afforested deep peat.

In the same vein, Goss, Williams and Howse highlight dissimilar rates of N-mineralization from different crop residues, dependent on plant species, and from contrasting management practices. Turnover is reduced by a growing crop, possibly due to the release of root exudates. In upland soils, particular importance is attached to the immobilization of applied N by SOM, a serious problem in pasture improvement, where liming and reseeding can increase SOM turnover rates, nullifying the effects of added fertilizer-N. In forest soils, enhanced N-mineralization under mixed species may be related to the activities of enchytraeid worms. Nutrient release following forest harvesting is related to the alteration in SOM turnover, allowing the accumulation of potentially-leachable mineral-N.

Nelson, Oades and Cotsaris have examined the release of dissolved organic matter (DOM) into streams. DOM is important in both the ecology of surface waters and in their potential for potable water supplies. In addition to mineralization releasing nutrients which lead to eutrophication, disinfection practices can lead to toxic products, e.g. trihalomethanes, from humic precursors. Land use practices have been blamed for high DOM in certain catchments in Australia, but Nelson et al. have shown that catchment outputs depend rather on properties of the soils related to parent geology. DOM concentrations in streams and clay contents of catchment soils show an inverse relationship, complicated by hydrology, where clay B horizons could either be permeable and thus filter out DOM or impermeable, causing flow through sandy A horizons incapable of retaining organic matter.

Recently, the importance of various novel mechanisms has been advanced concerning the partition of synthetic organic molecules between solid and liquid phases in soils. In a consideration of these theories, especially with regard to non-ionic pollutants and pesticides, Malcolm discussed the work of Chiou and his co-workers[1]. Retardation of these molecules by SOM is governed by the size and charge on both the humic substance and the solute molecule, the water solubility of the latter, the physical and chemical composition of the soil matrix, and the kinetics of interactions between the various phases. Malcolm discussed the relevant sorption mechanisms and their relative importance in the retention of pollutants. Dissimilarities in composition between bulk SOM and DOM were stressed, but similar sorption mechanisms can lead to significant transport of pollutants into surface and ground waters. Chemical partitioning can lead to the movement of supposedly insoluble components within the bulk matrix, by mechanisms distinct from those involved in chemisorption. Since this work has appeared in detail in several recent publications, it is felt that a formal presentation in the present volume would constitute unnecessary duplication. Interested readers are referred to the bibliography contained in a recent review article.

Kögel-Knabner, Deschauer and Knabner invoke a three compartment model (bulk soil - soil water - DOM) in discussing the behaviour in soil of 3-methyl-7-chloro-8-carboxy-quinoline, a new acidic herbicide. The behaviour of DOM and the herbicide depended upon the cation composition of the soil solution; composition and pH exerted major influences on sorption processes, both in the bulk and the DOM phases. Effective sorption isotherms, describing interactions with the soil matrix, can be calculated which take into account sorption both to the bulk soil and to DOM.

Even where pesticides have been used in the officially-approved manner, subsequent release into water-courses is not infrequent. Harrod, Carter and Hollis describe an investigation into the accumulation of dieldrin in eels in Cornish streams. Although erosion control in agriculture has been aided by increased aggregate stability due to enhanced SOM contents, such increases also lead to greater amounts of the insecticide aldrin being retained by sorption to SOM. Comparison of the colluvial deposits with source soils has shown higher SOM, clay and/or silt contents in the former, reflecting either depositional conditions or preferential erosion. This leads to concentrations of aldrin, and the derived dieldrin, significantly larger than those applied on the field.

The areas of research covered in these papers serve to stress the importance of the need to adopt a holistic approach to studying soil-water interactions, especially where interactions are influenced by both soluble and bulk phases. Recent advances in our knowledge of soil organic matter allow a better understanding of the reaction rates and mechanisms which are governing processes resulting from current land use practices.

1. C.T. Chiou. "Roles of organic matter, minerals and moisture in sorption of nonionic compounds and pesticides in soil", in "Humic Substances in Soil and Crop Sciences; Selected Readings" (P. MacCarthy, C.E. Clapp, R.L. Malcolm and P.R. Bloom, eds.), Amer. Soc. Agron. and Soil Sci. Soc. Amer., Madison, 1990, 111-159.

Organic Nitrogen in Soils and Associated Surface Waters

H.A. Anderson, M. Stewart, J.D. Miller, and A. Hepburn
DIVISION OF SOILS AND SOIL MICROBIOLOGY, MACAULAY LAND USE RESEARCH INSTITUTE, CRAIGIEBUCKLER, ABERDEEN, AB9 2QJ, UK

1 INTRODUCTION

The pivotal role of nitrogen in plant nutrition has been highlighted in numerous publications and the interplay between "plant-available" and "immobilized" forms is now understood more clearly.[1] In global terms, the chemistry of nitrogen regulates the biosynthesis of cellular amides and the heterocyclic compounds which, in the form of enzymes and nucleic acids, then control the production of major structural components such as proteins, polysaccharides and lignins.

All the biological forms of organic N have been shown to be present in soils and are regarded as components of the soil N cycle, with its various compartments or pools of organic and mineral N. The N cycle is unquestionably "leaky", and the aim of this paper is to use specific field-based examples to illustrate several of the points of leakage from such a cycle, with special regard to the forms of soil-N involved and to the processes giving rise to them.

Humified and Non-humified Soil Organic Nitrogen. Organic materials in soils and waters are conveniently categorized as humic and non-humic substances, by operational definitions. Humic substances are organic materials derived principally from decaying plant remains,[2] but with the normal plant components considerably altered by the soil animal and microbial populations. Abiological chemical reactions are also possible, the entire process giving rise to a complex mixture of macromolecules whose composition is site-dependent, especially on vegetation and pedogenic processes.[3]

Fractionation of Humic Substances. Since Achard's first extraction of humic substances in 1786, the use of different isolation and fractionation techniques led to considerable confusion until, as recently as 1982, the International Humic Substances Society proposed related methodologies for the extraction and fractionation of humic substances from both soils and waters.[3] These methods have become generally accepted, and have been used in the work described here. Briefly, soils are extracted with alkali under nitrogen and the acidification of the clarified extract precipitates the soil humic acid; acidification of filtered water samples precipitates the aquatic analogue. The soluble "fulvic acid fraction" in either instance contains both humic and non-humic substances, and these are separated by hydrophobic sorption of the humified "fulvic acid" per se, from a hydrophilic acid fraction which contains most of the soluble mineral components and non-humified organic materials, especially those which include polysaccharides and proteins and/or peptides.

Such a procedure allows better control of the fractionation when compared with dialysis which has been used in the past, and recovers more N.[4] This last point is critical where attempts are being made to detail pathways and quantify processes, and where losses must be minimised.

Immobilized N and Humic Substances. The numerically-insignificant fraction of total global N which embodies biosphere N can be sub-divided further (Table 1), with so-called "dead organic matter" being the major sub-component.[5,6] Hydrolysis of complete soils merely integrates the products from various components, organic and mineral, living and dead. However, hydrolysis of isolated humic and non-humic fractions gives broadly similar results, suggesting that the easily-characterized products arise from co-extracted materials which are evenly distributed, possibly as a consequence of the aggressive reagents used in the extraction process, and also of the absorptive capacity, shape and surface activity of the humic macromolecules.[10] There remains a large unidentified organic N fraction, which possibly results from interactions between amino acids and phenols[7] or sugars;[8] the effect of strong alkali on these reactions is largely unknown.

These aggressive reagents can be largely avoided where the humic substances are already soluble or dispersed in drainage and

Table 1. Percentage distribution of forms of N from global, through terrestrial contents to organic fractions in acid hydrolysates.[4,5,8]

% Distribution of

	Global N		Terrestrial Biosphere N	
	Lithosphere	82.6	Plant	1.32
	Atmosphere	17.4	Animal	0.02
	Biosphere	0.01	Inorganic	15.4
			Dead organic	83.3

% Distribution in 6M HCl hydrolysates*

	Arable Soils	Humic Acid	Fulvic Acid	Hydro. Acid
1)	15±6	12±3	6±2	15±8
2)	19±6	30±10	38±10	18±5
3)	21±5	20±5	18±6	50±10
4)	40±7	40±10	35±5	35±15
5)	7±2	5±3	3±2	45±15
6)	0.7±0.3	0.7±0.3	1±0.8	ND

* Humic, fulvic and hydrophilic acids extracted and fractionated by the I.H.S.S. procedure.[3] Forms of N - 1) Unhydrolysed, 2) Hydrolysed, unidentified, 3) Ammonium, 4) Amino acid, 5) Amino-sugar, 6) Nucleic acid. ND = not determined.

surface waters. Although alkaline solutions are used at later stages of the IHSS water fractionation procedure,[3] the non-humified fractions have already been removed. Comparisons can then be made between the aquatic substances and their contiguous soil counterparts.

One possible means of circumventing these problems is by using ^{15}N-enriched mineral N as a fertilizer, or ^{15}N-enriched organic N as a probe, and following the fate of the heavy isotope in a soil plant system, in classical isotopic tracing methods.

Isoptic-enrichment techniques are less useful in aquatic ecosystems, mainly for reasons of cost, but recent advances in isotope analysis have allowed detection of differences at natural abundance levels, permitting the full flexibility of field experimentation to be realized. An aid to the analysis of many soil-water systems is the existence of a slight ^{15}N enrichment due to selective loss of the lighter isotope during bacterial denitrification.[11]

While laboratory incubations and glasshouse pot experiments, under strictly-controlled conditions, can be of great value in mechanistic studies using labelling techniques, field experimentation is likely to provide more realistic data. Such a field experiment will be discussed below.

2 IMMOBILIZATION AND MINERALIZATION OF N

In a recent field study of N-partitioning within the potato plant in relation to N-supply,[9] ^{15}N-labelled fertilizer was applied as calcium nitrate to a crop growing in a sandy loam Countesswells Series soil. Soil organic fractions were isolated after exchangeable ammonium and nitrate had been extracted, and the patterns of isotope immobilization and mineralization were followed prior to fertilizer application (at natural abundance levels of ^{15}N), through the planting, emergence and growth of the crop. Prior to planting, the natural ^{15}N-enrichment data showed that no meaningful isotopic fractionation had occurred in the top 30 cm of the soil profile, the isotopic distribution being the same as that of the total N contents of the extracted humic, fulvic and hydrophilic acid fractions (Table 2). The peaks and troughs shown in the ^{15}N distribution after fertilizer application (Figure 1) were readily interpreted in terms of both plant growth

Table 2. Soil profile nitrogen distribution (%) in humic substance fractions from potato ^{15}N field experiment; upper line - total N, lower - ^{15}N

Horizon	Humic Acid	Fulvic Acid	Hydro. Acid
Ap	24.9±0.8	3.5±0.3	3.7±0.4
	23.8±1.3	3.7±0.5	3.8±0.3
B$_2$	9.0±3.5	4.5±2.2	6.9±3.5
	18.9±4.8	7.4±2.5	11.0±3.7

AP n = 32; B$_2$ n = 15

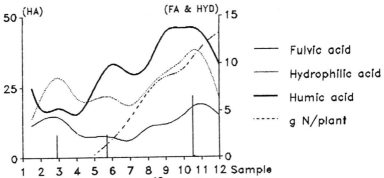

FIGURE 1 % DISTRIBUTION OF ^{15}N IN 0-5 CM SOIL SAMPLING AT 14-DAY INTERVALS, PLANTING AT #2, EMERGENCE AT #4; VERT. BARS – RELATIVE RAINFALL

and soil moisture influencing N-immobilization and -mineralization.

Much argument has centred on the relative age of the humic and fulvic acids. In the data shown here, the greater incorporation of ^{15}N label into the humic acid fraction obviously does not reflect age, but rather the ability of the humic substances to adsorb non-humified material.[10,13] Peptides and their copolymers with carbohydrates, e.g. mucopeptides, are readily soluble in alkali and co-precipitate with humic substances; similar sorption reactions can take place within soils.[11] The present tracer experiment allows the determination of the relative rates of synthesis of the adsorbed components, and these are reflected in the slopes of the curves in Figure 1.

The growing season (1986) was typical for NE Scotland, with low rainfall augmented by periodic storm inputs. At sample #3, the first apparent peak of ^{15}N incorporation corresponds to increasing immobilization being reversed by the high rainfall prior to the peak. A similar feature is seen, after crop emergence at #4, around #6; the general trend between #4-10. for increasing immobilization, parallels N-uptake by the crop,[9] (Figure 1). This time period was one of extended drought, which was broken by rainfall between #10-11. Plant uptake of N increased noticeably,[9] coinciding with a release of available-N by mineralization of the immobilized fertilizer N.

The humic acid contains the major part of the extracted ^{15}N

Table 3. Potato field experiment - unextracted N expressed as %age of total soil N

Sample No.	2	5	8	11
% Unextracted N	38.4±1.5	40.5±1.5	42.5±1.2	41.9±3.5
% Unextracted ^{15}N	31.0±2.3	38.0±3.6	60.2±8.3	43.5±2.4

n = 32

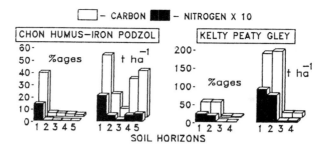

FIGURE 2 CONVERSION OF %C & N INTO TONNES HA^{-1}
SOIL HORIZONS — CHON 1—LFH, 2—E, 3—Bh, 4—Bs,
5—BC; KELTY 1—O1, 2—O2, 3—Bg, 4—BCg

(Figure 1). Between 30-45% of the total soil N was not extracted and the amount of non-extracted ^{15}N label was broadly similar (Table 3). This similarity, and the lack of any obvious isotopic fractionation between the humic, fulvic and hydrophilic acids, suggest that the organic N is distributed by factors other than soil processes, possibly by the extraction and isolation techniques. A similar distribution of extracted and unextracted soil N has been reported previously.[4]

3 SOIL N STORES AND N IN SURFACE WATERS

The calculation of soil N pools for different forms of organic N is also necessary in attempts to relate the N pools in plants, soils and surface waters. Using bulk densities, stone and gravel contributions, and horizon thicknesses, the normal soil profile analytical data, expressed in units per dry wt. of soil, have to be converted into the absolute sizes of soil parameter stores. Two such conversions are shown in Figure 2 for forested hillslope soils at Loch Chon and Kelty Water, studied intensively during the Surface Water Acidification Programme (SWAP).[12] In these, %C and %N for the soil profile, to a nominal depth of 1 m, have been converted to a tonnes ha^{-1} basis.

The hydrologies of the two sites are similar, in having

Table 4. Soil stores of amino-acid N (t ha^{-1}) in humic substances in two soils at Loch Ard, Central Scotland.

Site & vegetation	Soil	Horizon	Humic acid	Fulvic acid	% Total soil N
Loch Chon; Norway spruce;	Humus -iron podzol	LFH	0.56	0.05	28
		E	0.08	0.01	18
		Bsh	0.18	0.02	40
		BC	0.12	0.02	23
Kelty Water; Sitka spruce;	Peaty gley	O_1	1.11	0.03	13
		O_2	0.69	0.01	10
		AHg	0.16	0.002	41
		BCg	0.06	0.004	13

episodic flow predominantly at depth.[12] Thus the input of organic N to the soil drainage waters is dictated, not necessarily by the high concentrations of soluble material from the upper organic horizons, but by more dilute solutions, draining from the lower mineral layers, which are hydrochemical integrals of processes upslope. These processes will possibly incorporate "return flow", in which lateral water movement includes water flow routed from the lower mineral horizons into upper organic layers downslope.

Soil stores of amino acid N present in humic substances are shown in Table 4; the predominant fraction is in the humic acids. When the soil hydrology is taken into account, the small amounts of fulvic acid N in the basal horizons at Kelty are obviously the result of intense leaching of the gleyed soil, a similar, though less intense process probably also accounts for the smaller stores in the BC horizon at Chon. The increased fulvic acid stores at Chon compared with Kelty are examples of the ability of spodic horizons to retain organic N.[14] Comparison of seasonal amino acid N throughputs at both sites, for the LFH and BC horizons, and for the stream outputs are shown in Figure 3. Humic acid outputs are negligible compared with the fulvic and hydrophilic acid fractions. Although a decrease in hydrophilic acids in the LF horizons is evident during the winter period (Nov-Apr), as expected with decreased biological activity during colder soil regimes, subsoil throughputs show a systematic increase during the winter period. One difficulty arises in formal comparison of the output data, with LFH horizon throughflow being easily expressed on a simple area basis similar to rainfall, but the lower soil horizon output being expressed as a contour throughflow,[12] and the stream output appearing as absolute gravimetric values. However, the data serve to highlight the large amounts of N being transferred through these afforested systems into surface waters. Nitrate concentrations are small ($<10^{-6}$M) in these systems which are predominantly ammonifying, and when the data are converted into distribution diagrams as in Figure 4, the extent of the so-called "unknown" forms of N^4 becomes apparent. The subsequent fate of these different forms of N in the output streams is varied, with mineralization and uptake by stream biota an obvious

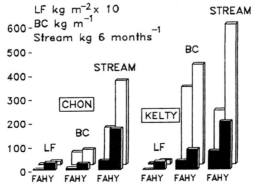

FIGURE 3 SEASONAL OUTPUTS AT TWO FORESTED SITES IN CENTRAL SCOTLAND; FRONT — SUMMER OUTPUT, REAR WINTER; FA = FULVIC, HY = HYDROPHILIC ACIDS

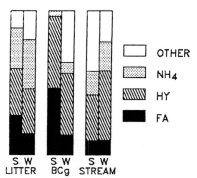

FIGURE 4 KELTY WATER 1988–89; DISTRIBUTION CHART FOR FORMS OF N AS % TOTAL N THROUGHPUTS; S=SUMMER, W=WINTER

route. However, in most streams with large organic inputs, streambed gravel is coated by coloured gelatinous precipitates. At Chon, such coatings are much in evidence, and although the fulvic acid components show a greater surface area coverage on the gravel, the now-evident humic acids contain a greater proportion of the amino acid N (Table 5).

4 DISTURBANCE OF SOILS DURING CHANGE OF LAND USE

Given the present over-production in EC agriculture, greater emphasis is being placed on the afforestation of less-productive areas. In addition to other impacts which occur during afforestation,[15] soil disturbance during site preparation and harvesting can be a major factor leading to the loss of organic N, either directly, in the form of organic matter losses in drainage or by erosion, or indirectly, through a change in the nitrogen-cycling in the soil system. The release of nitrate from clearfelled sites was explained by Tamm[16] as a consequence of an ammonium surplus, not being taken up by the missing trees, followed by increasing nitrification.

In a catchment clearfell study in the Loch Ard forest, there was the expected increase in nitrate in the output stream at harvesting, compared with that of a neighbouring control catchment (Table 6). Using ion selective electrodes (ISE), rapid measurements of both nitrate and ammonium were made in the field, on water samples from streams, drains and soil solutions from tension lysimeters (Anderson & Stewart, unpublished).

Although laboratory incubation experiments using standard

Table 5. Amino acid N contents of stream bed gravel coatings at Loch Chon

	g m^{-2}	mg AA-N m^{-2}
Humic Acid	2.2	50
Fulvic Acid	8.5	12

Table 6. Nitrate concentrations (mg l^{-1}) at specific points in Loch Ard clearfell sampled quarterly 1989-90.

Control stream (outlet)	0.04-0.46
Main stream (outlet)	2.04-2.70
(undisturbed source area)	0.10-0.15
Blackwater drainage (under discarded brash)	0.19-0.76
Bank seepage at organic-mineral soil interfaces	0.12-3.68
Soil solutions (lysimeters)	0.03-1.85
Disturbed organic soil drainage	45.0 -1.33

techniques,[17] had indicated the possibility of the catchment soils releasing nitrate (and ammonium) by mineralization of organic N, none of the soil solution samples showed large nitrate concentrations. Although most of the soils showed some gleying, i.e. impeded drainage, the stream is very reactive to rainfall, suggesting that drainage is rapid.

During harvesting, grid-sampling of water throughout the catchment, in a series of streams and drains, established that nitrate production was not common to each block. It was largely limited to small areas, often less than 1 ha, in which disturbance of deep organic soils and peats had taken place (Table 6).

Disturbance of peatland during afforestation is presently shown at its height in the so-called "Flow Country" of Caithness and Sutherland. Indeed, the perceived effects of afforestation in this area have led to a moratorium on planting being imposed. The Forestry Commission have established an experiment at Bad a'Cheo in Rumster Forest (Caithness), to examine the effects of current forestry practice on water quality and the peat formation.[18] The peat is 4-5 m deep and is situated in a pristine area, where atmospheric inputs are dominated by marine salts rather than anthropogenic pollution, this blanket bog containing 53% C, 2% N and 1.4% S. One obvious cause for concern regarding drainage water quality lies in the high nitrogen and sulphur contents, which, if mineralized, could lead to increased nitrate and sulphate in outputs to locally-important fisheries.

Using the ISE techniques mentioned above, nitrate production was confirmed by in vitro incubation, but no increase was seen in the field. Tamm[16] concluded that nitrification was inhibited by acidity in soil solutions, and this explanation must apply at Bad a' Cheo. The pH of the drainage waters is 4.4-4.5, dictated by the high concentrations of organic acidity. Average concentrations of N during August 1989 included 35-50 μmol l^{-1} ammonium, 12-19 humic acid N, 2-3 fulvic acid N and 18-26 hydrophilic acid N. Thus the concentrations of mineral and organic forms are approximately similar. Data shown in Figure 5

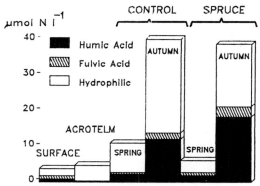

FIGURE 5 ORGANIC N IN PEATLAND DRAINAGE BEFORE
AND AFTER SITE CULTIVATION FOR AFFORESTATION
CONTROL=PERIMETER DRAIN; SPRUCE=SAME+30 CM
DMB PLOUGHING AT 4.0 M INTERVALS

illustrate the major effects of the initial stages of afforestation on the concentration of organic N in drainage water. Under undisturbed conditions, water collecting in surface pools is dominated by non-humified hydrophilic acid N, which is also the major component in the acrotelm solution (upper 20 cm peat). Cultivation, either by a simple perimeter drain, or more emphatically by ploughing for tree-planting, leads to a large increase in the concentration of humic substance N, especially the humic acid fraction. This fraction will include particulate material <0.45 µm, the widely-accepted lower value for the size of suspended solids in water samples. Colloidal organic matter, with a high N (and S) content is very important in the output from the Bad a' Cheo plots.

5 CONCLUSIONS

Modern field experimental procedures have widened the scope for the application of ^{15}N isotope techniques in studying N inputs, throughputs and outputs from plant-soil systems. In most of these scenarios, organic nitrogen associated with humic substances can be adsorbed rapidly to the surface of the organic macromolecules. Soil drainage contains predominantly fulvic and hydrophilic acids which can be precipitated in streams in the form of N-rich coatings on gravel. The disturbance of soil organic N can lead to the loss of either mineral or organic forms of N from the plant-soil system, which can have serious nutrient consequences for the growing crop.

In the contemporary world, with its greater interest in the environmental consequences of land use, such losses are perceived as avoidable pollutant inputs into water courses. As our knowledge of organic N behaviour increases, so we may become more adept in managing this important resource.

ACKNOWLEDGEMENTS

We should like to thank the sponsors of SWAP for financing some of the work described above, and the Forestry Commission for permission to use the forested sites at Loch Ard and Rumster Forests.

REFERENCES

1. Recent reviews of this subject include:-
   F.J. Stevenson, 'Humus Chemistry - Genesis, Composition, Reactions', Wiley & Sons, New York, 1983; H.A. Anderson & D. Vaughan, in 'Soil Organic Matter and Biological Activity' (D. Vaughan & R.E. Malcolm, eds.), Nijhoff/Junk, Dordrecht, 1985; H.A. Anderson et al., in 'Humic Substances II - In Search of Structure' (M.H.B. Hayes, P. MacCarthy, R.L. Malcolm & R.S. Swift, eds.), Wiley & Sons, Chichester, 1989.
2. F.E. Allison, 'Soil Organic Matter and Its Role in Crop Production', Elsevier, New York, 1973.
3. G.R. Aiken, D.M. McKnight, R.L. Wershaw & P. MacCarthy (eds.), 'Humic Substances in Soil, Sediment & Water', Wiley & Sons, New York, 1985.
4. H.A. Anderson in ref. 1.
5. C.C. Delwiche, Sci. Am., 1970, 137.
6. C.G. Kowalenko, in 'Soil Organic Matter' (M. Schnitzer & S.U. Khan, eds.), Elsevier, New York, 1978.
7. R.L.M. Synge, Phytochemistry, 1975, 14, 859.
8. C. Eriksson, Prog. Food Nutr. Sci., 1981, 5, 1.
9. P. Millard, D. Robinson & L.A. Mackie-Dawson, Ann. Bot., 1989, 63, 289.
10. R.S. Swift, in 'Humic Substances II - In Search of Structure', loc. cit.
11. F.J. Stevenson, in 'Humus Chemistry', loc. cit.
12. J.D. Miller, H.A. Anderson, R.C. Ferrier & T.A.B. Walker, Forestry, 1990, 63, 251 and 311.
13. M.V. Cheshire, P.A. Cranwell & R.D. Haworth, Tetrahedron, 1968, 24, 5155.
14. H.A. Anderson, A.R. Fraser, A. Hepburn & J.D. Russell, J. Soil Sci., 1977, 28, 623.
15. P.S. Maitland, M.D. Newson & G.A. Best, 'The impact of afforestation and forestry practice on freshwater habitats', Nature Conservancy Council, Peterborough, 1990.
16. C.O. Tamm, H. Holmen, B. Popovic & G. Wiklander, Ambio, 1974, 3, 211.
17. B.L. Williams, J. Soil Sci., 1983, 34, 113.
18. Macaulay Land Use Research Institute, Annual Report 1988-89, MLURI, Craigiebuckler, Aberdeen.

Organic Matter Turnover and Nitrate Leaching

M.J. Goss[1]*, B.L. Williams[1], and K.R. Howse[2]

[1] MACAULAY LAND USE RESEARCH INSTITUTE, CRAIGIEBUCKLER, ABERDEEN, AB9 2QJ, UK
[2] ROTHAMSTED EXPERIMENTAL STATION, AFRC INSTITUTE OF ARABLE CROPS RESEARCH, HARPENDEN, HERTS, AL5 2JQ, UK

1 INTRODUCTION

Nitrate leaching in soils depends on the relative rates of plant uptake and several nitrogen transforming processes, as well as on rainfall, drainage and fertilizer inputs. Organic matter turnover is important because more than 90 per cent of mineral nitrogen in soil after the harvest of arable crops has been mineralized from organic nitrogen[1]. Similar accumulations of mineral nitrogen develop in intensively utilised grassland[2]. In upland areas which are managed with lower fertiliser inputs, leaching losses are generally small[3]. Nevertheless, ploughing pastures can stimulate organic matter decomposition and increase nitrate losses to streamwater[3]. Beneath trees, nitrification and nitrate leaching are extremely variable between sites[4] and fluxes of nitrate through the soil profile following clearfelling of coniferous stands may be either immediate or delayed[5]. In each situation, the potential for nitrification and nitrate leaching may be altered by the activities of other nitrogen transforming processes. These include ammonification, immobilisation of ammonium and nitrate in the soil microbial biomass and organic matter fraction and denitrification of nitrate under anoxic conditions. However, the factors that control the extent and duration of these processes in each situation and how they interact have not been elucidated.

In this paper, we review field and laboratory experiments relevant to nitrification and nitrate leaching in soils under different cropping practices and land uses. In particular the effects of organic matter turnover and different nitrogen transformations on nitrification and leaching are considered

* Present address: Department of Land Resource Science, University of Guelph, Ontario, Canada

and discussed.

2 NITRATE LEACHING IN ARABLE SOILS

Nitrate is the principal form of mineral nitrogen in arable soils. When there is little uptake by plants during winter, leaching loss can be appreciable and cause contamination of water resources. Agricultural practices such as tillage, cropping sequence and crop residue management change the rate of turnover of soil organic matter and consequently may modify the rate of nitrogen mineralization. At the experimental site at Brimstone farm, Oxfordshire, some of these factors are being quantified. Compared with direct-drilling, tillage to 20 cm depth stimulated organic matter turnover by an average of 7 kg N ha^{-1}, i.e. the cumulative difference over 9 years was 63 kg N ha^{-1}. Ultimately this increased the release of nitrate

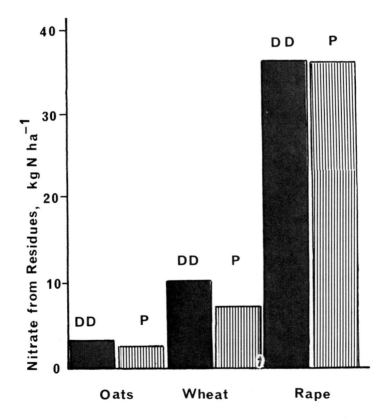

Figure 1   Contribution to nitrate leaching under winter wheat from the residues of the previous crop. DD-direct drilled, P-ploughed

Table 1   Winter losses of nitrate, kg N ha^{-1} leached from plots at Brimstone, Oxfordshire

Cropping system	Leaching loss
Winter fallow/spring wheat	62.2
Winter cereal (oats), ploughed straw burnt	25.6
Winter cereal (oats), ploughed straw incorporated	17.3

nitrate into the soil and contributed to 27% more being lost by leaching from ploughed than from direct drilled plots. The leaching loss between the harvest of one crop and the application of nitrogen fertilizer as top-dressing to the next was very dependent on the nature of the previous crop[6]. The contributions from the plant residues to nitrate leaching can be calculated (Figure 1) from mean annual decreases of soil organic nitrogen of 15 and 8 kg N ha^{-1} yr^{-1} in ploughed and direct drilled plots, respectively. Thus if the previous crop were winter oats, and straw and chaff were removed by burning, 80% of the nitrate loss from ploughed soil over the period was derived from older organic matter. The corresponding values for winter wheat and oilseed rape were 68% and 29%.

A growing crop results in some uptake of mineral N, and in addition the presence of plant roots limits mineralization of soil organic matter[7]. In consequence winter fallow shows the maximum release of nitrate (Table 1). A crop decreases the nitrogen available for leaching while incorporation of straw residues with a high C:N ratio encourages further immobilization of mineral nitrogen by assimilation into the microbial biomass (Table 1).

The incorporation of straw residues can also increase the potential for denitrification. Both immobilization and denitrification can occur simultaneously if microbial activity is intense, and there is an adequate supply of oxygen[8].

3   NITRIFICATION IN PEAT SOILS

The nutrient content of peat increases with increasing eutrophic conditions[9]; poorer peats have slower rates of organic matter turnover and greater proportions of undecomposed plant residues than peats from mineral-rich areas[10]. The fibrous peats also have high C:N ratios and a greater tendency for net immobilisation of nitrogen and other nutrients. In a comparison of

Table 2 Herbage-N and accumulated net changes in mineral-N contents, (∆-Min N) kg N ha^{-1} of reseeded peat, 0 to 10 cms depth, incubated *in situ* during the growing season at Forsinard, Highland Region

	∆-Min N*			Herbage-N		
	PK	NPK	SE	PK	NPK	SE
1985	16	-3	15.7	14	63	8.9
1986	41	89	20.8	19	68	12.5
1987	-20	-27	11.7	49	98	4.4

* Negative values indicate net immobilisation of nitrogen
NPK - 250 kg N, 60 kg P and 60 kg K ha^{-1}yr^{-1}
SE - standard error of the mean (n = 3)

samples from different peatland areas[11] only peats with pH (CaCl$_2$) > 3.8 and an ash content > 20 per cent showed active nitrification. The type of peat is important for management of nitrogen nutrition when areas of peatland are limed for reseeding. At a blanket bog site in Forsinard (Highland Region), liming and reseeding raised the pH (CaCl$_2$) of the peat from 3.4 to 5.0, and increased the numbers of ammonifying bacteria by a factor of between 10 and 100[12] compared with values obtained for a comparable unfertilized blanket bog[13]. Immobilization of mineral nitrogen predominated during the spring in this peat and then over the growing season as herbage production increased (Table 2). Five years after reseeding no ammonium oxidizing bacteria were detected[14] and only trace amounts of nitrate could be extracted. Denitrification of added nitrate was very active in this reseeded peat[15], with rates exceeding 6 kg N ha^{-1} day^{-1} at 10°C during April and May. A combination of poor drainage and enhanced organic matter turnover seems to stimulate both denitrification and immobilisation in this peat, but gaseous loss of nitrogen only occurred when nitrate was added. The source of the substrate for decomposition could be either plant residues from the original vegetation or fresh litter and exudates released by the grass sward. Whatever the substrate, the constraints on the management of nitrogen nutrition at the site are significant. In contrast, profiles of reseeded peatland on the Isle of Lewis (Highland Region) contained appreciable amounts of nitrate whereas peat from adjacent unimproved areas contained none (Table 3). On incubation, only peat from the surface 10 cm of the reseeds produced additional amounts of nitrate so the nitrate at greater depths had been leached from the surface. The sites on Lewis had been improved by traditional methods employing calcareous shell sands to neutralise acidity and little or no fertiliser inputs after the first two years[15]. The nitrifying peat was still very acid after

Table 3 Amounts of NH_4^+ and NO_3^--N, kg ha^{-1}, in the profiles of natural and 8 year old reseeded peatland at Barvas, Lewis

Depth (cm)	NH_4^+ Natural	NH_4^+ Improved	SE	NO_3^- Natural	NO_3^- Improved	SE	Δ-NO_3^-*
0 - 10	124	118	18.7	0	168	48.1	18
10 - 20	133	137	19.9	0	0.5	2.8	0
20 - 30	103	96	4.4	0	2.3	3.7	0
30 - 40	106	62	10.2	0	1.1	2.4	0
40 - 50	82	91	8.8	0	0	-	0

* ΔNO_3^- -NO_3^- produced during aerobic incubation at 30°C
SE -standard error of the means (n = 3)

improvement (mean pH ($CaCl_2$) = 3.8), and the stimulation of nitrification may well have been the result of several interacting factors. For example, the mineral content of the surface peat horizons had been increased by the shell sand residues and these may have enabled microbial processes to occur by providing an inorganic or inert surface. Equally the quantities of lime and the time elapsed after application may have been important factors together with the forms of nitrogen returned in the excreta of grazing animals.

4 FOREST SOILS

There is considerable interest in the production of mineral nitrogen under forests, both because of problems associated with clear felling and because of the need to improve the long term management of nitrogen in nitrate sensitive areas. Mineral nitrogen production has been studied in forest floor and organic horizons under stands of Scots pine (*Pinus sylvestris,* L.) aged 45 and 110 years, Sitka spruce (*Picea sitchensis* (Bong.) Carr) and larch (*Larix kempferi* (Lamb. Carr) at Glen Tanar (Grampian Region). The trees were planted on humus iron podzols and alluvium derived from granitic parent materials. Twigs, cones and stones were removed by screening the litter and humus horizons through a 5 mm sieve and the humus returned to the forest floor for *in situ* incubation on pvc trays. Leachates and samples for extraction were taken at two-week intervals and values summed to obtain annual amounts of mineral nitrogen produced (Table 4). Similar quantities of mineral nitrogen were produced on an area basis for larch, spruce and the old pine stands, but the last contained appreciably more organic matter and total nitrogen than the younger crops. The mineral nitrogen fraction was therefore a smaller proportion of the total nitrogen than at the two pine sites (3 to 4 per cent), spruce (10 per cent) and larch (20 per cent). The more rapid turnover of nitrogen in

Table 4 Mineral-N, kg ha^{-1}, produced by LFH horizons beneath 4 conifer stands during incubation *in situ* for 12 months at Glen Tanar

Crop	NH$_4^+$	SE	NO$_3$	SE	Min-N	SE
Old Scots pine	35	11.1	2 (5)[a]	0.6	37.6(3)[b]	11.5
Young Scots pine	8	3.9	0.5(6)	0.1	9.1(2)	3.9
Sitka spruce	16	1.4	8 (33)	3.2	24.2(9)	2.2
Larch	20	4.2	13 (39)	1.9	33.1(20)	2.8

[a] Nitrate expressed as per cent of total mineral-N
[b] Mineral-N expressed as percentage of total-N
SE - standard error of the mean (n = 3)

the larch and spruce humus was consistent with their greater ash contents and pH (Table 5), but C:N ratio did not vary greatly. Microbial biomass carbon values in the humus, obtained by the method of Anderson and Domsch[16], were 3.59 and 1.89 g C kg^{-1} for larch and spruce, respectively, compared with 2.1 to 2.8 g C kg^{-1} for the two pine sites. Fungi comprised the predominant component of the biomass in these humus horizons, but beneath larch there was a greater proportion of bacterial biomass.

Nitrification of ammonia was less than 6 per cent of total mineral nitrogen in pine humus whereas this proportion reached between 30 and 40 per cent in the spruce and larch.

Total amounts of nitrate leached from spruce and larch humus, during one year at Glen Tanar, were 8 and 13 kg N ha^{-1}, respectively. These quantities are much smaller than the flux of 70 kg NO$_3$-N ha^{-1} yr^{-1} measured on clearfell sites in Wales cropped with Sitka spruce[5]. Although measurements of denitrification were not available this process would account for less than 10 kg N ha^{-1} yr^{-1} in the decomposing forest floor on clear fell sites[17]. Differences between sites will be affected by atmospheric inputs as well as by the nitrogen content of the soil and vegetation which in turn will influence the rate of immobilisation during the decomposition of harvest residues[18].

Table 5 Organic matter content, pH, loss on ignition and C:N ratios for F & H horizons beneath pine, spruce and larch at Glen Tanar

	Mg ha^{-1}	pH (CaCl$_2$)	%LOI	C:N
Old Scots Pine	97.1	2.7	73	29
Young Scots Pine	34.6	2.8	80	37
Sitka spruce	43.7	3.4	40	24
Larch	19.5	3.8	45	39

Table 6 Mean rates of production of mineral-N and NO_3-N between weeks 11 and 21, mg kg^{-1} dry litter week^{-1} in pure litters of Sitka spruce, Scots pine and a 1:1 mixture continuously leached with distilled H_2O at 20°C for 21 weeks

Litter	Net mineralisation	Nitrification
Sitka spruce	240.8(2.6)	245.6(2.7)
Scots pine	25.5(0.2)	10.0(0.1)
1:1 mixture	222.5(2.0)	130.8(1.2)

a Values in parentheses expressed as kg N ha^{-1} week^{-1}

Differences between species in the ammonification and nitrification rates have been measured in spruce and pine litters[19]. Furthermore when these litters were mixed, organic matter turnover and net mineralisation of nitrogen were enhanced but nitrification was not (Table 6). The explanation of these results was complex. Where the litters were mixed the greater fungal biomass in the pine needles became available to enchytraeid worms present in the spruce litter. The worms proliferated and bacterial action on the material ingested by the worms stimulated mineralization. Separate experiments to test the effect of worms on nutrient mobilisation in spruce litters[20] showed that they enhanced the leaching of all of the major nutrients and increased the proportion of mineral nitrogen present as nitrate.

5 CONCLUSIONS

Organic matter turnover in arable, forest and peaty soils occurs at different rates, the potential for nitrification and nitrate leaching being much greater in arable than in the more acid organic soils. Nevertheless, nitrification and leaching can occur during organic matter turnover in peaty and forest soils. Not all of the factors controlling nitrification in organic soils have been identified. Immobilisation of nitrogen in the soil microbial biomass during decomposition of fresh plant residues can alleviate problems of nitrate leaching in the short term, but remineralisation of this nitrogen may be problematic at a later date in all soils. The role and activities of secondary decomposers could be important factors controlling the duration of immobilisation in the microbial biomass.

6 REFERENCES

1. K.W.T. Goulding, D.W. Powlson, P.R. Poulton, C.P. Webster, A.J. Macdonald and M.J. Glendining. *J. Sci Food Agric.*, 1989, 48, 123.
2. J.C. Ryden, P.R. Vall and E.A. Garwood. *Nature* Lond. 1984, 311, 50.
3. A.M. Roberts, J.A. Hudson and G. Roberts. *Soil Use and Management*, 1989, 5, 174.
4. G.P. Robertson. *Phil. Trans. R. Soc. Lond*. 1981, B.296, 445.
5. P.A. Stevens and M. Hornung. *Soil Use and Management*, 1988, 4, 3.
6. M.J. Goss, K.R. Howse, P. Colbourn and G.L. Harris, 1988. 11th ISTRO Conf. Proceedings 679.
7. J.B. Reid and M.J. Goss. *J. Soil Sci.* 1982, 33, 387.
8. M. Gök and J.C.G. Ottow. *Biol. Fertil. Soil*, 1988, 5, 317.
9. D.G. Pyatt, M.M. Craven and B.L. Williams, 1979. 'Classification of Peat and Peatlands'. Int. Peat Soc. Helsinki.
10. B.L. Williams. *J. Soil Sci.*, 1983, 34, 113.
11. B.L. Williams. Proc. 7th Int. Peat Congr., 1984, 4, 410. Irish Peat Society, Dublin.
12. B.L. Williams and R.E. Wheatley. Submitted to *Biol. Fert. Soils*.
13. B.L. Williams and R.E. Wheatley. *Biol. Fertil. Soil*, 1988, 6, 141.
14. R.E. Wheatley and B.L. Williams. *Soil Biol. Biochem.*, 1989, 21, 355.
15. B.L. Williams, R. Boggie, J. Cooper and J.W. Mitchell. *Irish J. Agr. Res.*, 1985, 24, 229.
16. B.L. Williams and G.P. Sparling. *Soil Biol. Biochem.*, 1988, 20, 579.
17. J. Dutch, PhD Thesis, University of Aberdeen, 1989.
18. M.F. Proe and B.L. Williams. Proc. Worksp. on 'Research Strategies for long-term site Productivity' IEA/BE A3 Report No 8 1989, 187. Forest Research Institute, New Zealand.
19. B.L. Williams and C.E. Alexander. *Soil Biol. Biochem.*, 1990, 23, 71.
20. B.L. Williams and B.S. Griffiths. *Soil Biol. Biochem.*, 1989, 21, 183.

The Influence of Soils on Organic Matter in Water

P.N. Nelson[1], J.M. Oades[1], and E. Cotsaris[2]

AUSTRALIAN CENTRE FOR WATER TREATMENT AND WATER RESEARCH
[1] DEPARTMENT OF SOIL SCIENCE, WAITE AGRICULTURAL RESEARCH INSTITUTE, GLEN OSMOND, SOUTH AUSTRALIA 5064
[2] STATE WATER LABORATORIES, SALISBURY, SOUTH AUSTRALIA 5108

1 INTRODUCTION

Dissolved organic carbon (DOC) is of major importance in the ecology of freshwater bodies. It also causes problems with water treatment which are difficult and expensive to control[1]. Alum (flocculant) and chlorine (disinfectant) dose rates increase in proportion to the concentration of DOC, as does the production of disinfectant byproducts such as trihalomethanes. Little is known about the factors which influence concentrations of DOC in streams, and geology, soils, topography, vegetation, land use and management may all be involved. A recent study[1] in the Mt. Lofty Ranges in South Australia found that the concentration of DOC in streams varied from 2-26 mg l^{-1}. The Mt. Lofty Ranges serve as catchment areas for the Adelaide reservoir system, and the concentrations in some Adelaide reservoirs (up to 20 mg l^{-1}) are high by international standards. Although land-use has a large effect on water quality throughout the Mt. Lofty Ranges[2], variations in land-use do not appear to influence concentrations of DOC in streams. Therefore, the question was: Why is the concentration of DOC high in some streams and low in others?

Two pairs of small catchments (Fig.1) were chosen to investigate the origins of DOC. Each pair had similar land-use and climate, but different concentrations of DOC. Retreat catchment (1.2 km^2) and Lawless catchment (3.0 km^2) are situated in the Mt. Lofty Ranges. Both catchments consist of grazed pastures and have an annual rainfall of 800 mm. Despite the similarity of the two catchments, widespread sampling of streams of the Mt. Lofty Ranges in 1987 indicated that they had markedly different concentrations of DOC. The second pair of catchments are situated in the Otway Ranges in Victoria. Clearwater catchment (19.1 km^2) and Redwater catchment (7.1 km^2) both consist of native forest utilized for hardwood production and have an annual rainfall of 1500 mm. Despite the similarity of the catchments, the streams are known to have different DOC concentrations, as their names suggest.

Preliminary investigations indicated that a major diffence between the catchments in each pair was the types of soil present. Therefore, subsequent investigations concentrated on examining the relationship between soil properties and the quality of stream water. Although it is accepted that soil is a major source of DOC[3], it is not known which soil factors influence concentrations of DOC in streams. The work done in Retreat and Lawless catchments has been reported in detail by Nelson[4]. This paper presents an overview of our investigations into the relationships between soil properties and DOC in streams.

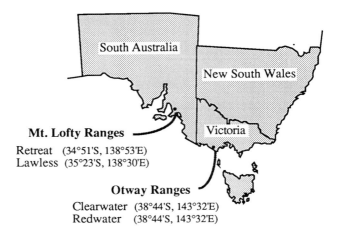

Fig.1 Location of the catchments in southeastern Australia

2 METHODS

<u>Water sampling and analysis</u>

Monitoring sampling stations were established in Retreat and Lawless catchments in 1988. The stations consisted of a flow control structure, stream depth probe, tipping bucket rain gauge, data logger and automatic sampler. Rainfall and stream depth were recorded continually during the latter half of 1988 and all of 1989. The data logger calculated discharge and triggered the sampler to take a sample at flow proportional intervals. Stream samples were taken by hand in Clearwater and Redwater catchments in early 1990. Sampling in Clearwater and Redwater catchments included water which had moved through the canopy and litter, but had not yet contacted the soil (throughfall). Samples were taken from ponds, springs and tributaries throughout all four catchments. Water samples were analysed for total dissolved and total suspended solid content, C, N and P. The DOC in water samples was characterised using ^{13}C CPMAS nuclear magnetic resonance spectroscopy of evaporated, freeze-dried samples[5]. Specific surface areas were determined by nitrogen adsorption.

<u>Soil sampling and analysis</u>

The soils of all the catchments were mapped and described[4]. The maps provided a basis for determining the importance of the various types of soil present, and for structuring the sampling program. Samples were taken at various depths from the major soil units. Analysis of the soil samples included particle size distribution, total and water extractable C content (1:5 soil:water extract), specific surface area, permeability and adsorption capacity for DOC.

3 RESULTS AND DISCUSSION

Although there were slight differences between the vegetation in the catchment pairs, the differences in C assimilation were not large enough to explain the observed differences in stream DOC. The concentrations of DOC in the streams (mg l^{-1}) were:

Retreat	Lawless	Clearwater	Redwater
6.5	21.9	3.8	32.0

The carbon contents of the soils (Table 1) were not related to the concentration of DOC in the streams. The C contents of the soils in the Otway Ranges were higher than those of the Mt. Lofty Ranges soils because of higher C assimilation, and longer turnover times due to cooler and wetter conditions.

Table 1 Soil properties

	Retreat	Lawless	Clearwater	Redwater
Carbon content (%)				
A horizon	4.5	2.6	7.4	9.3
B horizon	0.4	0.2	2.3	3.1
DOC in 1:5 soil:water extracts (mg l^{-1})				
A horizon	25.0	31.3	49.4	60.2
B horizon	3.0	3.0	13.3	24.8
Adsorption capacity (µg C g^{-1} soil)				
A horizon	153	71	108	31
B horizon	460	480	417	237

The concentration of water extractable C in the soils (Table 1) was related to the concentration of DOC in the streams in both of the catchment pairs. Lawless soils had higher water extractable C content than Retreat soil, and Redwater soils had higher water extractable C content than Clearwater soils. The difference between the Retreat and Lawless soils was accentuated when the DOC of water allowed to leach through undisturbed cores was measured indicating the importance of flow pathways through the soils. Thus, water moving through some soils would extract more C than in other soils. However, these differences in water extractable carbon were not large enough to fully explain the differences in the concentrations of DOC between the streams.

The DOC of water entering the soil (throughfall) was 100-200 mg l^{-1} in The Otway Ranges catchments, and the fate of this DOC upon entering the soil was examined by testing the ability of the soils to remove DOC from solution. The relative adsorption capacity of the soils for DOC was tested by mixing soil samples with throughfall and measuring the amount of DOC adsorbed (Table 1). Adsorption capacity in the A horizons was high in the catchments yielding water of low DOC concentrations (Retreat and Clearwater) and low in the catchments yielding water with high DOC concentrations (Lawless and Redwater). In the B horizons the situation was not as clearcut. In the paired catchments in the Mt. Lofty Ranges the adsorption capacity of the B horizons was the same in both catchments. However, when the permeability of the soils in Lawless catchment was measured, the B horizons were found to be 20 times less permeable than the corresponding A horizons. Therefore in the Lawless catchment water tends to move through the A horizons only and the properties of the B horizons may be disregarded. Although the permeability of the soils in Retreat Valley catchment was not measured, the B horizons are known to be reasonably permeable. The movement of water through the B horizon in this catchment would be expected to enhance the difference in DOC between the catchments due to the clay content of the A horizons. The differences in adsorption capacity of the catchments were large enough to explain the differences in water quality. The adsorption behaviour of dissolved organic N and P could be expected to parallel that of DOC. The results thus have significance for algal growth in reservoirs. Characterisation of the DOC by ^{13}C NMR showed that no specific structural group was adsorbed preferentially. The spectra from ^{13}C NMR of DOC from Redwater and Clearwater streams (Fig.2) are essentially similar except for the peak at 168 ppm in the Clearwater spectrum, which is most likely to be due to carboxyl carbon.

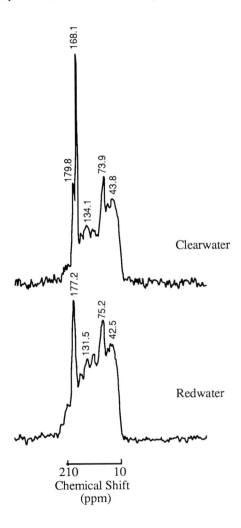

Fig.2 ^{13}C CPMAS NMR spectra of DOC from the Otway Ranges catchments

The next problem was to find out which soil properties determined adsorption capacity. Adsorption capacity was related to both clay content and specific surface area (Fig.3). The points on the graph represent samples which were taken from 0-10 cm and 60-70 cm depth in the catchment in the Otway

Ranges, and composites from the major horizons in the case of the catchments in the Mt. Lofty Ranges. With specific surface areas greater than 5 m^2g^{-1}, all samples showed net adsorption of DOC. With specific surface areas less than 5 m^2g^{-1} net adsorption or desorption occurred. Including water extractable C in the linear regression model improved the fit slightly because it helped to account for the amount of desorption that occurred at sites with soils having low specific surface areas. All of the soil samples from Redwater catchment had very low surface areas, and hence low adsorption capacities. In the other catchments, the samples with surface areas greater than 20 m^2g^{-1} are from the B horizons, while those with surface areas less than 20 m^2g^{-1} are from the A horizons. The soils from both the Mt. Lofty Ranges and the Otway Ranges followed the same relationship, despite their different natures, suggesting that the relationship is a fairly general one.

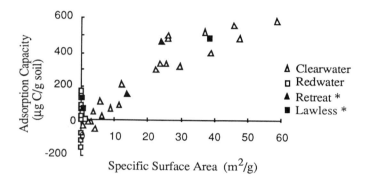

Fig. 3 Adsorption capacity and specific surface area (* horizon composites)

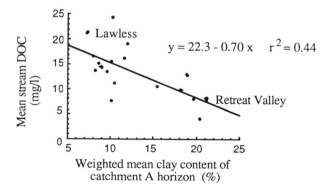

Fig. 4 Stream DOC and soil clay content in the Mt. Lofty Ranges

The relationship of DOC in streams with the clay content of soils was then tested on a wider scale. All the data on DOC in streams collected in the Mt. Lofty Ranges was collated and plotted against the weighted mean clay content of the A horizons of the corresponding catchments (Fig.4). The regression was significant, with clay content of the A horizon accounting for 44% of the variation in the DOC of streams.

4 CONCLUSIONS AND IMPLICATIONS

The concentration of DOC in streams is a function of the clay content of the soils in the catchments and the flow path of water through the soils. An implication is that many of the land use controls designed to improve water quality may not have any effect on one of the parameters most important for water treatment. It may be possible to reduce concentrations of DOC in streams by modifying the adsorption capacity or flow path of water through the soils.

REFERENCES

1. Hine, P.T. and Bursill, D.B. 1987. Seasonal trends of natural organics in South Australian waters and their effects on water treatment. AWRC Research Project 84/167, Final Report.
2. Wood, G. 1986. The Mt. Lofty Ranges watershed. Impact of land use on water quality and implications for reservoir water quality management. Engineering and Water Supply Department, South Australia. EWS 96/19.
3. Thurman, E.M. 1985. Organic Geochemistry of Natural Waters. (Martinus Nijhoft/Dr. W. Junk Publishers: Dordrecht).
4. Nelson, P.N., Cotsaris, E., Oades, J.M. and Bursill, D.B. 1990. The influence of soil clay content on dissolved organic matter in stream waters. Aust. J. Mar. Freshw. Res. (in press).
5. Oades, J.M., Vassallo, A.M., Waters, A.G. and Wilson, M.A. 1987 Characterization of organic matter in particle size and density fractions from a Red-brown earth by solid-state ^{13}C N.M.R. Aust. J. Soil Res. 25, 71-82.

Dissolved Organic Matter as Carrier for Exogenous Organic Chemicals in Soils

Ingrid Kögel-Knabner[1], Peter Knabner[2], and Helmut Deschauer[1]

[1] LEHRSTUHL FÜR BODENKUNDE UND BODENGEOGRAPHIE, UNIVERSITÄT BAYREUTH, POSTFACH 10 12 51, D-8580 BAYREUTH, GERMANY
[2] INSTITUT FÜR MATHEMATIK, UNIVERSITÄT AUGSBURG, UNIVERSITÄTSSTRASSE 8, D-8900 AUGSBURG, GERMANY

SUMMARY

Dissolved organic matter (DOM) can act as sorbent for exogenous organic chemicals in soils, possibly leading to enhanced leaching of these compounds. The distribution of exogenous organic chemicals in soils is considered here as an equilibrium of environmental chemicals in a three compartment system: dissolved, sorbed to dissolved macromolecules, and sorbed to the bulk soil matrix. In the present study the distribution of a new acidic herbicide in the three compartment system is investigated. Sorption isotherms are determined for the sorption of organic chemicals to bulk soil materials (Ap horizons) and to water-soluble soil organic matter. From these data an effective equilibrium sorption isotherm to bulk soil can be calculated which takes into account the sorption of organic chemicals to dissolved organic matter. A markable influence of DOM on sorption isotherms of the experimental herbicide is found only for high amounts of DOM (100 mg C L^{-1}).

1 INTRODUCTION

Sorption is one of the most important processes affecting the behaviour of exogenous organic chemicals in soils[1]. Most predictions and models for sorption and transport in soils are based on the assumption of a simple partitioning or reaction between the dissolved phase (A) and the phase sorbed to the bulk soil organic matter (B), which is considered as the main sorbent for organic chemicals in soils. The soil solution contains dissolved organic matter (DOM) that is also a potential sorbent and carrier for environmental organic chemicals[2-4]. The occurrence of sorption to this third phase can have significant impact on leaching of organic chemicals. Modeling sorption in complex systems, like soils, must therefore include binding to two types of materials, dissolved humic macromolecules and the bulk soil organic matter. Consequently, we consider sorption of chemicals in soils as an equilibrium in three phases: dissolved (A), sorbed to dissolved macromolecules (C), and sorbed to the bulk soil matrix (B)[5]. Sorption experiments in this respect have mainly been carried out with (model) humic acid preparations as a surrogate for DOM and neutral hydrophobic compounds[6]. In the present investigation we describe the sorption behavior of a new acidic herbicide in the three-phase-system and the influence of DOM on sorption and transport in a model. As described previously the composition of the soil solution has a decisive influence on the sorption of quinmerac to DOM[7]. Therefore water soluble soil organic matter (WSSOM) extracted from soil instead of (model) humic acid preparations is used for the sorption experiments.

2 MATERIALS AND METHODS

The sorbate used for the present investigation is the experimental herbicide quinmerac. Quinmerac is a substituted quinolinecarboxylic acid (7-chloro-3-methyl-quinoline-8-carboxylic acid)[8]. It is applied in suspension in a concentration of 0.25 kg ha^{-1}. Water solubility of quinmerac is 210 mg L^{-1}. For all laboratory sorption experiments quinmerac was used in a 50 % formulation (BAS 518H)[8].

The soil material used for sorption experiments and extraction of water-soluble soil organic matter was collected from the Ap horizon of a mollic Leptosol, a gleyic Cambisol, and a terric Histosol in Bavaria. Table 1 gives some chemical characteristics of the soil materials. A detailed description of extraction procedure for water-soluble soil organic matter (WSSOM) is found elsewhere[7]. Briefly, water-soluble soil organic matter was extracted from the air-dried (25 °C), sieved (< 2mm) soil material of the mollic Leptosol at a soil/solution ratio of 1/2.

Batch equilibrium sorption isotherms to bulk soil were determined at an adsorbent/solution ratio of 1/3 or 1/5[9]. Sorption to dissolved organic matter was investigated by ultrafiltration as described previously[7]. We used an Amicon ultrafiltration equipment with a molecular weight cut off 1000 filter membrane (YM-2). Sorption was determined in so-called continuous-flow wash-in experiments[10, 11]. Sorption isotherms for the sorption of quinmerac to bulk soil and to WSSOM are assumed to be of the Freundlich form $x/m = K_f C^{1/n}$, where x is the quantity adsorbed by mass m of soil or DOM in equilibrium with the solution concentration C, and K_f the sorption coefficient. K_f and $1/n$ are fitted to the experimental data by linear regression for the data in the logarithmic scale. For the coefficient $1/n$ near 1 the isotherm is approximately linear and gives the distribution coefficient K_d ($x/m = K_d C$).

Table 1: Chemical characteristics of the soil samples used for sorption experiments[9].

sample	C_{org}[a] %	N[b] %	pH (H_2O)	CEC[c] meq g^{-1}	clay %
I Ap	7.8	0.67	7.26	0.37	5
II Ap	19.0	1.46	7.33	0.63	10
III Ap	3.1	0.23	6.61	0.38	5

soils:
I	Mollic Leptosol	[a]C_{org}	organic carbon content
II	Terric Histosol	[b]N	total nitrogen content
III	Gleyic Cambisol	[c]CEC	cation exchange capacity

3 RESULTS AND DISCUSSION

<u>Sorption Experiments</u>

The sorption of quinmerac onto soils depends on soil pH, organic carbon content, and cation composition of the exchange complex[9]. At pH-values > 5.5 quinmerac is present as anion and bound via cation bridges to the exchange complex of the soil. The predominating cation in the agricultural soils used for the sorption experiments is Ca^{2+}. At pH < 5.5 the neutral form of quinmerac is dominating. Consequently it is supposed that the sorption mechanisms operating at these pH values are hydrophobic interaction or H-bonding.

Table 2 gives the equilibrium sorption data for the sorption of quinmerac onto different Ap horizons. Soil III did not show sorption at all, due to the low organic matter content. Some isotherms have a slope $1/n < 1$ and therefore show a decreasing percentage of quinmerac sorbed with increasing concentration in solution. Sorption is linear to a quinmerac concentration of about 20 g mL^{-1}. The K_f values range from 0.37 to 2.88. The data indicate that the sorption of quinmerac is low in neutral to slightly acidic agricultural soils. These sorption data are comparable to sorption data of other acidic pesticides, like 2,4-D or picloram[1, 12].

Table 2: Regression data for sorption isotherms according to Freundlich* (data from 7 and 9).

Soil	n	R	slope 1/n	K_f
soil/solution ratio 1/5				
I	27	0.836	0.98	0.37
II	41	0.961	0.62	2.88
soil/solution ratio 1/3				
I	17	0.976	0.86	0.81
II	23	0.988	0.80	2.51
WSSOM	36	0.958	2.20	1.32

* $\lg x/m = \lg K_f + 1/n \, c_s$
x/m   amount sorbed per mass of soil (g g^{-1})
c_s concentration in solution (g mL^{-1})
n   number of sorption data
R   regression coefficient for linear regression of logarithmic Freundlich equation.
1/n   slope of logarithmic Freundlich equation
K_f constant K_f of logarithmic Freundlich equation

Table 2 also gives the sorption isotherm for the sorption of quinmerac to WSSOM. The sorption to WSSOM results in distribution coefficients K_{oc} normalized to organic carbon from 1500 - 3000. They are much higher compared to the distribution coefficients obtained for sorption to the bulk soil matrix, where K_{oc} ranges from about 4 to 18. This has been described previously for the sorption of other herbicides to WSSOM[13].

Effective Sorption Isotherm

The influence of WSSOM on sorption of environmental chemicals is difficult to investigate in batch experiments. Therefore we want to calculate the sorption to bulk soil for different concentrations of WSSOM from the data known of the single sorption isotherms to bulk soil and to WSSOM. For this purpose we derive a transport model for the total concentration of dissolved organic chemicals in solution (A+C), denoted by $C_{(A+C)}$. It is based on the following assumptions:

The adsorption processes are in equilibrium described by adsorption isotherms as determined above, and the phase C does not undergo adsorption onto soil. The mass flux of dissolved phases (A) or (C) consists of a convective and a diffusive-dispersive part described in a uniform way with the same volumetric water flux and the same diffusion-dispersion coefficient. Additionally, the water content is taken to be the same for both phases (A) and (C). The consequences of omitting the last two assumptions will be addressed below.

Then mass conservation leads to the linear dispersion-convection equation for $C_{(A+C)}$ with a sink term due to the adsorption of $C_{(A)}$, the concentration of the phase (A), onto the soil matrix. For a specific value of $C_{(A+C)}$ there is a unique corresponding value of $C_{(A)}$, although in general there is no explicit expression for $C_{(A)}$. Therefore the transport equation can be transformed to a differential equation only in $C_{(A+C)}$. Thus $C_{(A+C)}$ fulfils the dispersion-convection equation with a sink term which can be interpreted as being the consequence of one adsorption process. This process is described by a functional relationship between $C_{(A+C)}$ and the concentration sorbed to bulk soil, which we call the effective isotherm. The concentration $C_{(WSSOM)}$ of WSSOM enters into the definition of the effective isotherm, which therefore varies with space or time, if the same applies for $C_{(WSSOM)}$. The value of the effective isotherm is less than the corresponding value of the sorption isotherm onto bulk soil. This reflects quantitatively the enhanced mobility of the organic chemical due to the adsorption to WSSOM. For constant concentration of WSSOM the effective isotherm is exactly the isotherm describing the adsorption process to bulk soil measured in batch experiments in the presence of WSSOM.

In general there is no explicit form for the effective isotherm, as such a formula is also lacking for the dependence of $C_{(A)}$ on $C_{(A+C)}$. But the value of the effective isotherm for some value of $C_{(A+C)}$ can be easily approximated numerically, if the two other isotherms are given by some closed form. In the special case of a linear sorption isotherm onto WSSOM, with a sorption coefficient $K_{f(WSSOM)}$, we have

$$C_{(A)}/C_{(A+C)} = 1/(1 + K_{f(WSSOM)} C_{(WSSOM)})$$

It is this factor less than 1, which enters the isotherm for sorption to bulk soil to produce the diminished effective isotherm. Thus if the sorption to the soil matrix is given in Freundlich form by $K_{f(soil)}$ and $1/n$, the effective isotherm has Freundlich form with the same coefficient $1/n$, but with a modified K_f-value $K_{f(eff)}$, obtained from

$$K_{f(eff)} = K_{f(soil)} 1/(1 + K_{f(WSSOM)} C_{(WSSOM)})^{1/n}.$$

If both isotherms are linear, i.e. also $1/n = 1$, the model is equivalent to the one developed by Enfield et al.[14]. In Fig. 1 the calculated sorption isotherms are shown for different concentrations of WSSOM. We use the sorption data for soil II for a soil/solution ratio 1/3, but we use a linear isotherm for the sorption to WSSOM with $K_f = 1627.1$. The presence of low concentrations of WSSOM has only a slight influence on the effective isotherm. Markable differences are found for high concentrations of WSSOM (100 mg L^{-1}) only.

The presence of WSSOM will influence the leaching of quinmerac under conditions when high concentrations of DOM occur in soils. These conditions are expected e.g. after application of compost, sewage sludge, or after accelerated release of DOM due to fertilization or liming.

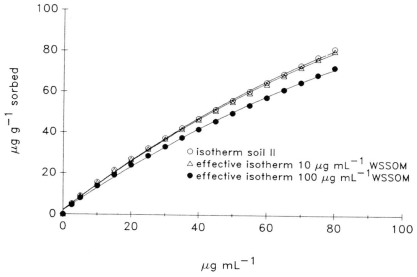

Fig. 1: Effective equilibrium sorption isotherms for the sorption of quinmerac onto soil II in the presence of different concentrations of WSSOM (g C mL^{-1}).

Depending on the relative strength of the two competing adsorption processes the effective isotherm can show new qualitative behaviour. Assume that both the isotherms for sorption to bulk soil and to WSSOM are of Freundlich form, given by $K_{f(soil)}$ and $1/n$, and $K_{f(WSSOM)}$ and $1/m$, respectively, with $1/n < 1$. Then the isotherm for sorption to bulk soil is of type (H) in the classification of Giles et al.[15], reflecting the strong sorption for low concentrations. If $1/m$ $1/n$, then the effective isotherm is also of the type (H), but it is of type (S) for $1/m < 1/n$.

As the WSSOM which is responsible for sorption of quinmerac is a macromolecule, it seems reasonable to use a reduced water content for the phase (C) reflecting the inaccessibility of small pores[14]. If we allow for this in the model, and also for different diffusion-dispersion coefficients, two additional terms appear in the differential equation for $C_{(A+C)}$. One is a source term, which can be interpreted as a further diminishing of the effective isotherm. The second can be interpreted as a reduction of diffusion. For the data obtained for quinmerac and an assumed reduction of water content of 10 %[14] these effects are negligible. Whether these effects may be prominent for other data sets has to be investigated further.

ACKNOWLEDGEMENTS

We thank Gisela Badewitz for able technical assistance. This work was financially supported by BASF Limburgerhof and by the Deutsche Forschungsgemeinschaft (Ko 1035/1-1).

REFERENCES

1. J.B. Weber and C.T. Miller, In B.L. Sawhney and K. Brown (Eds.), Reactions and movement of organic chemicals in soils, SSSA Spec. Publ. No. 22, Madison, Wisconsin 1989, 305.
2. T.M. Ballard, Soil Sci. Soc. Am. Proc., 1971, 35, 45.
3. G. Bengtsson, C.G. Enfield, and R. Lindqvist, Sci. Total Environ., 1987, 67, 159.
4. J.F. McCarthy and J.M. Zachara, Environ. Sci. Technol., 1989, 23, 496.
5. P.M. Gschwend and S. Wu, Environ. Sci. Technol., 1985, 19, 90.
6. D.C. Bouchard, C.G. Enfield, and M.D. Piwoni, In B.L. Sawhney and K. Brown (Eds.), Reactions and movement of organic chemicals in soils, SSSA Spec. Publ. No. 22, Madison, Wisconsin 1989, 349.
7. H. Deschauer and I. Kögel-Knabner, Sci. Total Environ., 1991, submitted.
8. B. Wuerzer, R. Berghaus, H. Hagen, R.D. Kohler and J. Merkert, Proc. Brit. Crop Prot. Conf. - Weeds, 1985, 1, 63.
9. H. Deschauer and I. Kögel-Knabner, Chemosphere, 1991, in press.
10. D.S. Gamble, M.I. Haniff, and R.H. Zienius, Anal. Chem., 1986, 54, 732.
11. R.E. Grice, M.H.B. Hayes, P.R. Lundie, and M.H. Cardew, Chem. Ind., 1973, 3, 233.
12. A. Moreale and R. Van Bladel, J. Environ. Qual., 1980, 9, 627.
13. Y.A. Madhun, J.L. Young, and V.H. Freed, J. Environ. Qual., 1986, 15, 64.
14. C.G. Enfield, G. Bengtsson, and R. Lindqvist, Environ. Sci. Technol., 1989, 23, 1278.
15. C.H. Giles, D. Smith, and A. Huiton, J. Colloid Surface Sci., 1974, 47, 755.

The Role of Soil Organic Matter in Pesticide Movement via Run-off, Soil Erosion, and Leaching

T.R. Harrod, Andrée D. Carter, and J.M. Hollis

SOIL SURVEY AND LAND RESEARCH CENTRE, SILSOE CAMPUS, SILSOE, BEDFORD, MK45 4DT, UK

1 INTRODUCTION

The movement of pesticides in soil is of concern following the discovery of residues of commonly used agrochemicals in ground and surface waters (Lees and McVeigh 1988)[1]. The E.C. Drinking Water Directive (European Community 1980)[2] sets maximum concentrations of single pesticides at 0.1 µg l^{-1} and total pesticides at 0.5 µg l^{-1}. Interest has focused on soil and environmental factors which, by interacting with inherent physiochemical properties of pesticides (Nicholls 1988)[3], may influence their movement. Organic matter directly influences soil adsorption of pesticides and it directly and indirectly increases water holding capacity and structural stability. Downward percolation and leaching of soluble pesticides are lessened by increased water holding capacity, while structural stability reduces erosion and loss of particles in run-off. This paper seeks to outline important processes and then report on a case study.

2 PESTICIDE ADSORPTION

Soil organic matter has a strong positive correlation with the physical adsorption of most pesticides [3]. Soil/water partition coefficients (K_D ml g^{-1}) can be used to derive comparative measure of compound mobility (K_{OC}) by correction for organic carbon (Gustafson 1989)[4]. Organic matter content varies according to clay content, moisture regime and land use (Table 1). It follows that a compound of a given K_D will be more mobile in freely draining sandy soils than in seasonally wet loamy ones and will be more mobile under continuous arable than under rotations with leys, due to the differences in organic matter content.

Mobility classes can be assigned to the range of K_{OC} values which when integrated with a rating of persistence (half life in soil) and classification of soil and substrate hydrologic properties, enable assessment of

TABLE 1. AVERAGE TOPSOIL ORGANIC CARBON CONTENT FOR TWO TEXTURES WITH DIFFERING WATER REGIME AND LAND USE

Texture and water regime	Arable	Ley	Long term grass	Semi-natural
Loamy sand, freely draining	1.1 (43)	1.8 (8)	2.3 (10)	2.5 (12)
Loamy sand, seasonally wet	1.8 (16)	2.5 (1)	2.8 (6)	2.6 (3)
Clay loam, freely draining	2.5 (128)	3.2 (68)	4.2 (160)	5.5 (30)
Clay loam, seasonally wet	2.5 (181)	3.1 (77)	4.1 (243)	5.7 (57)

Bracketed figures are numbers in sample. Based on sampling at 5 km intervals across England and Wales.

pollution risk to aquifers and surface waters (Hollis 1990)[5]. Greatest risk to shallow aquifers is assessed when persistent and highly mobile pesticides are applied to soils where structure encourages by-pass flow or where coarse textured topsoils form over gleyed profiles which indicate shallow water tables. Surface waters are most readily affected where impermeable or alluvial substrates are overlain by profiles with gleying close to the surface or having very small integrated air capacity (drainable porosity).

The type and age of organic matter strongly influence adsorption; for example dissolved organic matter can have a higher K_{OC} value than soil organic matter. This can enhance pesticide mobility and compounds may be found at greater depth in the soil than originally expected. The degree of hydration of organic matter can also influence adsorption. In dry soils there is less sorption and increased likelihood of persistence. Boesten and van den Linden[6] showed that the natural variability of the K_{OC} of atrazine was 68 ± 23 ml g^{-1}. Leaching models show K_{OC} to be a very sensitive parameter, with predictions of amounts leached varying up to 100 fold, according to input data. The critical nature of K_{OC} can be shown by the need for modelling and leaching studies in West Germany for registration of compounds when K_{OC} values are less than 500 ml g^{-1}, (BBA 1990)[7]. Sandy soils with under 1.5 % organic carbon are studied to represent the greatest risk of leaching of potentially mobile compounds.

3 SOIL STRUCTURE AND WATER RETENTION

The place of organic matter in maintaining soil structural stability is universally recognised, with organic carbon content of about 2% a critical value (Greenland, Rimmer and Payne 1975)[8]. This effect is acknowledged in advice on control of soil erosion by water, as in MAFF Leaflet 890 (1984)[9] and Colborne and Staines (1987)[10] and in many models of erosion risk, for example the recent U.K. Soil Protection Scheme (ITE 1989)[11]. In the latter a similar organic matter level is taken as a threshold value.

Organic matter increases soil water retention, particularly at suctions below 2 bars (Pembridge 1989)[12]. Most soil organic matter is hydrophilic and absorbs large quantities of water relative to mineral soil. However some may be inherently hydrophobic, or become so on dehydration. In many arable soils (Table 1) organic matter content is too small for significant direct effects on water holding, but its indirect improvement of soil structure can be important. With well developed structure soil water retained at low suctions is greater and is potentially mobile (Addiscott 1977)[13] and is easily displaced by rainfall, increasing the contamination risk of water bodies.

4. A CASE STUDY

A) Run-off and Water Erosion of Soil

Recent comparison of freshly deposited colluvium with the source soil in fields eroded by water in the south west of England have demonstrated (Table 2) how erosional and related processes can enhance or deplete organic matter levels, according to circumstances. In west Cornwall light or medium loamy soils generally contain more than 2% organic carbon, while on silty or very fine sandy soils in south Somerset amounts are nearer 1%. The Cornish soils are stony and their sand fraction ranges from very fine to coarse, while at the Somerset sites silt and very fine sand is dominant and they are without material coarser than 100 μm. Colluvium at the Cornish sites tends to have organic carbon amounts increased above that in the parent site, in south Somerset depletion appears more usual. There is a parallel enhancement of clay and silt in many of the Cornish samples, while in the Somerset colluvia clay amounts are reduced to 0.20 and 0.22 of the field contents.

Soil erosion by water is dependent on a number of interacting considerations. Precipitation clearly needs to be sufficient to induce run-off, but this is strongly modified by slope, soil textural and infiltration properties, as well as land use, including crop or vegetation cover and cultivation practices.

TABLE 2. ORGANIC CARBON % IN COLLUVIUM

Site	Field centre	Colluvium	Concentration factor*
a) Cornwall			
Titanic	4.1	5.2	1.27
Offdowns	4.3	7.5	1.74
Crowan	4.5	7.6	1.69
Bostrase	2.7	3.7	1.37
Tregembo	2.7	3.5	1.28
Relubbus	2.6	4.0	1.50
Godolphin Bridge	2.9	3.8	1.28
Mean			1.45
b) Somerset			
Ben Cross	1.1	0.5	0.45
Coleford Bridge	1.2	0.1	0.08

* Concentration factor = $\dfrac{\text{colluvium}}{\text{field centre}}$

The initial stages of soil erosion by water, particularly on relatively recently cultivated or structurally stable soils which are reasonably endowed with organic matter, may involve movement of structural aggregates or fragments, which are usually redeposited (after only minor abrasion, within short distances downslope, as at the end of a crop row). Such movement is unlikely to achieve any serious modification to soil particle-size distribution other than reduction of stone content.

On soils that have been wetted and exposed to rain impact there may be surface slaking and capping, particularly if organic matter amounts are small. Then there can be preferential movement of relatively fine material and organic matter which may be removed in any run-off, leaving behind the coarser fractions, without obvious rilling of the soil surface. The process also seals surface pores and fissures so that future run-off is increased.

With more vigorous erosion, involving transportation over greater distances, the aggregates are destroyed and the soil dispersed. Redeposition is then as laminated colluvium in which sorting into limited particle-size ranges is typical. In practice colluvium at the lower end of fields can contain laminae comprising rolled, aggregated soil as well as bands of sorted material. Again this form of deposition can occur without the production of rills upslope, but as the erosion intensifies the probability that such channels will be cut increases.

The amount and proportion of colluvium deposited within the field depends on local circumstances such as ground configuration and amounts and velocity of run-off. Where erosion has occurred, any run-off leaving the field, even after having contributed to deposition, is always turbid, containing suspended organic matter, finer soil particles and colloidal material. It will approach or enter watercourses, drains and ditches, inevitably introducing the risk of water pollution by both soil particles and by any chemicals adsorbed on them.

The harvesting of crops at times when the land is unsuitably wet results in compaction, puddling and slurrying of the soil surface by traffic. This gives rise to pools, ruts and puddles containing mud and turbid water and can lead to the redeposition of slurried soil as colluvium, as well as soil erosion and deposition, along with turbid run-off leaving the land via drains and watercourses. Such management of the soil is most often confined to land with high value crops harvested in the winter, although in wet seasons lifting of more normal crops, such as sugar beet, may bring about comparable damage.

B) Movement of Pesticide Residues

A national survey of organochlorine pesticide residues in eels (Milford 1989)[14] revealed high concentrations of dieldrin in samples from the Penzance area. As a result a study of pathways within catchments in West Cornwall was commissioned by National Rivers Authority, South West Region.

Although land cover in the west of Cornwall is dominated by grassland, both daffodils and potatoes are traditional crops. All 3 crops are grown on generally well drained light or medium loamy soils, in rotations which can also involve winter cauliflower (broccoli), cabbage and cereals. The insecticide aldrin has been applied for many years to the soil of bulb fields for the control of narcissus fly and to eradicate wireworm in pastures ploughed prior to potatoes.

On sloping land, potatoes, daffodils, broccoli and spring cabbage tend to be planted up and down the slope. Potatoes and daffodils are always planted in ridges; broccoli and cabbage may be banked up for protection in exposed sites. For cereals or new leys direction of working is less critical. However many fields have their long axes aligned with the slope, encouraging working that way. Soil erosion can take place if sufficient rain occurs before consolidation of the soil or development of the crop cover.

In early spring, harvesting daffodil flowers involves daily foot traffic between the ridges when the soil is moist or wet. This inevitably puddles and compacts the topsoil and restricts infiltration of rainwater. At the same time frequent tractor traffic along the headlands to collect the picked flowers ruts the ground and in wet weather creates a slurry of soil and rainwater. Downslope, and often streamward, movement of this slurried soil from headland wheelings can be considerable during heavy rain. On redeposition it forms fine, stoneless colluvium which may be laminated, as after erosion.

In wet weather removal of soil by small scale water movement down the flanks of the ridges and along furrows between them can be a very subdued process. It can proceed without clear scouring out of rills normally associated with soil erosion with only limited signs of water movement after events.

That run-off leaving fields where agricultural slurrying of the land or erosional processes have taken place can, by transporting soil material, introduce adsorbed pesticide residues to watercourses is clearly shown by Table 3. The dieldrin recorded is a metabolite of the pesticide aldrin. The half life of aldrin is given by Edwards (1973)[15] as 0.3 year, that of dieldrin 2.5 years. Rain on 14th March 1989 produced run-off across daffodil fields, entering the Newlyn River via the ditch in Titanic Field and the Trereife Stream. Percolation via land drains gave aldrin and dieldrin amounts (77 and 70 ng litre^{-1}) at the source of the Titanic Ditch, much greater than at low flows (typically about 2 and 21 ng litre^{-1}). However overland flow of water with a large suspended solid load raised the ditch water content to 650 and 670 ng litre^{-1} at the outfall, where low flow values are close to those at the source. Run-off leaving daffodil fields into the Trereife Stream during the same rainfall clearly had a similar effect. In the stream above the main bulb fields neither aldrin nor dieldrin were detectable when stream flow had fallen the next day, while at the confluence with the Newlyn River amounts fell to 9 and 15 ng litre^{-1} of aldrin and dieldrin respectively.

TABLE 3. PESTICIDE RESIDUES NEAR NEWLYN, MARCH 1989

	Aldrin µg kg^{-1}	Dieldrin µg kg^{-1}
Titanic Field		
a) Solids: Field soil	300	150
Colluvium	2474	580
Ditch sediment	149	59
b) Water: (14/3/89)	ng litre^{-1}	ng litre^{-1}
Ditch source	77	70
Ditch outfall	650	670
Overland flow depositing colluvium and entering ditch containing 1130 mg/litre^{-1} suspended solids	960	890
Trereife Stream		
Water: (14/3/89)		
Above main bulb fields	9	15
Run-off from bulb field entering stream	77	240
At confluence with Newlyn River	110	76

Table 3 illustrates the enhancement in the colluvium of aldrin (over 8 fold) and of dieldrin (nearly 4 fold) against field soil levels in Titanic Field. This mirrors, but strongly exceeds, the enhancement of colluvial organic matter, clay and silt (Tables 2 and 4). Similar levels of enhancement of aldrin (Table 5) were present at 6 out of 8 sites where colluvium had been deposited, while at 5 of the 8 locations relatively minor (on average 3 fold) enhancement of dieldrin had occurred.

Accounting for these enhancements, including the generally greater effect on aldrin than on dieldrin, must be a matter of some interest which has yet to be addressed. While the concentrations of organic carbon and silt in colluvium in the Cornish sites are almost always greater than that of clay (Tables 2 and 5), there are clearly no simple relationships with total organic matter. The comparative susceptibility of those organic complexes with greater affinity for the pesticide residues to erosional removal from the soil or to deposition within the colluvium may be a possible explanation.

TABLE 4. CLAY (<2 µm) AND SILT (2-60 µm) % IN COLLUVIUM

Site	Clay			Silt		
	Field centre	Colluvium	Concentration factor	Field centre	Colluvium	Concentration factor
a) Cornwall						
Titanic	13	17	1.31	29	50	1.72
Offdowns	15	16	1.07	42	72	1.71
Crowan	21	30	1.43	35	64	1.82
Bostrase	27	31	1.15	47	65	1.38
Tregembo	29	28	0.97	44	67	1.28
Relubbus	19	19	1.00	52	76	1.46
Godolphin Br.	28	21	0.75	56	68	1.33
Mean			1.09			1.63
b) Somerset						
Ben Cross	16	4	0.25	43	32	0.74
Coleford Br.	18	9	0.50	43	44	1.02

TABLE 5. ALDRIN AND DIELDRIN ($\mu g\ kg^{-1}$) IN COLLUVIUM

Site	Aldrin			Dieldrin		
	Field centre	Colluvium	Concentration* factor	Field centre	Colluvium	Concentration* factor
Titanic	300	2474	8.24	150	580	3.86
Offdowns	33	1499	45.4	74	600	3.44
Crowan	56	174	1.32	281	600	2.13
Bostrase	2	103	51.5	10	158	15.8
Trenear	<2	11	>5.5	66	150	2.27
Tredrea	<2	<2	1	34	20	0.59
Godolphin Bridge	258	20	0.08	400	43	0.11
Roskennals	56	1300	2.32	1100	710	0.64
Mean			14.42			3.61

*Concentration factor = $\dfrac{\text{Colluvium}}{\text{Field Centre}}$

An aldrin treated field on the River Hayle floodplain, now in cereals, is subject in the winter to temporary surface ponding of water when groundwater rises. Movement of the water was only evident between pools, where although very limited it became slightly turbid. The pools were undisturbed by livestock or agricultural activity and were not affected by direct entry of water from upslope. On the fall of groundwater in the spring a film of fine sediment remained on the site of the lower pools. This material (Table 6) showed concentration of aldrin, dieldrin and organic matter of a similar order to several sites in Tables 2, 4 and 5, with clay concentrated somewhat more (1.61 fold).

In view of the scale of concentration of aldrin and dieldrin in colluvium at so many of the Cornish sites, comparable effects involving other pesticides deserve consideration. From the range of pesticides used on daffodils analysis for simazine and carbendazim was undertaken, again comparing field amounts against fresh colluvium (Table 7). While the residues of simazine show no concentration, at 2 of the 4 sites sampled for carbendazim there were substantial (13 and 15 fold) enhancements in the colluvium.

5 DISCUSSION

The Cornish work demonstrates that both soil erosion and slurrying of the ground by ill-timed farming operations can mobilise pesticides associated with organic matter in run-off. This mobilisation can involve their concentration along with the organic matter and finer mineral fractions of the soil. Whether the resulting load is redeposited as colluvium or carried directly into watercourses depends on local circumstances. This has implications for all land where erosion is a risk or where the land is used when it is wet, producing turbid water and mud. At the sites in Cornwall the colluvium had retained substantial amounts of soil organic matter, although what proportion of the total remains to be determined. In the coarser textured soils sampled in South Somerset, which are more representative of land with high erosion risk under British agricultural conditions, organic matter (Table 2) and clay were substantially depleted in the colluvium and are most likely to have been carried into watercourses.

TABLE 6. SOIL AND PESTICIDE RESIDUES IN POOL SEDIMENT

	Clay % (<2µm)	Silt % (2-60µm)	Organic carbon %	Aldrin µg kg^{-1}	Dieldrin µg kg^{-1}
Field soil	26	55	3.4	<2	23
Pool sediment	42	58	4.6	11	62
Concentration factor*	1.61	1.05	1.34	>5.5	2.69

* Concentration factor = $\dfrac{\text{sediment}}{\text{field soil}}$

TABLE 7. SIMAZINE AND CARBENDAZIM IN COLLUVIUM

Site	Simazine µg kg^{-1}			Carbendazim µg kg^{-1}		
	Field	Colluvium	Concentration* factor	Field	Colluvium	Concentration* factor
Bostrase	72	74	1.03	95	1500	15.8
Tregembo	44	45	1.02	45	590	13.1
Relubbus	-	-	-	106	88	0.83
Godolphin Br.	-	-	-	680	650	0.95

*Concentration factor = $\dfrac{\text{colluvium}}{\text{field soil}}$

6 ACKNOWLEDGEMENTS

The work reported here was funded by the National Rivers Authority, South Western Region and the Ministry of Agriculture, Fisheries and Food. Pesticide analyses were undertaken by South West Water Authority, Exeter, the Water Research Centre, Medmenham and Wessex Water, Bath. Soil particle size and organic matter were analysed by the Soil Survey and Land Research Centre, Silsoe.

7 REFERENCES

1. LEES, A. and MCVEIGH, K. (1988). An investigation of pesticide pollution in drinking water in England and Wales. Friends of the Earth, London.
2. EUROPEAN COMMUNITY. (1980). Council Directive on the quality of water for human consumption. (80/778/E.E.C.). D. of E. Circular 20/82 H.M.S.O. 1982.
3. NICHOLLS, P.H. (1988). Factors influencing entry of pesticides into soil water. Pesticide Sci. **22**, 123-37.
4. GUSTAFSON, D.I. (1989). Groundwater ubiquity score; a simple method for assessing pesticide leachability. Environmental Toxicology and Chemistry. **8**. 339-57.
5. HOLLIS, J.M. (1990). A map of pesticide leaching risk. (Unpublished). Soil Survey and Land Research Centre, Silsoe.
6. BOESTEN, J.J.T.I. and VAN DEN LINDEN, A.M.A. (in press). Model calculations on the influence of sorption and transformation on pesticide leaching to groundwater and persistence in the plough layer. J.Env. Quality. **1**.
7. B.B.A. (FEDERAL BIOLOGICAL RESEARCH CENTRE FOR AGRICULTURE AND FORESTRY). (1990). Guidelines for the testing of agrochemicals. Part 4-3. Lysimeter tests to establish the mobility of agrochemicals in the subsoil. Braunschweig.
8. GREENLAND, D.J., RIMMER, D. and PAYNE, D. (1975). Determination of the structural stability of English and Welsh soils, using a water coherence test. J. Soil Sci. **26**. 294-303.
9. M.A.F.F. (1984). Soil erosion by water. Leaflet 890.
10. COLBORNE, G.J.N. and STAINES, S.J. (1987). Soils in Somerset 1. Sheet ST41/51 (Yeovil). Soil Surv. Rec. No.111.
11. I.T.E. (1989). An assessment of the principles of soil protection in the U.K. Institute of Terrestrial Ecology Report PECD 7/2/45.
12. BEMBRIDGE, CHRISTINE. (1989). Water retention properties of organic soils and the problems associated with laboratory measurements. Soil Survey and Land Research Centre Report R3805.
13. ADDISCOTT, T.M. (1977). A simple model for leaching in structured soils. J. Soil Sci. **28**. 554-63.
14. MILFORD, B. (1989). Organochlorine pesticide residues in freshwater eels. South West Water Environmental Protection Report EP/WZ/89/2.
15. EDWARDS, C.A. (1973). Persistent pesticides in the environment. CRC Press, Ohio.

Section 3

Organic Matter and Soil Structure

Introductory Comments

by D.A. Rose, University of Newcastle upon Tyne

The interaction between physics and soil organic matter is in two directions. First, physical techniques are used with increasing success to identify components of soil organic matter and to quantify the chemical and biological properties of these components (see Section 1 of this volume), often non-destructively. Second, organic matter affects the physical properties of the soil, in particular by being associated with an improvement in both the structure of the soil and the persistence or stability of such structure, however imperfect or ambiguous our definitions of structure and structural stability. Such improvements in structure and stability occur because chemical and biological agents in the soil organic matter increase both the degree of aggregation of the primary mineral particles and the strength of the bonds between these mineral particles.

The papers in this section fall into three distinct subject groups.

(i) Two papers consider the effect of organic matter in creating and improving soil structure. Waters and Oades use scanning electron microscopy to confirm that soil aggregates are not random assemblages of smaller particles but that soil particles are stabilized in increasingly larger units by different organic binding agents, supporting the hypothesis of aggregate hierarchy in soils. Rose demonstrates that long-continued applications of farmyard manure to arable soils have beneficial effects on soil structure by increasing both the amount of pore space within the aggregates and the quantity of water available to plants.

(ii) Three papers discuss the stability of soil structure. Gabriels and Michiels review the role of organic matter in protecting soils from erosion by water. Swift suggests that soils contain aggregates whose stabilities to wetting differ widely, and that change in the average value of a stability index is caused by a change in the proportion of unstable aggregates rather than a general change across all aggregates. Rogers, Cook and Burns demonstrate that extracellular polysaccharide-producing phototropic microorganisms used as soil conditioners can slow or reverse the degradation of soil structure.

(iii) Three papers assess the effect of soil physical conditions or agronomic practice on soil organic matter. Howson shows that the rate of breakdown of cellulose in soil can be inferred from the decomposition of a strip of cotton cloth, relates this breakdown rate to the annual potential evaporation and demonstrates the utility of the technique in ecological studies. Kaiser, Walenzik and Heinemeyer conclude that soil compaction causes both microbial biomass and organic carbon in the soil to decrease. They suggest that this decrease is due to poor soil aeration resulting from an increase in the percentage of the water-filled pore space of the soil. Mendonça, Moura Filho and Costa relate the degradation of organic matter by oxidation to soil structure and crop type, though their results may have been confounded by soil differences because the clay content of the soil under grass (59%) was considerably higher than under natural forest or rubber trees (c.42%).

Finally, there is an agenda for the future. The papers in this section demonstrate three needs which must be addressed if progress is to continue. First, the authors, from different scientific disciplines, use a plethora of undefined, ill-defined and/or contradictory terms and unsatisfactory units to describe aspects of soil structure. Some consistency in definitions and terminology must be imposed, particularly to aid multi-disciplinary work, similar to that commissioned on soil water by the International Society of Soil Science and published in 1976. Second, our work is descriptive, and treated only by statistical methods which merely redescribe the data or confirm significant differences between observations. There is an urgent need to develop quantitative, mechanistic models of the creation and degradation of soil structure which will act as an intellectual focus for our experimental studies. These experiments should be critical tests of mechanistic hypotheses, rather than observations of structural phenomena. We also need to refine our measures of structural stability and to inter-relate these measures for stability to different stresses - wetting/drying, freezing/thawing, salinity, mechanical pressure. Third, there is the problem of the hidden or unmeasured, yet controlling, variable which we often overlook. For example, Kaiser et al. conclude that the effect of compaction on soil microorganisms is due to a lack of aeration, but fail to estimate or measure the air-filled pore space. Using the information in their paper and assuming a soil density of 2.60 mg m^{-3}, the air-filled porosities arising from compactive treatments I, II and III (their Table 1) are 0.22, 0.14 and 0.07 m^3(air) m^{-3} respectively, clearly demonstrating the point they wish to make. Similarly, air-filled porosities (their Table 2) decrease from 0.31 to 0.23 (no traffic) and from 0.22 to 0.15 (high traffic) m^3(air) m^{-3} between November and December 1988. The percent water-filled pore space is irrelevant. How much of what we do is not relevant to the problems we seek to solve?

Soil Organic Matter and Water Erosion Processes

D. Gabriels and P. Michiels

NATIONAL FUND FOR SCIENTIFIC RESEARCH, BRUSSELS, BELGIUM AND STATE UNIVERSITY GENT, FACULTY OF AGRICULTURAL SCIENCES, DEPARTMENT OF SOIL PHYSICS, COUPURE LINKS 653, B-9000 GENT, BELGIUM

Introduction

Soil erosion is a combination of three processes : (1) sediment detachment by rainfall; (2) sediment entrainment by runoff water; and (3) sediment deposition. Hence, a clear distinction should be made between "on-site" and "off-site" erosion processes and problems. Crust formation, soil and nutrient losses cause a yield reduction "on-site" which can eventually lead to a complete degradation of the land. Flooding and sedimentation are major problems "off-site". Dissolved and sorbed nutrients cause eutrophication "off-site", which can lead to chemical pollution of surface waters.

In this paper, the effect of organic matter on water erosion is analyzed. When dealing with erosion processes, three questions arise from the viewpoint of soil technology.

1) where does organic matter originate
2) how does organic matter affect soil properties and
3) what is the effect of soil properties on erosion ?

Where does organic matter originate ?

Organic material can be stored either "above the soil", "on the soil" or "in the soil".

The organic carbon distribution "in the soil" due to native vegetation is essentially the same in tropical and temperate regions. Forested areas show a marked organic matter accumulation in the topsoil as a result of litter fall and superficial tree roots. Savannas and prairies generally produce more carbon in the subsoil because of the decomposition of deep grass roots.

The organic carbon content of a soil _in equilibrium_ with its natural vegetation is a function of the annual addition and decomposition of soil organic carbon. The following formulae explain the relationships between synthesis and decomposition under steady state conditions:

$$C = b.m/k \tag{1}$$

and

$$a = b.m = C.k \tag{2}$$

where C = the amount or percentage of organic carbon stored in the soil at equilibrium (tonnes ha^{-1});
b = the annual amount of fresh organic matter added to the soil (tonnes ha^{-1});
m = the conversion rate of fresh organic matter into soil organic carbon (humification coefficient) (percent);
a = the annual addition of soil organic carbon (tonnes ha^{-1}); and
k = the annual decomposition rate of soil organic carbon (mineralisation coefficient) (percent).

Table 1 summarizes the magnitude of the parameters used in equation (1) and (2) for several natural vegetations.

Table 1 Annual additions (b) and conversion rates (m) of fresh organic matter and annual additions (a), decomposition rates (k) and equilibrium levels (C) of topsoil organic carbon in some tropical and temperate locations[1] (recalculated data[2]).

	Tropical forests	Temperate forests	Tropical savannas	Temperate prairies
b (tonnes ha^{-1})	3.8-6.0	0.8-1.7	0.4-1.4	1.4
m (%)	47-51	47-52	43-50	37
a (tonnes ha^{-1})	2.0-2.9	0.4-0.9	0.2-0.7	0.5
k (%)	2.0-5.2	0.4-1.0	1.2-1.3	0.4
C (tonnes ha^{-1})	55-400	87	16-55	134
C (%)	1.2-9.0	2.0	0.4-1.2	3.0

The annual addition of fresh organic matter such as litter, branches, and dead roots (b) is of the order of 5 tonnes ha^{-1} of dry matter in tropical forests, and about 1 tonnes ha^{-1} in temperate forests. The actual ranges in the literature are from 3 to 15 tonnes ha^{-1} in tropical forests and 1 to 8 tonnes ha^{-1} in temperate forests. Tropical savannas add from 0.5 to 1.5 tonnes/ha^{-1}, and temperate prairies about 1.5 tonnes ha^{-1}. Due to faster growing, tropical forests add about 5 times more raw organic matter to the soil as do temperate forests. Fresh organic matter additions in grasslands are similar for tropical and temperate regions.

The conversion rate (m) of fresh organic matter into soil organic carbon is of the order of 30 to 50 percent per year. These rates are nearly the same for tropical and temperate vegetations.

Therefore, the annual additions of soil organic carbon (a) are about four times higher in tropical than in temperate forests. For tropical and temperate grasslands, these annual additions are fairly similar.

The annual decomposition rates (k) of soil organic carbon vary considerably. These rates are 2 to 5 percent in tropical forests, but 0.4 to 1 percent in temperate forests, probably as a result of temperature limitations. The annual decomposition rate of soil organic carbon in tropical savannas averages 1.2 percent, or three times that of temperate prairies.

Tropical rain forests produce about five times as much biomass and soil organic matter per year as do temperate forests. The rate of organic matter decomposition is also about five times greater than in temperate forests. Thus the organic matter contents stored "in the soil" at equilibrium are similar for tropical and temperate forests. The organic matter contents stored "above the soil" and "on the soil" are up to five times higher for tropical forests than for temperate forests.

The commonly held view is that the organic matter contents of tropical soils is low compared to temperate soils because of the high temperatures and fast decomposition rates. However, contrary to commonly held views, the organic matter contents in tropical soils are not very different from those in the temperate regions. The traditional way to assess the organic carbon stored "in the soil" is by the soil colour. The absence of a direct relationship between soil colour and organic matter content is one of the explanations why tropical soils are higher in organic matter content than is generally believed. Many red Oxisols can have higher carbon contents than black Vertisols. The apparent colour of the soil is affected more by its moisture content and by its amount of oxides than by its amount of organic matter.

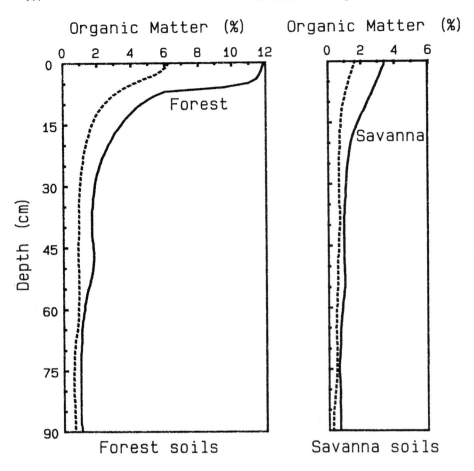

Figure 1   Organic matter distribution in typical West African forest and savanna soil profiles before (solid line) and after (broken line) cultivation[3].

The annual decomposition rate (k) of soil organic carbon increases with increasing temperature, aeration, cultivation, and denitrification. Annual organic carbon additions are drastically decreased when forests are brought under cultivation; crop residues usually provide only a fraction of the 5 tonnes ha^{-1} of dry matter previously supplied by the forest. Since the annual decomposition rate (k) increases with temperature and aeration, cultivation accelerates organic carbon decomposition. In the savannas, exposure and plowing results in a fourfold increase in k relative to the equilibrium values. Figure 1 shows the general effects of cultivation on the organic carbon contents in two West African soil profiles.

Inorganic fertilization can increase the annual addition of soil organic carbon by a large amount because of its effect on plant growth, and increase in the amounts of residues added to the soil. In properly managed systems, it can also reduce the decomposition constant. In the tropics the organic matter can be kept at high levels with good management practices.

Soil aggregation, aggregate stability or how does organic matter affect soil properties ?

It is generally believed that the chemical and physical soil properties are influenced by a modification of the organic matter content. Soil aggregation and aggregate stability are the main physical properties of the soil which affect its susceptibility against the impact of raindrops, runoff water and consequently soil erosion.

Organic matter contributes to the binding of soil particles and hence the formation of aggregates. The stability of these aggregates depends on the organic binding agents involved and can be classified as:

(1) <u>transient</u> binding agents, or organic materials which are decomposed rapidly by microorganisms. Polysaccharides are the most important group of these binding agents. They are produced rapidly and decomposed rapidly. The aggregation of this kind of organic matter is associated with large (>250 µm diameter) transiently stable aggregates[4];

(2) <u>temporary</u> binding agents such as thin roots and fungal hyphae, build up in the soil within weeks or months, and they persist for months or years, they also stabilize macroaggregates. Macroaggregation is controlled by soil management, i.e. crop rotations[5], because such management influences the growth of plant roots and the oxidation of organic carbon;

(3) <u>persistent</u> organic binding agents which are characteristic of the soil, controlling the water stability of micro-aggregates. These binding agents are resistant components associated with polyvalent metal cations, and strongly sorbed polymers. Organo-mineral associations function as binding agents in stable aggregates, particularly for those less than 250 µm in diameter[6,7,8].

In the surface layers of many agricultural soils, organic matter plays an important role in binding soil particles to aggregates, enabling the soil to withstand stresses caused by rapid wetting and by the impact of raindrops. In this respect, suitable laboratory methods have been developed to assess the aggregate stability or the soil structure building by organic matter. The wet-sieving methods[9,10] and the water-drop techniques[11,12] are the most widely used.

Hartmann and De Boodt[13] found that a sandy soil with moderate to high organic matter content (4.33 %) shows the same aggregation pattern than a sandy loam soil. The correlations between the organic carbon content in soils and the water-stability of aggregates have not always been significant because only part of the organic matter may be responsible for the water-stable aggregates, or the organic carbon may not be the only binding agent, or the stability may be more related to the disposition or the free organic material rather than to the type or amount of the organic matter[5].

Soil erodibility, or what is the effect of soil properties on erosion processes ?

One of the factors involved in the on-site erosion problem is the susceptibility of the soil towards erosion, and in particular its vulnerability against the impact of raindrops and runoff water. In terms of the Universal Soil Loss Equation[14] this susceptibility is called "erodibility".

The meaning of the term "soil erodibility" is distinctly different from that of the term "soil erosion". The rate of soil erosion on any area may be influenced more by land slope, rainstorm characteristics, cover, and management, than by the properties of the soil itself.

Some soils erode more readily than others, even when the slope, rainfall, cover, and management are the same. This difference, due to properties of the soil itself, is referred to as the soil erodibility.

Factors which affect the erodibility of the soil include, (1) those that influence the infiltration rate, permeability, and the total water capacity, (2) those that resist the rainfall forces which lead to splashing, dispersion and slaking and (3) those that resist the abrasion and transport of particles by runoff water.

A soil erodibility nomograph for farmland and construction sites as outlined in Figure 2, was developed[15]. Data were obtained from a detailed study which included laboratory determinations of soil physical and chemical properties, rainfall simulation tests, field plot experiments, and measurements of soil erosion from fallow plots under natural rainfall conditions. The nomograph (Figure 2) indicates how, and to what extent each of various properties of a soil affects its erodibility.

Usually a soil type becomes less erodible as the silt content decreases, regardless of whether the corresponding increase is in the sand fraction or in the clay fraction. Overall, the organic matter content is ranked next to particle-size distribution as an indicator of erodibility. However, a soil's erodibility is a function of the complex interactions of a substantial number of its physical and chemical properties, and it often varies within a standard texture class.

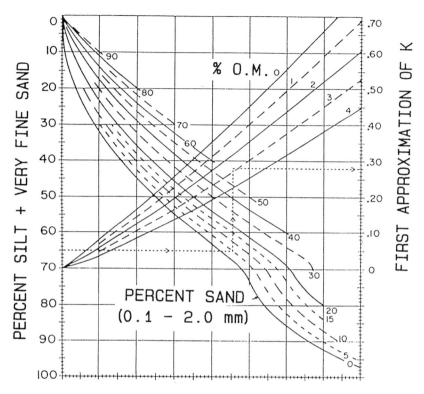

Figure 2   The soil erodibility nomograph[15].

Table 2   Erodibility K as determined from the nomograph (Figure 2) for different organic matter (O.M.) and sand contents.

sand %	O.M. %	erodibility K	erosivity R	soil loss ton/acre/yr
5	3	0.28	100	28
	1	0.36	100	36
20	3	0.35	100	35
	1	0.47	100	47
30	3	0.40	100	40
	1	0.53	100	53

An example of the effect of different contents of organic matter (1 percent and 3 percent) on the soil erodibility K, is given in Table 2 for soils containing 65 percent silt + very fine sand but with different contents of sand namely 5 percent, 20 percent and 30 percent. The nomograph is used to determine K (first approximation).

For a "unit" erosion plot, with a uniform standard slope, in continuous fallow, tilled up and down the slope, the soil loss is 100 K for 100 rainfall erosivity-index units per year. The soil loss corresponding to the 100 rainfall erosivity-index is given for each soil in the right-hand column of Table 3. It can be seen that the higher the sand content the greater the amount of organic matter is needed to decrease the soil losses.

Soil losses and losses of organic matter.

Loss of soil by erosion is a serious problem all over the world. Loss of topsoil by erosion has been shown to decrease the contents of organic matter, together with the fine clays. In this way, the sorbed nutrients such as nitrogen and phosphorus are also lost on-site. Available water holding capacity, plant rooting depth, soil productivity, and crop yields are also decreased[16,17].

Organic carbon losses from bare-fallow plots with slopes of 1, 5, 10 and 15 percent on an Alfisol in Western-Nigeria[18] varied from 54 to 3080 kg/ha^{-1}.

Table 3 Weight distribution of particles of different sizes and of organic matter in soils in an eroded slope[19].

Particle size class	% in initial field (1)	% in erosion base (2)	% in soil on meadow (3)	% in soil in field (4)
0-2 μm	4.1	1.4	6.5	0.6
2-20 μm	8.8	0.9	5.2	1.2
20-50 μm	16.4	11.2	35.4	5.5
50-100 μm	34.8	45.1	34.7	24.7
100-200 μm	29.7	34.6	15.5	54.8
200-500 μm	6.1	6.7	2.6	13.1
500-2000 μm	0.1	0.1	0.1	0.1
% O.M.	2.1	0.3	2.6	0.3

Rill and gully erosion was measured on a sloping arable field in Belgium[19], having a sandy loam topsoil according to the USDA classification[20]. Eight gullies in a 1.4 hectare field showed a total soil loss of 18.4 tonnes, or 1.3 mm of denudation when 213 mm of rain fell. Table 3 reports the results of the physico-chemical analyses of the soil at different positions on the gully: (1) soil in the initial field before gully development, (2) soil deposited at the erosion base, (3) soil deposited 10 metre below the erosion base (on the meadow), (4) soil left in the field after erosion.

A comparison between the texture of the soil left in the field and the texture of the soil in the initial field (Table 3) illustrates an alarming situation of erosion of sensitive soils, namely a selective impoverishment of the colloidal complex (0-50 µm and O.M.). The very fine sand (50-100 µm) and fine sand fraction (100-200 µm) are deposited on the erosion base; the clay and organic matter are carried with the runoff water further downslope on to the meadows where the water can infiltrate.

Addition of organic matter

The traditional way to increase organic matter in cultivated areas is to add undecomposed raw materials in the form of animal manures, composts, or plant materials incorporated as green manure.

Organic farming differs from conventional farming mainly in tillage methods, crop rotations, fertilizer applications, and pest control methods. Conventional farming systems depend on chemical fertilizers and pesticides while organic farming systems largely exclude their use and rely upon crop rotations, manuring, mechanical cultivation, organic fertilizers, and biological pest control to maintain soil productivity, supply plant nutrients, and control pests.

Incorporation of green manures is often effective only for the next subsequent crop, and not for complete rotation schemes since the organic matter contents are not maintained for the duration of the crop rotation[1].

Perhaps the most valuable approach to maintaining soil organic matter with cropping is to provide a surface mulch that will lower soil temperatures enough to prevent larger increases in decomposition rates and to provide protection against erosion. The uses of mulches may be applicable to a wider area than are animal or green manures.

The effect of organic mulches in reducing erosion can be attributed to reducing the direct impact of raindrops, maintaining maximum soil infiltrability, and decreasing the quantity, the velocity, and the transport capacity of runoff water[18].

One practical way to obtain mulch is by crop-residue management, using no-tillage or mulch-tillage techniques. Lal[18] found that mulching with straw at a rate of 4 to 6 tonnes ha^{-1} effectively prevented runoff and soil loss from slopes ranging from 1 to 15 percent, and the effectiveness of no-tillage treatments in preventing runoff and erosion was comparable to applying 4 to 6 tonnes ha^{-1} of straw mulch. Straw or hay should be anchored so that it will not blow or wash away. Cutback asphalt and asphalt emulsions were used to bind individual pieces of mulch to each other and to the soil surface[21,22]. A latex-emulsion was also effective in binding pieces of wood-bark to the soil surface, and hence in decreasing soil losses[23].

Gabriels[24] and Gabriels et al.[25] incorporated industrial composts (paper mill sludge mixed with bark, and household waste) at a rate of 30 tonnes per hectare in the upper 10 cm of a silt loam soil layer (USDA classification[20]) to improve soil structure building and sugar beet production. An increase of 1 percent in the organic matter content was found after one year, and there was a 7 to 14 percent increase in the dry weight of the beets. However, care should be taken of successive additions of those kinds of composts since this could result in an accumulation of heavy metals in the soil.

When economical mechanization and fertilizer practices are feasible, the maintenance of organic matter content can be considered as a minor management goal. In fact, adequate fertilizer practices can increase the organic matter content because of increased plant, stubble, or residue and root production. However, in poorly aggregated soils, susceptible to surface sealing, crusting and compaction, the maintenance and application of organic matter and mulching techniques are essential.

Conclusions

Organic matter affects the soil susceptibility towards "on-site" erosion through the effects of surface aggregate stability, surface sealing and crusting, and compaction.

Organic matter contributes to the formation and the stability of soil aggregates, although the correlation between the content of organic soil carbon and the water-stability of aggregates has not always been significant. Organic carbon is ranked next to particle size distribution as a mean indicator of erodibility.

Traditional ways to increase organic matter in cultivated areas include the use of animal manures, composts or green manures. Maintenance of soil organic matter is achieved through the use of surface mulches, such as straw, hay or industrial composts.

If there is not enough organic matter present, then the use of fertilizers and mulches should be considered as major management practices.

References

1. P.A. Sanchez (1976). Soil Organic matter. In Properties and management of soils in the tropics. ed. P.Sanchez, 162-183.
2. D.J. Greenland and P.H. Nye (1959). Increases in carbon and nitrogen contents of tropical soils under natural fallows. J. Soil Sci., 9, 284-299.

3. E.A. Brams. (1972). Cation exchange as related to the management of tropical soils. Mimeographed lecture presented at the Tropical Soils Institute, University of Puerto Rico. Prairie View. Texas A&M University, Texas.
4. A. Guckert, T. Chone and F. Jacquin (1975). Microflore et stabilité structurale des sols. Revue d'Ecologie et Biologie du Sol, 12, 211-223.
5. J.M. Tisdall, J.M. and Oades (1982). Organic matter and water-stable aggregates in soils. J. of Soil Sci., 33, 141-163.
6. A.P. Edwards and J.M. Bremner, J.M. (1967). Microaggregates in soils. J. of Soil Sci., 18, 64-73.
7. A.P. Hamblin (1977). Structural features of aggregates in some East Anglian silt soils. J. of Soil Sci., 28, 23-28.
8. L.W. Turchenek and J.M. Oades (1978). Organo-mineral particles in soils. In Modification of Soil Structure (eds W.W. Emerson, R.D.Bond and A.R.Dexter), 137-144, London, Wiley.
9. L. De Leenheer and M. De Boodt (1959). Determination of aggregate stability by the change in mean weight diameter. Int. Symp. Soil Structure. Meded. Landbouwhogeschool, Gent, 24, 290-300.
10. S. Hénin, G. Monnier and A. Combeau (1958). Méthode pour l'étude de la stabilité structurale des sols. Annales Agronomiques, Paris, 9, 71-90.
11. T.M. McCalla (1944). Waterdrop method for determining stability of soil structure. Soil Sci., 58, 117.
12. E. Bruce-Okine and R. Lal (1975). Erodibility as determined by raindrop technique. Soil Sci., 119, 149-157.
13. R. Hartmann and M. De Boodt (1974). The influence of the moisture content, texture and organic matter on the aggregation of sandy and loamy soils. Geoderma, 11, 53-62.
14. W.H. Wischmeier and D.D. Smith. (1978). Predicting rainfall erosion losses - A guide to conservation planning. Agr. Handbook No. 537. U.S.D.A. Washington, D.C. 58 pp.
15. W.H. Wischmeier, C.B. Johnson and B.V. Cross (1971). A soil erodibility nomograph for farmland and construction sites. Journal of Soil and Water Conservation, 26, 189-193.
16. R.E. Uhland (1949). Crop yields lowered by erosion. USDA-SCS-Tech. Paper-75, USDA Soil Conservation Service, Washington, D.C.
17. McDaniel, T.A. and Hajek, B.F. (1985). In Erosion and Soil Productivity ed. McCool 1985, 48-58, American Society of Agricultural Engineers, St. Joseph, Michigan.
18. R. Lal (1976). Soil erosion problems on an alfisol in Western Nigeria and their control. IITA Monograph No 1.
19. D. Gabriels, J.M. Pauwels and M. De Boodt (1977). A quantitative rill erosion study on a loamy sand in

the hilly region of Flanders. <u>Earth Surface Processes</u>, <u>2</u>, 257-259.
20. U.S.D.A. (1975). Soil Taxonomy. Agricultural Handbook no. 436. Soil Conservation Service U.S.D.A.
21. W.S. Chepil, N.O. Woodruff, F.H. Siddoway and L. Lyles (1960). Anchoring vegetative mulches. <u>Agric. Eng.</u>, <u>41</u>, 754.
22. W.S. Chepil, N.P. Woodruff, F.H. Siddoway, D.W. Fryrear and D.V. Armbrust (1963). Vegetative and nonvegetative materials to control wind and water erosion. <u>Soil Sci. Soc. Am. Proc.</u>, <u>27</u>, 86.
23. D. Gabriels. (1975). Preventing erosion with butadiene-styrene copolymer or polyurethane. Proc. Third Intern. Symp. Soil Conditioning, Gent, Belgium.
24. D. Gabriels (1988). Use of organic waste materials for soil structurization and crop production: initial field experiment. <u>Soil Technology</u>, <u>1</u>, 89-92.
25. D. Gabriels, P. Michiels, W. Cadron and J. Rejman (1989). The use of organic waste materials and plant cover for soil surface protection and crop production. Soil Technology Series 1, 133-137.

Effects of Humic Substances and Polysaccharides on Soil Aggregation

R.S. Swift

DEPARTMENT OF SOIL SCIENCE, UNIVERSITY OF READING, READING RG1 5AQ, UK

1 INTRODUCTION

Many observations and experiments carried out under both field and laboratory conditions have confirmed the importance of soil organic matter in the formation and stabilisation of soil aggregates[1,2]. At the same time it is clear that the observed effects are not due to the soil organic matter as a whole but to specific components of the organic matter. Two of the components which have been shown to be involved in the aggregation processes are the soil polysaccharides and soil humic substances[3,4,5]. This paper explores the relative ability of these two components to promote the formation of aggregates and the persistence of their stabilisation effects. In order to achieve this, studies have been carried out on natural and reformed soil aggregates. Further studies were carried out on natural soils to assess the importance of various organic components in field situations, particularly those in which the stability of aggregates is changing quite rapidly in response to changing cropping systems.

2 MATERIALS AND METHODS

Studies of natural soil aggregates were carried out using various soils from Scotland[1] and New Zealand[6,7]. Studies on reformed aggregates were carried out using samples from Stirling series soils which had been maintained under either long-term pasture or continuous arable production systems[4,5]. For studies on reformed aggregates additions of glucose or of polysaccharide were made following the destruction of natural aggregates by careful mechanical crushing or as a result of washing procedures carried out in order to produce mono-ionic soils[4]. All aggregate reformation studies involved incubation at 30°C for various periods of time following the addition of the particular treatment plus a soil inoculum. The polysaccharides used were two microbially produced exocellular polysaccharides, xanthan gum and alginate. Xanthan gum consists of an α-1,4 D-glucose backbone substituted on alternate residues with trisaccharide side chains. Alginate has a linear α-1,4 linked structure composed of D-mannuronate and L-guluronate residues. Additions of adsorbed humic substances also required the temporary adjustment of soil pH before returning the soil to the Ca^{2+} saturated form at pH = 7[5]. Full details of experimental procedures are given elsewhere[1,4,5,6,7].

Aggregate stability was measured by wet-sieving of rewetted aggregates through a nest of sieves[1]. The results are expressed as Mean Weight Diameter values (MWD) where;

MWD = Σ (percentage of sample on sieve X mean interseive size).

For the data presented the lower limit is 25 (for completely unstable aggregates and 240 (for stable aggregates).

3 RESULTS AND DISCUSSION

For soils with similar texture and mineralogical composition, it is usual to find a good correlation between organic matter content and aggregate stability. Good examples of such correlations (a) for a group of silt and clay loams and (b) for soils from individual soil series have been presented elsewhere by Swift and co-workers[1,7].

A significant portion of the organic matter in soils exists as discrete particles of decomposing plant remains, or as well-humified material. Large particles of such materials are unable to interact with the surfaces of mineral particles, and therefore cannot be directly involved in the formation and stabilisation of soil aggregates. Interaction with mineral particle surfaces occurs at the molecular level and must therefore involve the colloidal components of the organic matter.

In addition, it has frequently been observed that aggregate stability can change substantially, even though there is little change in the overall soil organic matter levels. It is apparent from observations such as these that particular fractions or components of the soil organic matter, such as soil polysaccharides or humic substances, are responsible to a large extent for the aggregation processes. It has not been possible to determine which of these components are most important by using correlation procedures. In general, the amounts of total organic matter, carbohydrate and humic substances are all highly correlated with aggregate stability. This is illustrated by the data given in Table 1 for a number of sandy clay loams and silty clay loam soils which tend to exhibit structural problems when organic matter levels decline.

Table 1   Correlation coefficients between aggregate stability (MWD) and various organic matter constituents for groups of medium and heavy textured soils.

Soil Group and (No.)	Soil Component		
	Organic Matter	Carbohydrate	Humic Substances
Kilmarnock (13)	0.7439**	0.8811***	0.6900**
Humbie (14)	0.7743***	0.7106**	0.7463**
Stirling (27)	0.8674***	0.7916***	0.7642***
Winton (21)	0.7057***	0.8906***	0.6136***
Miscellaneous (18)	0.8264***	0.7552***	0.8431***

* Significant at 5% level; ** significant at the 1% level; *** significant at the 0.1% level.

It is necessary to adopt other techniques to overcome this problem. Possible approaches are to make known additions of aggregating agents to soils, or to generate aggregating agents *in situ*. This work is carried out more effectively using reformed aggregates rather than existing natural aggregates due to the more effective mixing and homogenisation that can be achieved using the latter approach. An alternative approach is to selectively extract and identify the active aggregating agents from a variety of field soils. Both of these approaches have been used in the work described here.

Studies of Reformed Aggregates

Table 2 shows the results obtained following the incubation of soils after the addition of glucose, alginate or xanthan. All of these treatments were effective in producing stabilised, reformed aggregates from the crushed soil. Whereas the stabilising effects of the polysaccharides are due directly to the binding action of these compounds on the mineral soil particles, the results obtained from the glucose treatment are not attributable to the direct effects of the glucose but are due to the production of exocellular polysaccharides by micro-organisms as a result of metabolising the added glucose. The effects observed as a result of additions to the crushed soils were similar, irrespective of whether the polysaccharide was added directly or was formed *in situ* following the addition of glucose. The relatively stable aggregates were produced within a week and reached a maximum value within a period of 1 to 4 weeks.

For both arable and pasture soils, the glucose treatment produced the most stable aggregates. This indicates that the soil polysaccharides produced *in situ*

Table 2 Changes in stability with time of reformed aggregates (MWD values) following additions of glucose, alginate or xanthan to Stirling series soil.

Soil and treatment[1]	Aggregate Stability (MWD)				Natural Aggregates[2]
	Incubation period (weeks)				
	1	4	8	12	
Crushed samples					
Pasture + none	30	35	32	32	223
Pasture + glucose	204	172	128	90	
Pasture + alginate	196	165	61	51	
Pasture + xanthan	177	90	45	45	
Arable + none	27	28	27	27	118
Arable + glucose	119	132	100	62	
Arable + alginate	77	108	70	49	
Arable + xanthan	132	115	49	40	
Mono-ionic (Ca^{2+})					
Arable + none	27	27	ND	ND	118
Arable + glucose	37	30	ND	ND	
Arable + alginate	49	32	ND	ND	

[1] Glucose was added at the rate of 0.5 per cent W/W and alginate and xanthan at the rate of 0.2 per cent W/W.

[2] MWD values for the original, natural soil aggregates are given for comparative purposes.

ND = Not determined.

were more effective than the pure polysaccharides added. The aggregate stability imparted by these treatments did not persist at the maximum value but declined steadily with time as the polysaccharides were decomposed. Even so, at the end of 12 weeks there was still some residual aggregate stability, most notably with the glucose treatments which received larger additions of metabolisable carbon. It should be noted that little or no aggregate stabilisation occurred in the control soils (Table 2) indicating the importance of the added or synthesised polysaccharides in the formation of stable macro-aggregates. It is also of interest to note that higher aggregate stability values were obtained for the pasture than the arable soils indicating the effects of the higher levels of indigenous organic matter in producing more stable aggregates.

In contrast, these carbohydrate treatments were

relatively ineffective in producing stable aggregates in the mono-ionic soil in which the original aggregate structure had been destroyed by ion-washing procedures (Table 2). Consequently, these experiments were not continued beyond 4 weeks by which time any small amounts of aggregation produced in the initial stages had disappeared. The mono-ionic soils were included in this study sequence as they were needed to study the additions of humic substances and, in particular, to allow differentiation between the effects of physically admixed and adsorbed humic substances on aggregate stabilisation.

The results obtained following the addition of humic substances either as a simple physical admixture or through carefully controlled surface adsorption processes, are shown in Table 3. The data presented are for Stirling arable soil converted into the Ca^{2+}-saturated, mono-ionic form, with and without glucose addition. Incubation following the physical addition of humic acid alone was ineffective and only marginally better when glucose was included. As this treatment produced little aggregation, the measurements were not continued beyond the first two weeks.

In contrast, incubation with adsorbed humic acid was very effective at reforming stable aggregates and the effect was significantly enhanced by addition of glucose (Table 3). Furthermore, the aggregate stability persisted for 8 weeks with only a relatively small decline. Given the inability of any of the carbohydrate treatments to produce stable aggregate in the Ca^{2+}-saturated, mono-ionic soil, these results indicate that adsorbed humic substances are remarkably effective in stabilising soil aggregate even in the absence of polysaccharide material.

From the results presented, we can conclude that polysaccharides are capable of producing stable aggregates, but the effect is transient and declines as the polysaccharides are decomposed. Adsorbed humic acid also produces stable aggregates and the effect is more persistent. The most stable and persistent reformed aggregates were produced by the combined action of polysaccharides and humic acid. No doubt the increased persistence is due to the relative resistance of humic acid to biological decomposition in the soil compared to polysaccharides. In all instances the involvement of soil micro-organisms is an essential ingredient of the aggregate formation and stabilisation process.

Studies of Field Soils

Soils under permanent pasture tend to have high organic matter contents and high aggregate stability values, whereas the values for soils under continuous cultivation are much lower. In both cases the values for organic matter content and aggregate stability tend to

Table 3 Stability of reformed aggregates (MWD) with time following additions humic acid (2%) to Ca^{2+}-saturated Stirling arable soil.

Humic acid treatment	Glucose Addition	Aggregate Stability (MWD) Incubation period (weeks)	
		2	8
Physical addition	none	27	ND
Physical addition	+ 0.5%	35	ND
Adsorbed	none	134	127
Adsorbed	+ 0.5%	181	175

ND = Not determined.

stabilise after several years under the respective system of land management as the system reaches equilibrium.

When soils are changed from pasture to arable use (or vice versa) then the overall soil organic matter levels change steadily over a number of years. In contrast, it has been observed that when such changes in land use occur the stability of aggregates changes more rapidly than the overall organic matter contents[6,7,8]. For example, in the data presented in Table 4 the pasture soil had been under continuous grassland for 25 years, the arable soil had been continuously cultivated for 15 years, and the regrassed soil had been returned to pasture for two years after 13 years of continuous cultivation. Although the organic matter content of the regrassed soil had not yet risen significantly, the aggregate stability had shown a substantial increase (Table 4). These observations are indicative of the presence of an 'active' component whose content is changing more rapidly than the overall organic matter levels. In order to investigate these changes and the identity of the 'active' component it was necessary to study soil types on which permanent pasture, permanent arable, and arable/pasture rotation systems were practiced.

In addition, from observations made during this study, it became apparent that the more stable aggregates which survived the initial period of the wet-sieving process tended to remain intact for long periods of time. On the other hand most of the unstable aggregates broke down rapidly within the first few minutes of the wet-sieving procedure. Thus the single value Mean Weight Diameter value used in this study to represent the overall aggregate stability is not a mean value of a

Table 4 Comparison of the organic matter composition of stable and unstable aggregates from Wakanui soils with different cropping history.

Soil and aggregate fraction	Aggregate stability (MWD)	Total N (%)	Organic C (%)	Hydrolysable carbohydrate (%)
Pasture	213			
Stable		0.30	4.2	0.43
Unstable		0.24***	3.5***	0.40***
Arable	104			
Stable		0.16	2.1	0.22
Unstable		0.13***	1.7***	0.16***
Regrassed	200			
Stable		0.16	2.0	0.18
Unstable		0.15ns	1.9ns	0.16*

Table 4 (Continued)

Soil and aggregate fraction	Extractable carbohydrate ($\mu g\ C\ g^{-1}$)			
	Cold water (20°C)	Hot water (80°C)	0.1M HCl	0.5M NaOH
Pasture				
Stable	76	301	262	1408
Unstable	65***	252***	226***	1100*
Arable				
Stable	49	136	169	784
Unstable	38***	92***	145**	732*
Regrassed				
Stable	49	134	170	824
Unstable	44***	102***	165ns	780*

+ Significance of difference between stable and unstable aggregates shown: * $P<0.05$; ** $P<0.01$; *** $P<0.001$; ns = not significant.

closely grouped Gaussian distribution of aggregate stability values, but it is the mean of widely-spaced values with significantly large numbers of values grouped at the extremes of the distribution range. It therefore became apparent that any properties or components responsible for changes in the stability of aggregates would be more readily observable by comparing the composition of the most stable aggregates with the least stable aggregates. In this case 'unstable' aggregates were defined as those which had passed completely through the nest of sieves within 1 min and 'stable' aggregates were defined as those which still remained on the top sieve after 15 min[6]. The soil samples obtained in this way were analysed for Total N, Organic C, Hydrolysable carbohydrate and carbohydrate extractable with cold water ($20°C$), hot water ($80°C$), 0.1 M HCl and 0.5M NaOH. The results obtained are shown in Table 4.

Most of these parameters showed significant differences between stable and unstable aggregates from long-term pasture and arable sites, whereas only the cold and hot water-extractable carbohydrate showed highly significant differences between stable and unstable aggregates in the soil which had been regrassed for two years after a long period of continuous cultivation. This is indicative that the rapid changes in the stability of these aggregates are probably due to differential changes in polysaccharide components which comprise a small proportion of the total organic matter.

At the same time it should be remembered that the indigenous, adsorbed humic substances play an important role in the aggregate stabilisation process, particularly with respect to maintaining the longer term stability of aggregates. This is particularly exemplified by the differences between Total N and Organic C in the pasture and arable sites as well as the differences between stable and unstable aggregates at those sites.

To further test the significance of the various organic matter fractions and, in particular, to investigate the carbohydrate components in more detail than shown in Table 1, the same organic matter components listed in Table 4 were determined for a number of whole soils with a range of cropping histories. The results obtained are shown in Table 5 as a series of correlation coefficients. All of the organic matter parameters show significant to very highly significant correlations and, of the various carbohydrate fractions examined, the hot water-extractable carbohydrate content consistently gave the highest correlation coefficients. These results provide further evidence for the overall importance of soil organic matter and its fractions in the stabilisation of soil aggregates. In terms of the carbohydrate components, the results indicate particular fractions, such as those components extracted by hot water, are more active than others. Indeed it has been

Table 5 Linear correlation coefficients (R) between aggregate stability and various organic matter components

Soil Organic Component	Soil Type		
	Barrhill sandy Loam	Lismore silt loam	Temuka clay loam
Total N	0.68**	0.78***	0.67*
Organic C	0.66**	0.77***	0.72**
Carbohydrate content			
Acid hydrolysable	0.67**	0.75**	0.76***
Cold water-extractable	0.72**	0.77***	0.74**
Hot water extractable	0.83***	0.84***	0.79***
HCl-extractable	0.56ns	0.76**	0.60ns
NaOH-extractable	0.68*	0.72**	0.72**

[1]Barrhill, n = 27; Lismore, n = 26; Temuka, n = 27.

[2]Statistical significance shown: * $P < 0.05$, ** $P < 0.01$, *** $P < 0.001$, ns = not significant.

shown that certain polysaccharide fractions are more strongly and more selectively adsorbed by clay minerals than other fractions[10] and could well be active in particle aggregation.

Although statistical correlations, such as those presented in Tables 1 and 5, are helpful in identifying components which warrants further investigation, they provide no information about the composition of these organic components or of their mode of action. The more detailed, controlled addition studies presented in this paper clearly indicate the importance of certain fractions of the soil polysaccharides and humic substances in the aggregate stabilisation process. However, if we are to understand properly the nature of these effects it will be necessary to elucidate the composition and structure of the organic macromolecules involved and to investigate the mechanisms and extent of their interactions with soil mineral particles. These aspects require a considerable amount of further work.

REFERENCES

1. K. Chaney and R.S. Swift, J. Soil Sci. 1984, 35, 223
2. J.M. Tisdall and J.M. Oades, J. Soil Sci. 1982, 33, 141.

3. M.V. Cheshire, 'Nature and Origin of Carbohydrates in Soil', Academic Press, New York, 1979.
4. K. Chaney and R.S. Swift, J. Soil Sci. 1986, 37, 329.
5. K. Chaney and R.S. Swift, J. Soil Sci. 1986, 37, 337.
6. R.J. Haynes and R.S. Swift, J. Soil Sci. 1990, 41, 73.
7. R.J. Haynes, R.S. Swift and R.C. Stephen, Soil and Tillage Research, 1991, 19, 77.
8. A.J. Low, J. Soil Sci. 1972, 23, 363.
9. J.M. Oades, Plant Soil, 1984, 76, 319.
10. M.H.B. Hayes and R.S. Swift, 'Chemistry of Soil Organic Colloids' in 'The Chemistry of Soil Constituents', Wiley, Chichester, 1978.

Organic Matter in Water-stable Aggregates

A.G. Waters and J.M. Oades

DEPARTMENT OF SOIL SCIENCE, WAITE AGRICULTURAL RESEARCH INSTITUTE, THE UNIVERSITY OF ADELAIDE, GLEN OSMOND, SOUTH AUSTRALIA 5064

1 INTRODUCTION

Water-stable aggregates are fundamental to good structure for plant growth, as their size distribution controls pore-size distribution and consequently the physical and chemical processes in the soil. The relationship between organic matter and structural stability in soils is well established. Cultivation decreases organic matter content and degrades soil structure, usually with an associated decrease in the number of water-stable aggregates, particularly the larger water-stable aggregates. Aggregates contain particles and pores which range in scale over several orders of magnitude (Fig.1) and it is important to define the scale of the soil structure under study. Failure to do so may result in contradictory results and confusion.

Tisdall and Oades[1] proposed an hierarchical conceptual model for soil aggregate structure in which various organic binding agents operated at different stages in the structural organization of aggregates.
Organic binding agents were classified into 3 groups:
  a) transient, mainly polysaccharides;
  b) temporary, roots and fungal hyphae; and
  c) persistent, resistant aromatic components associated with polyvalent metal cations, and strongly sorbed polymers.

The aim of this study was to determine whether an aggregate hierarchy as shown in the 'Aggregations' column of Fig.1 does exist, and the role of organic matter (OM) within such a hierarchy.

2 MATERIALS AND METHODS

Air-dry samples of 4 soils, an Alfisol after 60 yrs of wheat-fallow rotation (WF) and after 40 years under a mixed grass-legume pasture (PP), a Mollisol and an Oxisol were separated into size fractions.

Air-dry samples were subjected to fast wetting by immersing the soil in water, and slow wetting on filter paper on a wet sand bed before wet sieving to obtain the size fractions: 2.0-4.0 mm, 1.0-2.0 mm, 0.5-1.0 mm, 250-500 μm, 150-250 μm, 90-150 μm and 53-90 μm. The <53 μm fraction was then separated into 20-53, 2-20 and <2 μm fractions by sedimentation. Sonic dispersion was followed by wet sieving and sedimentation to obtain the size fractions >250, 53-250, 2-20, 0.2-2.0 and <0.2 μm.

Total carbon was determined by dry combustion using a Leco CR-12 carbon system with a furnace temperature of 1200°C[2]. Total N was determined by the Kjeldahl digestion method.

For SEM examination samples were gold-coated and examined with a Cambridge Stereoscan S250 electron microscope.

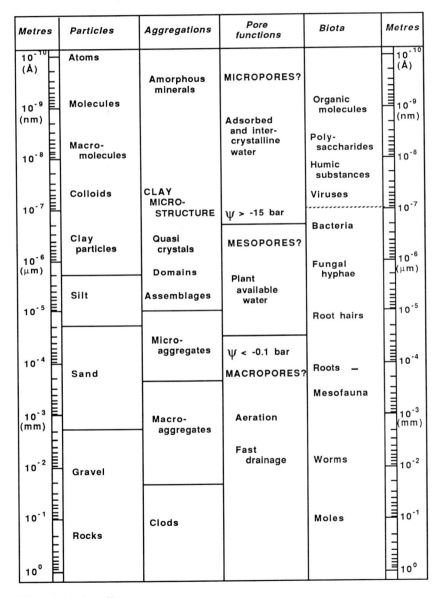

Fig.1 Scale in soil structure.

3 RESULTS AND DISCUSSION

There was no discernible difference by scanning electron microscopy between the aggregates obtained after slow and fast wetting pretreatments and further distinction between these treatments will not be made.

The C contents and C:N ratios of the various size fractions greater than 2 μm did not differ greatly (Table 1). This is in contrast with size fractions obtained after sonic dispersion of the soil between which there were substantial differences (Table 2).

Table 1 Carbon and nitrogen contents of particle size fractions after wet sieving (*PP - old pasture, WF -wheat fallow alternately)

	Alfisol-PP*		Alfisol-WF*		Mollisol		Oxisol	
	%C	C/N	%C	C/N	%C	C/N	%C	C/N
2-4 mm	3.6	12.3	1.2	10.6	-	-	-	-
1-2	3.0	11.4	1.3	13.3	5.8	11.4	5.3	12.6
0.5-1	2.7	11.5	1.3	13.3	5.8	11.4	5.3	12.6
250-500 μm	2.9	12.1	1.2	13.7	5.5	10.5	5.2	20.6
150-250	3.0	9.7	1.0	10.7	5.1	10.7	5.2	18.4
90-150	2.4	12.1	0.7	11.2	5.4	10.0	5.4	17.8
53-90	1.6	10.3	0.4	10.9	4.2	10.3	5.5	19.1
20-53	1.3	11.4	0.2	9.0	2.0	10.3	5.4	18.5
2-20	3.5	12.6	1.5	13.9	6.0	10.2	6.1	18.5
≤2	4.6	7.8	2.8	9.1	9.3	7.5	8.0	10.9
Total Soil	2.9	11.3	1.2	14.2	5.4	9.9	5.6	12.1

Table 2 Carbon and nitrogen contents of particle size fractions after ultrasonic dispersion (* data from Turchenek[3])

	Alfisol-PP		Alfisol-WF		Mollisol*		Oxisol	
	%C	C/N	%C	C/N	%C	C/N	%C	C/N
>250	4.6	28.9	1.9	23.4	10.0		11.3	-
53-250	1.3	14.9	0.6	20.7		15.5	10.5	26.5
20-53	0.4	11.8	0.2	11.7	1.8		3.5	25.6
2-20	2.6	13.5	1.3	14.6	7.8	13.3	7.5	14.3
≤2	4.9	9.8	3.3	10.5	7.1	10.2	4.0	11.6
	2.3	11.5	1.2	12.8	6.3	11.8	5.5	14.8

In particular the >250 μm fraction had higher C content and C:N ratio after sonic dispersion for all soils, indicating its plant like character and partial separation from inorganic components. In comparison the fractions greater than 250 μm obtained after wet sieving had C contents and C:N ratios similar to those for the whole soil, indicating a bulk composition of the aggregates similar to that of whole soil. On this basis it was concluded that further detailed chemical characterization was unlikely to yield information relevant to the stabilization of aggregates. Examination of the water-stable aggregates by scanning electron microscopy supported the concept of an aggregate hierarchy. On the basis of these observations 4 main stages were identified:

<20 μm → 20-90 μm → 90-250 μm → >250 μm.

Macroaggregates >250 μm

Root systems and VA mycorrhizal hyphae associated with the roots stabilise soil macroaggregates >250 μm diameter. Fig. 2a (from Tisdall and Oades[4]) shows an extensive network of hyphae from a pot culture of VA mycorrhizal fungi with soil particles firmly adhered, confirming observations on natural water stable aggregates of the binding effect of hyphae.

More detail of this stabilizing mechanism can be seen in Fig.2b which shows a fine wheat root, from the Alfisol under wheat-fallow rotation, coated with inorganic soil particles. The mucilages responsible for the adhesion are probably of both plant and microbial origin. Numerous electron micrographs show collapsed hyphae within and on surfaces of water-stable macroaggregates, for example Fig.2c.

The stability of macroaggregates is related to the growth of roots and hyphae which is controlled by agricultural practices. Cultivation causes the breakdown of macroaggregates as it decreases the length of roots and hyphae in the soil[1]. Miller and Jastrow[5] developed a conceptual model of the interrelationships among root size classes, mycorrhizal fungi and aggregate size distributon. They found that extraradical hyphal length and fine (0.2-1 mm diiameter) root length had the strongest direct effect on the geometrical mean diameter of water stable aggregates. Grasses appear to be the most beneficial plants in stabilizing macroaggregates of roots and hyphae[4].

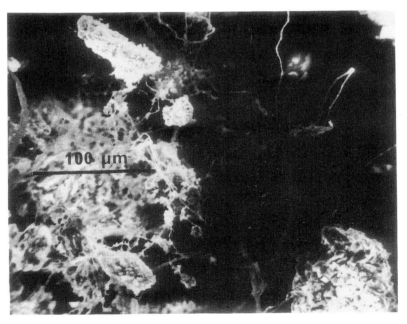

Fig.2a Macroaggregates >250 μm. Hyphae of VA mycorrhizal fungi binding soil particles into water stable aggregates, from Tisdall and Oades[4].

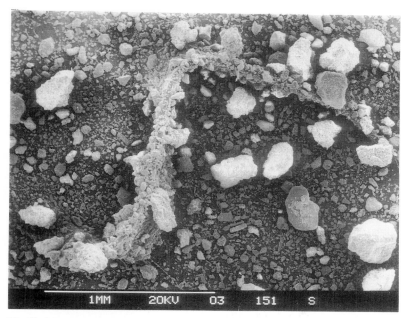

Fig.2b Macroaggregates >250 μm; Alfisol WF. Fine wheat root coated with inorganic soil particles.

Fig.2c Macroaggregates >250 μm; Alfisol PP. Collapsed hyphae on and within aggregate.

Microaggregates 90-250 μm

The next stage in the hierarchy comprises microaggregates 90-250 μm. These microaggregates have a nucleus of plant debris encrusted with inorganic components (Fig.3a). The debris represents fragments of roots and stems showing distinct cellular anatomy (Fig.3b) and probably represents plant materials from the previous few seasons. In cases where encrustation is complete aggregate shape can indicate an organic nucleus. Fig.3c shows a broken aggregate with internal remnants surrounded by inorganic particles. The age of such material is a matter of conjecture.

Microaggregates 20-90 μm

Few distinct organic entities were observed. In microaggregates of 20-90 μm diameter there were only occasional fragments of plant debris. The aggregates often contained voids (Figs. 4a and 4b) some of which showed remnants of plant anatomy. The voids were presumably left after biological oxidation of plant and microbial debris such as remnants of roots and hyphae. Fig. 4c shows a general view of the 53-90 μm fraction of the Alfisol from the pasture soil. In addition to aggregates of various morphology there were numerous quartz particles and phytoliths. One of the aggregates possibly contains a lignin coil, a rare example of plant debris in microaggregates of this size.

The microaggregates at this stage in the hierarchy began to reflect the clay morphology of the particular system. The Alfisol aggregates were more open than the denser Oxisol aggregates. The soil contains 70% fine sand and silt which explains why the C content of this size fraction is low (Tables 1 and 2). Microaggregates 20-53 μm diameter were very rare in the Alfisol, particularly under the WF rotation. This fraction contained dominantly quartz grains and phytoliths.

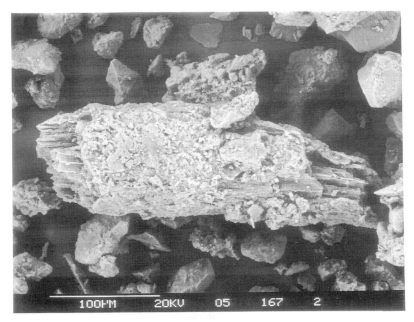

Fig.3a Microaggregates 90-250 μm; Mollisol. Bundle of elongate plant cells encrusted with inorganics.

Organic Matter in Water-stable Aggregates

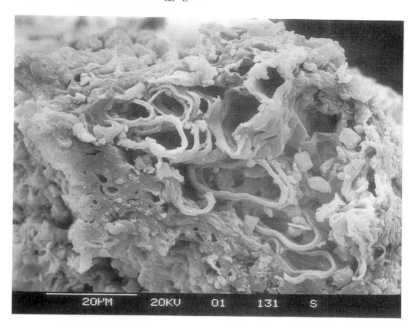

Fig.3b Microaggregates 90-250 μm; Alfisol PP. Partly decayed vascular bundle surrounded by inorganics.

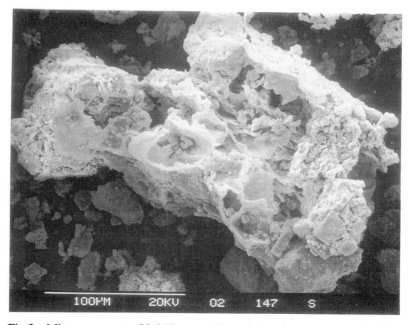

Fig.3c Microaggregates 90-250 μm; Alfisol PP. Broken aggregate showing internal cellular remnants surrounded by a layer of inorganics.

Fig.4a Microaggregates 20-90 μm; Oxisol. Large void showing remnants of plant anatomy (?) completely coated with inorganics.

Fig.4b Microaggregates 20-90 μm; Alfisol PP. Elongate void running from top left to bottom right of aggregate; no remnants of plant anatomy are apparent.

Organic Matter in Water-stable Aggregates 171

Fig.4c Microaggregates 20-90 μm; Alfisol PP. General view of aggregates, A, quartz particles, Q, and phytoliths, P.

Clay microstructure <20 μm

Below 20 μm diameter clay microstructure became dominant. Biological entities were not readily observed by scanning electron microscopy. However Foster et al.[6] have observed bacteria and mucilages by 'staining' and transmission electron microscopy of thin sections at this scale.

The structure of the clay aggregates observed by SEM agreed with the schematic representation of Oades[7]. The Mollisol was composed of quasicrystals of flexible sheets (Fig.5a). There has been no unequivocal evidence of organic components between the lamellae except perhaps in some sediments[8]. The Alfisol was made up of domains of rigid, platy particles, characteristic of illitic materials (Fig.5b). The Oxisol comprised assemblages of blocky plates characteristic of kaolinitic materials (Fig.5c). The kaolinite of this sample was very fine, having a surface area measured by nitrogen adsorption of >50 m^2/g.

4 GENERAL DISCUSSION

Recent studies into aggregate stability have focussed on the role of roots and fungal hyphae in binding microaggregates into macroaggregates[5,9,10], and have supported and extended the conclusions of Tisdall and Oades[11]. The OM associated with macroaggregates is qualitatively more labile, less highly processed and is more readily mineralized than that associated with microaggregates, and hence the OM binding microaggregates into macroaggregates is the primary source of nutrients released when OM is lost on cultivation.

Fig.5a Clay Microstructure < 20 μm; Mollisol. Quasicrystals of flexible sheets.

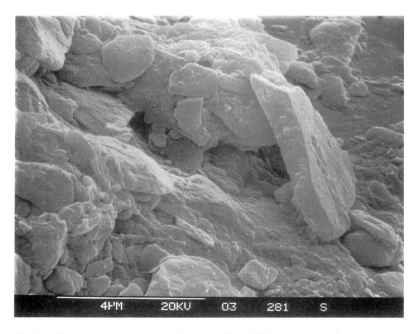

Fig.5b Clay Microstructure < 20 μm; Alfisol WF. Domains of clay plates, characteristic of illitic materials.

Organic Matter in Water-stable Aggregates 173

Fig.5c Clay Microstructure < 20 μm; Oxisol. Assemblages of blocky plates, characteristic of kaolinitic materials.

The importance of roots and hyphae in stabilizing larger aggregates explains why aggregate stability declines at a rate similar to the decomposition rate of plant materials added to soil. At any one time macroaggregate stability is a result of rates of root and hyphae growth and the rates of their decomposition.

The binding mechanisms of microaggregates are less well understood. Tisdall and Oades[11] found aggregates 20-250 μm to be stable to rapid wetting and not destroyed by agriculture, but destroyed by ultrasonic vibration. Several types of binding agents including persistent organic materials, crystalline oxides and highly disordered aluminosilicates combine to give the aggregates this high stability[12].

This study has identified encrustation of plant debris as an important mechanism in microaggregate formation and stabilization. This encrustation affords protection to the debris from further decomposition and may exist for years. Encrusted plant remnants are readily observed in the 90-250 μm aggregates, and voids in aggregates 20-90 μm are probably due to biological oxidation of this debris. The degradation of the organic material has rendered the aggregate less stable and thus this feature is characteristic of the lower level of the aggregate hierarchy.

5 CONCLUSION

It has been shown by scanning electron microscopy that soil aggregates are not completely random accumulations of smaller soil particles but that particles are stabilized in increasingly larger aggregates by different binding agents, supporting the concept of aggregate hierarchy in soils. Encrustation of plant debris is indicated to be an important feature of many stable aggregates and is a major mechanism responsible for protecting the embalmed plant debris.

REFERENCES

1. Tisdall, J.M. and Oades, J.M. 1980. The effect of crop rotation on aggregation in a red-brown earth. Aust. J. Soil Res. 18, 423-433.
2. Merry, R.H. and Spouncer, L.R. 1988. The measurement of carbon in soils using a microprocessor-controlled resistance furnace. Commun. Soil. Sci. Plant. Anal. 19, 707-720.
3. Turchenek, L.W. and Oades, J.M. 1979. Fractionation of organo-mineral complexes by sedimentation and density techniques. Geoderma 21, 311-343.
4. Tisdall, J.M. and Oades, J.M. 1979. Stabilization of soil aggregates by the root systems of ryegrass. Aust. J. Soil Res. 17, 429-441.
5. Miller, R.M. and Jastrow, J.D. 1990. Hierarchy of root and mycorrhizal fungal interactions with soil aggregation. Soil Biol. Biochem. 22, 579-584.
6. Foster, R.C., Rovira, A.D. and Cock, T.W. 1982. 'Ultrastructure of the root-soil interface'. Am. Phytopath Soc., U.S.A..
7. Oades, J.M. 1986. Associations of colloidal materials in soils. Proc. 13th Congr. Int. Soil Sci., Hamburg VI, 660-674.
8. Theng, B.K.G., Churchman, G.J. and Newman, R.H. 1986. The occurrence of interlayer clay-organic complexes in two New Zealand soils. Soil Sci. 142, 262-266.
9. Elliott, E.T. 1986. Aggregate structure and carbon, nitrogen and phosphorus in native and cultivated soils. Soil Sci. Soc. Am. J. 50, 627-633.
10. Gupta, V.V.S.R. and Germida, J.J. 1988. Distribution of microbial biomass and its activity in different soil aggregate size classes as affected by cultivation. Soil Biol. Biochem. 20, 777-786.
11. Tisdall, J.M. and Oades, J.M. 1982. Organic matter and water-stable aggregates in soils. J. Soil Sci. 33, 141-163.
12. Muneer, M. and Oades, J.M. 1989. The role of Ca-organic interactions in soil aggregate stability. 3. Mechanisms and models. Aust. J. Soil Res. 27, 411-423.

Microalgal and Cyanobacterial Soil Inoculants and Their Effect on Soil Aggregate Stability

S.L. Rogers[1], K.A. Cook[2], and R.G. Burns[1]

[1] BIOLOGICAL LABORATORY, UNIVERSITY OF KENT, CANTERBURY, KENT CT2 7NJ, UK
[2] SHELL RESEARCH CENTRE, SITTINGBOURNE, KENT ME9 8AG, UK

1 INTRODUCTION

A well-structured soil is comprised of a matrix of aggregates: inorganic and organic soil components bound together with a corresponding system of inter-aggregate pores[1]. It is important for the maintenance of soil structure that aggregates are stable and do not disintegrate when subjected to rapid wetting[2] or following mechanical disruption associated with intensive agriculture[3]. The degradation of soils leads to a number of serious problems such as slumping and hardsetting[4] and increases in erosion[5]. Ultimately soil aggregate destruction results in a reduction in fertility[6,7].

The role of soil organic matter in the formation and maintenance of soil aggregate structure is well documented[8-11], and microbial and plant polysaccharides in particular are of great significance in the adhesion of soil particulates[12-15]. There is evidence to suggest that the inoculation of soil surfaces with extracellular polysaccharide (EPS)-producing heterotrophic microorganisms[16,17] or EPS-producing phototrophic microorganisms[18-20] can increase the resistance of aggregates to further disruption and may even reverse soil structural degradation.

The study reported here has been undertaken in order to assess the feasibility of using EPS producing phototrophic microbial inocula as soil conditioners in order to remediate degraded soils.

2 MATERIALS AND METHODS

Primary screen of candidate microorganisms

Nine exopolysaccharidic phototrophic edaphic soil microorganisms were selected for evaluation in a primary screen (Table 1). In order to choose the most promising species to use as a soil stabilizer the following characteristics were monitored: survival of phototroph inocula; in-situ EPS production; and changes in aggregate stability. For this purpose a model unstable poor quality soil was constructed (70% agricultural silt loam + 30% fine sand-50-70 mesh) giving a soil textural analysis of 47.4% sand, 31.4% silt and 21.2% clay. The pH was 6.3 (2:5 soil to water mix). The model soil was used in preference to a natural poorly aggregated field soil in the primary screen in order that a wide range of physical and chemical variations could be applied in other experiments.

Five grammes of air dried and heat sterilized (121°C for 60 min. on three consecutive occasions) model soil was placed in each of 24 wells of a 25 well perspex plate (Sterilin Ltd. U.K.). The central well was filled with sterile water to prevent soil desiccation. Soils were held at 60% moisture holding capacity (M.H.C), i.e. 0.285ml H_2O g^{-1} soil. Exponentially growing cultures were harvested by centrifugation (10,000g x 10 min), washed and resuspended in sterile distilled water. Soils were inoculated at the rate of 0.44µg chlorophyll a g^{-1} soil dry weight (d.wt) (Chlorophyceae and Cyanophyceae), and 0.057µg chlorophyll a g^{-1} soil (d.wt) for *Porphyridium purpureum*. Twelve wells in each plate were inoculated and 12 left uninoculated (controls). Plates were duplicated and incubated at 20°C under cool white fluorescent light with a 16/8 hour light/dark photo-regime for 56 days. Soils were sampled every 7 days, air dried to a constant weight and stored at 4°C prior to analysis.

Determination of biomass, EPS and aggregate stability

Phototroph biomass in soil was determined spectrophotometrically (LKB 4050 spectrophotometer) following cold acetone extraction of chlorophyll a [21]. The production of polysaccharides by the inocula was assessed as total soil reducing sugars using the anthrone assay [22] after preparation of soil hydrolysates according to Cheshire [23]. Results are expressed as glucose equivalent carbohydrate carbon. Aggregate stability, which measures the resistance of air-dried soil aggregates to physical disruption in water, was determined using the turbidometric technique of Sparling et al., [24]. Readings range from 50% for an unstable soil to 90% for a stable well aggregated soil and standard errors are in the order of ±1-2%.

Table 1   Phototrophic Microorganisms Used in the Primary Screen.

Organism	Source	Division(Class)
Chlamydomonas mexicana	CCAP 11/55A	Chlorophyceae
UKC88E (Chlorella sp.)	Soil isolate	Chlorophyceae
Porphyridium purpureum	CCAP 1380/1A	Rhodophyceae
Scytonema hofmanii	UTEX 2349	Cyanophyceae
Scytonema ocellatum	CCAP 1473/2	Cyanophyceae
Synecocystis PCC 6714	UTEX 2470	Cyanophyceae
Nostoc muscorum	CCAP 1453/12	Cyanophyceae
Westiellopsis prolifica	UKC	Cyanophyceae
UKC ACS5 (Oscillatoria sp.)	Soil isolate	Cyanophyceae

Survival of inoculated species in natural soil

To study survival in natural soil the species identified in the primary screen (*C. mexicana, N. muscorum, P. purpureum*) were inoculated onto a poorly structured agricultural soil from the Deal area of Kent, namely a Hamble series Brick earth (pH 7.3, 22.6% sand, 54.2% silt, 23.6% clay) suffering from slumping, hardsetting and a reduction in fertility. Soil (2Kg d.wt) was contained in a seed tray (38x24x5cm) and in order to assess the effect of inoculum size on survival and stability, two inoculum rates were used: *C. mexicana* 7.76×10^5 cells g^{-1} soil and 3.09×10^5 cells g^{-1} soil; *N. muscorum* 4.04×10^5 cells g^{-1} soil and 1.61×10^5 cells g^{-1} and *P. purpureum* 1.76×10^5 cells g^{-1} soil and 7.34×10^4 g^{-1} soil These two rates are approximately equivalent to 2Kg and 5Kg dry weight inocula ha^{-1} Soils were maintained at 60% M.H.C. throughout the experiment and incubated under standard conditions. Survival was measured using dilution plate counts on selective media[25]: *C.mexicana*, green algal medium No.4; *N.muscorum*, Jaworskys media; *P. purpureum*, artificial seawater media.

Plant bioassays

In order to assess the effect of structural stability on soil fertility, experiments involving the emergence of spring wheat (*Triticum aestivum* var. Axona) were carried out. Percent emergence of wheat seedlings was calculated up to 5 days after planting in the poorly structured Hamble series Brick earth (aggregate stability 67.3±1.0%), a well structured non-agricultural calcareous grassland soil (aggregate stability 80.6±1.3%, 42% sand, 32% silt, 26% clay pH6.9), and a second agricultural silt loam soil (aggregate stability, 76.5±1.1%, 39% sand, 41% silt, 20% clay, pH7.0). Soils (500g) held in seed trays (21x8x5cm) were planted with the wheat seeds at a depth of 1.5cm (30 seeds per tray) and watered to 60% M.H.C. Watering the soil surface simulated the conditions under which surface aggregate disruption occurs during rainfall. Soils were then left to dry in order to mimic the conditions under which slumping and hardsetting occurs.

3   RESULTS

Primary screen

All species studied in the primary screen proliferated on the sterilized model soil, as measured by an increase in soil chlorophyll a content (Table 2). However *Porphyridium purpureum*, *Nostoc muscorum* and *Chlamydomonas mexicana* were identified as the most successful colonisers in the primary screen as shown in Figure 1. From an inoculation of $0.057\mu g$ chlorophyll a g^{-1} soil, *P. purpureum* chlorophyll a increased to $1.57\mu g$ g^{-1} at 56 days post inoculation. Soils inoculated with *N. muscorum* and *C. mexicana* ($0.44\mu g$ g^{-1}) contained $8.70\mu g$ g^{-1} and $11.8\mu g$ g^{-1} chlorophyll a, respectively at day 56. These three organisms also expressed appreciably higher levels of EPS production (> 40% in some cases) than the other species (Table 3) as revealed by increases in the glucose equivalent carbohydrate content of inoculated soils at day 56 (Figure 2).

Improvements in aggregate stability in inoculated soils ranged from little or no improvement (e.g UKCACS 5 showed no improvement in aggregate stability at 56 days, *Scytonema ocellatum* a non significant 3.3% increase) to a 20% increase in stable aggregates in soils inoculated with *C. mexicana*. Soils inoculated with *P. purpureum* and *N. muscorum* showed significant increases in water stable aggregates of 18% and 10%, respectively (Figure 3).

Table 2   Chlorophyll a In Inoculated Soils at Day 56

Organism	μg chlorophyll a g^{-1} soil (d.wt.)
Chlamydomonas mexicana	11.80
UKC88E (Chlorella spp.)	4.19
Porphyridium purpureum	1.57
Scytonema hofmanii	0.66
Scytonema ocellatum	2.50
Synecocystis PCC 6714	5.71
Nostoc muscorum	8.70
Westiellopsis prolifica	3.50
UKCACS5 (Oscillatoria sp.)	2.41

Mean S.E. ± 0.87

Table 3   Exopolysaccharide in Inoculated Soils at Day 56

Organism	mg carbohydrate carbon g^{-1} soil (d.wt.)
Chlamydomonas mexicana	9.95
UKC88E (Chlorella spp.)	6.94
Porphyridium purpureum	8.42
Scytonema hofmanii	6.02
Scytonema ocellatum	4.35
Synecocystis PCC 6714	5.67
Nostoc muscorum	9.60
Westiellopsis prolifica	5.98
UKCACS5 (Oscillatoria spp.)	5.89

Mean S.E. ± 0.46

Microalgal and Cyanobacterial Soil Inoculants

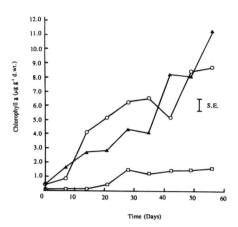

Figure 1   Phototroph colonisation of model sterilized soil. Initial inoculum densities: *C. mexicana* ▲, 0.44µg chlorophyll a g^{-1} soil (d.wt); N. muscorum ○, 0.44µg g^{-1}; *P. purpureum* □, 0.057µg g^{-1}. Mean S.E. ±0.87.

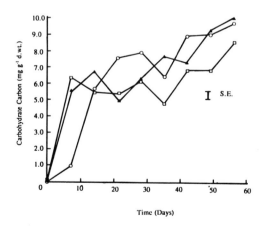

Figure 2   Phototroph exopolysaccharide production in inoculated sterilized model soil. *C. mexicana* ▲, *N. muscorum* ○, *P. purpureum* □. Mean S.E. ±0.46.

As a result of the primary screen *Chlamydomonas mexicana*, *Porphyridium purpureum* and *Nostoc muscorum* were chosen for more detailed study.

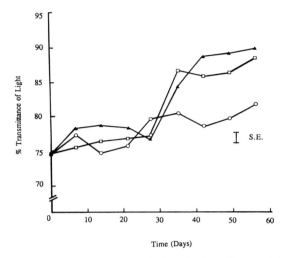

Figure 3 Aggregate stability in inoculated sterilized model soil. Initial stability 74.5%. *C. mexicana* ▲, *N. muscorum* ○, *P. purpureum* □. Mean S.E. ± 1.75.

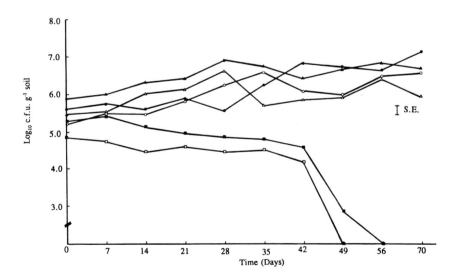

Figure 4 Survival of *C. mexicana* ▲, *N. muscorum* ○ and *P. purpureum* □ in non-sterilized poorly aggregated field soil. High rate inoculum (closed symbols), low rate inoculum (open symbols). Mean S.E. ± 0.25.

Establishment of inoculants in field soil

The establishment of *C. mexicana*, *N. muscorum* and *P. purpureum* in non-sterile field soil is reported in Figure 4. An increase in numbers of *C. mexicana* at both the low and high inoculation rates was recorded, with values of 8.14×10^5 g^{-1} soil and 4.42×10^6 g^{-1} soil, respectively after 70 days. Values for *N. muscorum* in non-sterile soil also increased with comparable numbers of 3.49×10^6 g^{-1} (low rate) and 1.70×10^7 g^{-1} (high rate). A green veneer was observed on *C. mexicana* and *N. muscorum* inoculated plots from 14 days onwards and this covered approximately 70-80% of the soil surface by day 70. In contrast *P. purpureum* inocula did not become established at either inoculation rate, with numbers declining steadily from day 1 and falling below the limits of detection (2.5×10^2) at day 49 (low rate) and day 56 (high rate).

Seedling Emergence

The results of the seedling emergence study are reported in Figure 5. The percentage emergence (at day 5) was highest in the well aggregated grassland soil, with 82.47% of seeds emerging by day 5 (no further emergence was observed after this time). In contrast 42.40% of seeds planted in the second agricultural soil emerged and in the Brick earth emergence was very poor at 29.12%. The Brick earth formed a hard seedling-resistant crust when allowed to dry out following watering, whereas the other two soils retained their surface aggregate structure. This resistance to surface aggregate disruption was most noticeable in the non-agricultural grassland soil.

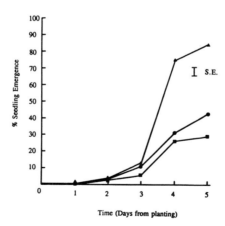

Figure 5 Percentage seedling emergence of spring wheat (*Triticum aestivum* var. Axona) in three soils with different structural stabilities. Grassland soil ▲, agricultural soil ●, Brick earth ■. Mean S.E. ± 5.0.

4   DISCUSSION

Unicellular green algae (Chlorophyceae) have been used successfully as soil stabilizers on irrigated temperate soils under field conditions[26]. In our study inoculation of sterilized model soil in the primary screen with Chlorophyceae, Cyanophyceae and Rhodophyceae, lead to a significant increase in soil aggregate stability, matched by an increase in the soil carbohydrate carbon (polysaccharide) content. Kroen[27] showed that unicellular green algae produced polysaccharides in soil and Foster[28] highlighted the role of polysaccharides in the process of aggregate stabilization on an agricultural scale. The successful colonisation of *N. muscorum* and the production of EPS may be assisted by its fixation of atmospheric nitrogen. In fact Cyanophyceae inoculation of temperate soils has been suggested as an alternative to inorganic nitrogen fertilizer[29], and has long been used in rice fields to reduce N-fertilizer amendments[30]. The successful establishment of *C. mexicana* and *N. muscorum* inocula on non-sterile field soil in our experiments is encouraging. However, there was a notable difference between the overall increase of viable cells in high rate and low rate inoculated soils. *C. mexicana* high rate increased by 5.7-fold whereas low rate inoculated soils only increased by 2.6-fold by day 70, *N. muscorum* high rate increased by 38.6-fold whereas the low rate inocula increased by 21.7-fold. At this stage it is too early to define what effect the inocula is having on the aggregate stability and polysaccharide content in the soil, however previous studies with Chlorophyceae and Cyanophyceae inoculation of soils have shown marked improvements in aggregate stability coupled with increases in soil polysaccharide content[19,31].

The reasons for the poor performance of *P.purpureum* inocula in non-sterile field soils in contrast to model soil in the primary screen are not clear. However, current studies investigating the effect of N and P fertilizer amendments on inocula establishment and polysaccharide production suggest that nutrient limitation may be a contributing factor. Poorly aggregated field soil has been amended with the equivalent of 67Kg ha^{-1} N and 12Kg ha^{-1} P (as NH_4NO_3 and K_2HPO_4). Viable cell counts of *P.purpureum* in soil inoculated with the high rate inocula show 1.01×10^6 cells g^{-1} soil at 14 days compared to 1.34×10^5 cells g^{-1} soil in non-fertilized soil. By 21 days a *P.purpureum* mat was observed on the surface of inoculated fertilized soil.

The influence of aggregate stability, on seed germination and seedling emergence has been investigated previously[32-33]. In this study the difference in emergence rates between the poorly structured agricultural soil and well aggregated non-agricultural soil confirm previous reports. As the poorly structured soil is from an intensive agricultural operation, it has been subject to agricultural rates of inorganic nutrient input (i.e. 500Kg 15:15:20 P:K:N fertilizer ha^{-1} per anum). Therefore plants are unlikely to be nutrient limited and the formation of a surface cap on the soil after a wetting/drying sequence is assumed to be the dominant factor in suppressing seedling germination[4].

Acknowledgement   We thank the Ministry of Agriculture Fisheries and Food for supporting this research.

REFERENCES

1. J.M. Oades, Plant and Soil. 1984, 76, 319.

2. J.M. Tisdall and J.M. Oades, J. Soil Sci., 1982, 33, 141.

3. V.V.S.R. Gupta and J.J. Germida, Soil Biol. Biochem., 1988, 20, 777.

4. C.E. Mullins, D.A. MacLeod, K.H. Northcote, J.M. Tisdall, I.M. Young, 'Advances in Soil Science', R.Lal, B.A. Stewart, Springer-Verlag New York Inc., 1990, Vol. 11, p. 38.

5. R.P. Harris, G. Chesters, O.N. Allen, Soil Sci. Soc. Am. Proc., 1966, 30, 205.

6. C.E. Mullins, I.M. Young, A.G. Bengough, G.J. Ley, Soil Use Man., 1987, 3, 79.

7. H.O. Rubido, M.K. Wood, M. Cardenas, B.A. Buchanan, Soil Sci., 1989, 148, 355.

8. J.M. Tisdall, B. Cockroft, N.C. Uren, Aust. J. Soil Res. 1978, 16, 9.

9. M.H.B. Hayes, 'The Role of Organic Matter in Modern Agriculture', Y. Chen, Y. Avnimelech, Martinus Nijhoff. Netherlands, 1986, p. 183.

10. E.T. Elliott, Soil Sci. Soc. Am. J., 1986, 50, 627.

11. R.J. Haynes and R.S. Swift, J. Soil Sci., 1990, 41, 73.

12 G.D. Swicer, J.M. Oades, D.J. Greenland, Aust. J. Soil Res., 1968, 6, 211.

13. L.E. Lowe, "Soil Organic Matter", M. Schnitzer and S.U. Kahn, Elsevier Science Press, 1978, Developments in soil science Vol.8, p. 65.

14. M.V. Cheshire, C.M. Mundie, J.M. Bracewell, G.W. Robertson, J.D. Russell, J. Soil Sci., 1983, 34, 539.

15. R.G. Burns and J.A. Davies, Biol. Agri. Hort., 1986, 3, 95.

16. K. Moavad, I.P. Babyeva, S.Y. Gorin, Soviet Soil Sci., 1976, 9, 65.

17. J. Lynch, J. Gen. Mic., 1981, 126, 371.

18. W.R. Barclay and R.A. Lewin, Plant and Soil, 1985, 88, 159.

19. B. Metting, Soil Science, 1987, 143, 139.

20. B. Metting, 'The Rhizosphere', J.M. Lynch, Wiley & Son, New York, 1990, p. 355.

21. E. Hansmann, 'Phycological Methods', J.R. Stein, Cambridge University Press, 1973, p. 359.

22. R.H. Brink, P. Dubach, D.L. Lynch, Soil Science, 1960, 89, 157.

23. M.V. Cheshire, 'Nature and Origins of Carbohydrates in Soils', Academic Press, New York, 1979.

24. G.P. Sparling, G. Gaskin, M.V. Cheshire, Commun. Soil Sci. Plant Anal., 1985, 16, 1219.

25. A.S. Thompson, J.C. Rhodes, I. Pettman, 'Culture Collection of Algae and Protozoa, Catalogue of Strains", Natural Environmental Research Council, 1988.

26. B. Metting and W.R. Rayburn, Soil Sci. Soc. Am. J., 1983, 47, 682.

27. W.K. Kroen, J. Phycol., 1984, 20, 616.

28. R.C. Foster, Science, 1981, 214, 60.

29. P.A. Reynaud and B. Metting, Biol. Agri. Hort., 1988, 5, 197.

30. B.A. Whitton and P.A. Roger, 'Microbial Inocula of Crop Plants', R.Campbell and R.M. Macdonald, IRL Press, Oxford, 1989, SGM Special Publication Vol. 25, p. 89.

31. D.L.N. Rao and R.G. Burns, Biol. Fert. Soils, 1990, 9, 239.

32. M.V. Braunack and A.R. Dexter, Soil and Til. Res., 1989a, 14, 259.

33. M.V. Braunack and A.R. Dexter, Soil and Til. Res., 1989b, 14, 280.

Organic Matter and Chemical Characteristics of Aggregates from a Red-Yellow Latosol under Natural Forest, Rubber Plant, and Grass in Brazil

E.deS. Mendonça[1], W. Moura Filho, and L.M. Costa

DEPARTMENTO DE SOLOS, UNIVERSIDADE FEDERAL DE VIÇOSA, VIÇOSA, MINAS GERAIS, BRASIL

1 ABSTRACT

The aim of this work was to study the organic matter and the chemical characteristics of four classes of aggregates from a Clayey Alic Red-Yellow Latosol under natural forest (Floresta Ombrófila Aberta), rubber plant and grass (Brachiaria humidicula). The work was carried out with aggregates larger than 4 mm, of 4-2 mm, 2-0.2 mm and 0.2-0.05 mm, from A1 and A3 horizons. The samples were subjected to one of two treatments to oxidize the organic matter: a) with hydrogen peroxide (H_2O_2 30% vol); and b) at room temperature, 100, 200, 250, 300 and 400°C. The organic matter percentage, humic substances, amorphous-Fe, exchangeable-Al, effective-CEC and available-P were measured. The results showed that the resistance of the organic matter to oxidation is related to the soil structure, changing between classes of aggregates and depth, and being influenced by soil cultivation. The quantities of organic matter, Fe, Al, P and humic substances vary between classes of aggregates and are influenced by soil cultivation. The macro- and microaggregates differ in organic matter-Fe complexation, which is influenced by the kind of organic matter. We found it is necessary to develop or adapt a methodology to give a better understanding of the clay-organic complex nature of the Brazilian soils. Humic substances, particularly the humin fraction, of the soils from the Amazonian Basin should be better characterized, with the objective of understanding their low resistance to oxidation.

2 INTRODUCTION

The productivity of soils from tropical and subtropical regions, rich in Fe and Al oxides, is directly related to organic matter, as well as to the clay-organic complex, which has an influence on the physical, chemical and biological soil properties. In these hot and humid regions, the soil reactions are very intensive. The substitution of natural vegetation for agricultural use may cause an imbalance in the ecosystem, and the new vegetative cover may have strong influence on the characteristics of

---

[1]Present Address: Department of Soil Science, University of Reading, London Road, Reading RG1 5AQ, England.

some soil properties, such as organic matter, the clay-organic complex and CEC[1,2].

Soil organic matter has been characterized in different ecosystems. Working with a Red-Yellow Latosol under grass, Longo[3] observed more fulvic than humic acids and their quantity decreased with depth until 60 cm. On the other hand, Arya et al[4] verified large quantities of fulvic acid with depth. Many researchers have studied the association between organic matter and inorganic constituents of soil[5-8]. In water-stable aggregates the soil organic matter is strongly complexed by inorganic colloids, being oxidized with more difficulty by H_2O_2 than the organic matter of aggregates with low water stability[9]. One of the main factors that controls the formation of stable aggregates is the wetting and drying cycle of the soil, which is related to the climate and soil management[10,11]. These factors together with the vegetal species and soil characteristics (texture, structure and type of clay) are the main factors that have an influence on the soil organic matter condition[12]. In the burned areas of forest soils, the hydrophobic organic materials may protect aggregates against the action of water[13]. The specific action of organic substances in the aggregates is to behave as an agent for the cohesion of clay fractions through H-bridges, covalent linkages, complex coordination with metallic cations, physical linkages by Van der Waal's forces, which can cover the surface of soil particles, and soil particles that can be joined by physical linkages of hyphae of fungi and microscopic roots of plants.

The microaggregate structure of Oxisols is a result of the activity of soil mesofauna[6]. The fulvic acids associated with amorphous oxi-hydroxides can be an important structural element in the process of microaggregate stabilization[8]. Barriuso et al[7] verified that the organic-ferric and metallic colloids are the result of soil organic matter preventing the solubilization of Al and Fe hydroxides. Some researchers[14-16] have found a decrease in the amount of soil organic matter with the size reduction of aggregates, whereas others[17-19] have shown that the amount of organic matter changed inversely with the size of aggregates.

Although many studies have been done with soil orgnic matter, many problems continue to exist without a solution at present. One of these is the effect of the association of organic matter with soil mineral constituents on the mechanisms of stabilization and mineralization of soil organic matter[20]. In tropical soils, where Fe and Al oxi-hydroxides and low activity clays are predominant, the understanding of dynamic clay-organic interactions, which act on the formation and stabilization of aggregates, is very important for a better management of these soils. The aim of this work was to study the organic matter and chemical characteristics of four classes of aggregates from a Red-Yellow Latosol under different type of cultivation.

3 MATERIALS AND METHODS

Characterization of the Area

The material of study was sampled in selected areas of an Alic Red-Yellow Latosol, a heavy clay soil, under natural forest (f), rubber plant (r) and grass (g), from Fazenda Barão, State of Mato Grosso - Brazil (Tables 1, 2 and 3). The region is 400 metres above sea level, at an approximate latitude of 12 °S and longitude of 56 °W. The relief is flat or gently undulating, and the annual average temperature and precipitation are 24 °C and 2200 mm, respectively. The natural vegetation is represented by Ombrófila Aberta Forest, which is a transition area between Amazon

Table 1. Physical and Chemical Properties of the Soil under Natural Forest

Horizon	Depth cm	Clay %	pH (1:2.5) H_2O	Extractable acidity Al^{3+}	H^+	CEC	O.C.	N	Total Elements SiO_2	Al_2O_3	Fe_2O_3
				--- meq/100g ---					-------- % --------		
A1	0-10	40	4.1	0.80	5.80	6.73	2.50	1.05	-	-	-
A3	10-14	43	4.4	0.40	3.56	4.05	1.55	0.69	13.76	16.83	5.10
B11	42-81	47	5.0	0.30	2.34	2.70	0.90	0.33	-	-	-
B12	81-138+	52	4.4	0.10	1.55	1.71	0.58	0.24	16.14	19.63	6.10

Table 2. Physical and Chemical Properties of the Soil under Rubber Plant

Horizon	Depth cm	Clay %	pH (1:2.5) H_2O	Extractable acidity Al^{3+}	H^+	CEC	O.C.	N	Total Elements SiO_2	Al_2O_3	Fe_2O_3
				--- meq/100g ---					-------- % --------		
A1	0-13	41	4.4	0.40	4.55	5.64	2.04	0.97	-	-	-
A3	13-44	41	4.6	0.40	2.57	3.43	1.29	0.47	11.62	15.30	4.40
B11	44-66	42	4.8	0.20	1.78	2.04	0.76	0.28	-	-	-
B12	60-103+	47	4.6	0.10	2.21	2.37	0.54	0.25	14.32	18.36	5.20

Table 3. Physical and Chemical Properties of the Soil under Grass

Horizon	Depth cm	Clay %	pH (1:2.5) H_2O	Extractable acidity Al^{3+}	H^+	CEC	O.C.	N	Total Elements SiO_2	Al_2O_3	Fe_2O_3
				--- meq/100g ---					-------- % --------		
A3	0-48	58	4.3	0.20	4.09	4.48	2.12	0.40	17.26	26.26	5.60
B11	48-75	59	4.7	0.10	4.52	4.68	1.29	0.29	-	-	-
B12	75-143+	59	4.4	0.10	7.49	7.66	0.94	0.20	19.84	29.07	5.70

Forest and Cerrado[21]. The rubber was planted four years before sampling and three years after being planted it was cultivated one cycle of rice between the rows. Limestone at 1000 Kg/ha and 4-20-10 (N-P-K) at 300 Kg/ha were applied at the begining of the growing season. The area of grazing was cultivated with Brachiaria humidicula four years before sampling. The A1 horizon had been eroded and the grass was in very bad condition.

Characterization of Studied Materials

The sample material was collected with an undisturbed structure. After air drying, the materials were separated into macroaggregates, aggregates larger than 2 mm (>4 and 4-2 mm), and microaggregates, aggregates smaller than 2 mm (2-0.2 and 0.2-0.05 mm). Macroaggregates were separated by dry-sieving and microaggregates by wet-sieving after horizontal shaking of 30 g of fine earth with 100 ml of H_2O for four hours.

Treatments and Chemical Analyses

In order to evaluate the influence of organic matter on the chemical characteristics of the soil materials, 0.5 ml of H_2O_2 at 30 % of volume per gram of material was applied in accordance with a previous test (Table 4), to oxidize soil organic matter. The 0.5 ml was chosen to avoid excess H_2O_2 and to be sure that most of the soil organic matter would be oxidized. The samples were subjected to environmental temperature, 100, 200, 250, 300 and 400 °C for one hour to study the resistance of soil organic matter to the oxidation.

Organic carbon was determined by the methodology of Walkley-Black, described by Jackson[22]. The extraction of humic substances was made with a mixture of 0.1N NaOH and 0.1N $Na_4P_2O_7$ at pH 12[23]. The humic substances were fractionated by chemical precipitation, when the pH was adjusted to a range of 2.0-2.5 with concentrated H_2SO_4. The partially separated humic and fulvic acids were then filtered, the humic acid being retained on the filter paper. The amount of organic carbon in the extracts was analyzed as described above.

Amorphous-Fe was extracted by 0.2N acidic ammonium oxalate[24] and exchangeable-Ca and Mg were extracted by 1N KCl, prior determination by atomic absorption spectrophotometry. The determination of available-K and Na, extracted by Mehlich-1, was made by flame photometry. Exchangeable-Al was extracted by 1N KCl and measured by titration with NaOH[25]. Phosphate was extracted by the method of Mehlich-1[25] and measured with molybdate by colorimetry. Effective-CEC was obtained by the addition of exchangeable cations[26].

Statistical Analyses

Horizon, aggregate and peroxide-treated or non-treated samples were combined in a 5x4x2 factorial arrangement. Horizon, aggregate and temperature treatments were arranged in a 5x4x6 factorial arrangement. All treatments were of a complete randomized design, with three replications. The effect of treatment with hydrogen peroxide was tested by F-test, at a 5 % probability level. The effects of the temperature treatments were tested by regression analyses for linearized, cubic, exponential and quadratic models.

Table 4. Organic Carbon Oxidized with Different Levels of Hydrogen Peroxide in fA3 and rA3 Horizons. Averages of Four Replications

Horizon	Total Organic Carbon	Organic Carbon Levels of H_2O_2 (ml/g of soil)			
		0.25	0.50	1.00	2.00
	-------------	%	-------------------		
fA3	1.55	0.19	0.19	0.16	0.16
rA3	1.29	0.21	0.20	0.21	0.21

Table 5. Levels of Organic Carbon (O.C.), Exchangeable-Al, ECEC, Amorphous-Fe, and Available-P in Macro and Microaggregates from a Red-Yellow Latosol Under Different Cultivations. Average of Three Replications

Vegetation	Horizon	Aggregate	Treatment H_2O_2	O.C. %	Al	ECEC meq/100g	Fe	P ppm
Forest	A1	macro	with	0.49	5.63	5.81	5293	10.31
			without	2.63	0.38	0.56	3684	1.77
		micro	with	0.36	7.23	7.31	2397	21.27
			without	2.79	0.60	0.67	3117	3.42
	A3	macro	with	0.56	3.90	4.00	5658	6.69
			without	2.05	0.20	0.31	3620	0.83
		micro	with	0.34	8.04	8.11	2690	18.96
			without	2.13	0.33	0.39	3052	1.39
Rubber Plant	A1	macro	with	0.23	4.93	5.27	4176	7.79
			without	2.11	0.19	0.65	2758	1.77
		micro	with	0.39	3.84	4.29	2417	8.07
			without	1.74	0.16	0.33	2992	0.75
	A3	macro	with	0.23	3.36	3.46	3505	4.50
			without	1.63	0.27	0.35	2711	0.56
		micro	with	0.47	1.48	1.64	2233	6.17
			without	1.33	0.17	0.48	2885	0.79
Grass	A3	macro	with	0.76	3.58	3.71	3106	3.02
			without	2.26	0.18	0.35	3121	0.40
		micro	with	0.53	3.11	3.19	2441	4.16
			without	2.10	0.16	0.22	3018	0.57

4 RESULTS AND DISCUSSION

For the soil under forest, the percentage of organic carbon in the microaggregates was larger than in the macroaggregates, without soil organic matter oxidation, but when the soil organic matter was oxidized the macroaggregates had more organic carbon (Table 5). The soil under rubber plant displayed an opposite trend, suggesting that soil cultivation changed the characteristics of the soil organic matter.

Before soil organic matter oxidation, it was observed that the amount of organic carbon in the soil sample decreased with depth and soil cultivation in the classes of aggregates studied (Table 5). But, as was expected, the soil organic matter in the A3 horizons was more strongly complexed with the mineral fraction than the soil organic matter from A1 horizons, and the soil cultivation in relation to the rubber had a negative effect on the stability of the complexation of the organic matter-mineral fraction in the macroaggregates (Table 6). This was also observed by Tisdall and Oades[26] and Oades[27]. In the microaggregates of soil under rubber, the soil organic matter forms a more stable complex with the inorganic soil fraction. The influence of grass on the soil organic matter resistance to the oxidation suggests that burning may have affected the physical-chemical properties of the humic substances[28], because of changes to the soil microorganisms. These changes may cause the formation of hydrophobic organic material, which can protect the aggregates against the action of water and so facilitate the formation of organic compounds that are more stable and more resistant to the oxidation[13].

In the soil aggregates larger than 4 mm under forest (fA1 and fA3) and rubber plant (rA1), and in the groups of aggregates of 0.2-0.05 mm under surface horizons of these soils (fA3 and rA3), it was found that the amount of organic carbon increased at 100 °C (Table 7). This behaviour suggests that the soil organic matter is strongly complexed within these classes of aggregates. The organic matter of aggregate classes of 0.2-0.05 mm is the most intensively complexed by the soil inorganic fraction, which can promote its protection against microbial and enzymatic degradation. In general, the cubic model gave the best fit of these data, where the temperature caused a strong decrease in the amount of organic carbon. In spite of this tendency, it may be said that in each situation the behaviour in the complexation of soil organic matter with the mineral fraction was different.

The percentage of organic carbon in the humin fraction comprises more than half of the total organic carbon, under any conditions (Table 8). In spite of the high amounts of humin, the soil organic matter had low resistance to oxidation with H_2O_2 (Table 4), which may indicate a different nature of this fraction, causing lowered stability and resistance to the oxidation. According to Duchaufour[5], one of the humification processes involves lignin slowly changing, to give origin to a humin, which forms a less stable linkage with the soil inorganic fraction. Another theory proposes that the size of the organic molecule is an important factor in determining the organic-metal complex, large organic complexes linking to a metal in a very weak way. These complexes cannot form a ring structure around the metal[29]. On the other hand, the stable complex between humin and the soil mineral fraction may facilitate its degradation, when there is disturbance of this complexation[23]. Then, it is assumed that this fraction and the oxides are the main agents of the stabilization of the aggregates studied. One of these hypotheses may explain why the organic

Table 6. Percentage of Organic Carbon Oxidized in Four Classes of Aggregates from A1 and A3 Horizons of Soils under Natural Forest (f), Rubber Plant (r) and Grass (g). Average of Three Replications

Aggregate mm	fA1	fA3	rA1	rA3	gA3
			%		
Macro 1/	82.05a	73.34b	88.66a	85.54a	66.16a
Micro 2/	87.38a	83.86a	77.06b	68.52b	73.30a
>4	89.23a	76.80a	87.34a	84.55a	65.96a
4-2	74.88b	69.89a	89.99a	86.54a	66.37a
2-0.2	90.08a	84.52a	87.22a	81.48a	72.60a
0.2-0.05	84.69a	83.20a	66.90b	55.56b	74.01a

1/ Larger than 2 mm.   2/ Smaller than 2 mm.
The averages with the same s at the columns do not differ by the F test, at the level of 5% of probability.

Table 7. Percentage of Organic Carbon in Four Classes of Aggregates From A1 and A3 Horizons of Soils under Natural Forest (f), Rubber Plant (r) and Grass (g), at Six Temperatures. Averages of Three Replications

Horizon	Aggregate mm	Temperature (°C)					
		Room	100	200	250	300	400
				%			
fA1	>4	2.38	2.53	1.96	0.96	0.63	0.17
	4-2	2.89	2.59	1.33	1.16	0.73	0.13
	2-0.2	2.32	2.19	1.43	0.91	0.62	0.12
	0.2-0.05	3.27	2.93	2.03	1.75	0.63	0.39
fA3	>4	1.67	1.80	1.30	0.79	0.51	0.12
	4-2	2.41	1.99	1.91	1.04	0.96	0.13
	2-0.2	2.00	1.72	1.40	1.32	0.55	0.15
	0.2-0.05	2.26	2.56	2.16	1.09	0.75	0.31
rA1	>4	1.92	1.99	1.46	0.80	0.78	0.12
	4-2	2.30	2.32	1.90	1.60	1.27	0.19
	2-0.2	1.80	1.72	1.10	0.96	0.72	0.12
	0.2-0.05	1.69	1.43	1.44	0.83	0.52	0.28
rA3	>4	1.75	1.67	1.38	0.84	0.62	0.23
	4-2	1.51	1.43	1.04	0.80	0.63	0.19
	2-0.2	0.88	0.80	0.48	0.45	0.28	0.20
	0.2-0.05	1.78	2.34	1.93	1.21	0.90	0.32
gA3	>4	2.23	2.16	1.52	1.04	0.83	0.19
	4-2	2.30	2.21	1.87	1.24	0.91	0.20
	2-0.2	1.76	1.68	1.56	1.03	0.80	0.24
	0.2-0.05	2.27	2.18	1.90	1.56	1.11	0.36

matter of this soil is very easily oxidized.

The amount of humin decreased with depth in the soil under rubber. This behaviour appears to be related to the better aeration in the surface soil, the more frequent changes in the wetting and drying cycle, and also the alteration of microbial composition in the superficial horizon. These may have caused losses of more unstable organic material to the deeper layers or out of the system, producing an increase in the percentage of humin, as well intensifying the polymerization of organic compounds, thus helping the accumulation of humin.

As has been observed by other authors[3,30,31], fulvic acids contain three or four times more organic matter than humic acids in Brazilian Oxisols (Table 8). As the environment of these soils has good aeration, the mineralization of soil organic matter is very fast, which causes preferential formation of fulvic acids relative to humic acids[32]. The microaggregates have more humic acid and less fulvic acid than the macroaggregates. An increase of organic carbon in the fulvic acids with the decrease of total organic carbon in some classes of aggregates may indicate that the fulvic acids are forming weak complexes with metals[33]. However, it may also show that these fractions are a result of breaks in the structure of the humin fraction (Table 9).

The degree of Fe crystallization in the microaggregates is higher than in the macroaggregates, which have the lowest amount of amorphous-Fe, with less stabilization of the clay-organic complex. The amount of amorphous-Fe increases in the macroaggregates with the oxidation of soil organic matter, while in the microaggregates, it decreases (Table 5). This pattern is supported by the negative correlation ($r=0.993$) obtained between organic carbon and amorphous-Fe of macroaggregates and positive correlation ($r=0.975$) between the same components of microaggregates. Since the macroaggregates have more fulvic acids and less humic acids than microaggregates, it appears that the amorphous-Fe behaviour is related to these acids. Hence the macro- and microaggregates differ in organic matter-Fe complexation. The smaller density of plant cover in the soils under rubber and grass had a negative effect on the organic complexes of Fe, which originated from root exudates, products of microbial synthesis and substances liberated from soil organic matter decomposition. The amount of amorphous-Fe in these soils was reduced as a result. Another negative effect of soil cultivation would be the dehydration of amorphous-Fe gels, as a result of high temperatures[34].

In all situations, the direct positive effect of the levels of soil organic matter on the exchangeable-Al is observed (Table 5). Where there is more organic matter, the amount of Al is higher. There is more Al complexed with the organic matter in the soil under grass than rubber plant. As the levels of fulvic acids are higher at the soil under grass, we can suppose that this fraction was responsible for most of the Al complexing by soil organic matter. Because of the low levels of nutrients in these soils, the composition of effective-CEC is basically exchangeable-Al, with a correlation index of 0.995 between them.

The high correlations found between effective-CEC, Al, soil organic matter and Fe for the aggregate classes studied (data not shown), suggest that in virtue of the considerable flocculating power of Al and Fe, changes in the balance of charge caused by excessive liming may promote clay dispersion and, consequently, dispersion of aggregates of this soil. Bivalent cations have loss flocculating power

Table 8. Percentage of Organic Carbon of Humic Acids, Fulvic Acids and Humin of Aggregates from a Red-Yellow Latosol under Different Vegetations

Vegetation	Horizon	Humic Acid		Fulvic Acid		Humin	
		\multicolumn{6}{c}{Aggregate}					
		Macro	Micro	Macro	Micro	Macro	Micro
		\multicolumn{6}{c}{%}					
Forest	A1	5.4	9.2	30.0	23.5	64.6	67.3
	A3	6.2	11.0	30.4	20.5	63.4	68.5
Rubber Plant	A1	4.9	12.9	27.0	20.9	68.1	66.2
	A3	6.3	13.4	29.1	24.2	64.7	62.4
Grass	A3	4.3	9.6	31.7	26.0	64.0	64.4

Table 9. Percentage of Organic Carbon of Humic Acids, Fulvic Acids and Humin of Four Classes of Aggregates from Soils under Natural Forest (f), Rubber Plant (r) and Grass (g)

Horizon	Aggregate mm	Humic Acid			Fulvic Acid			Humin		
		with	without	△	with	without	△	with	without	△
		\multicolumn{9}{c}{% O.C.}								
fA1	>4	0.12	0.20	+0.08	0.81	0.81	0.00	1.46	1.52	+0.06
	4 - 2	0.16	0.21	+0.05	0.75	0.88	+0.13	1.97	1.50	-0.47
	2 - 0.2	0.12	0.13	+0.01	0.55	0.67	+0.12	1.53	1.39	-0.14
	0.2 - 0.05	0.44	0.34	-0.10	0.75	1.04	+0.29	2.07	1.54	-0.53
fA3	>4	0.12	0.13	+0.01	0.59	0.70	+0.11	0.97	0.97	0.00
	4 - 2	0.12	0.11	-0.01	0.62	0.82	+0.20	1.67	1.06	-0.61
	2 - 0.2	0.12	0.13	+0.01	0.48	0.47	-0.01	1.40	1.12	-0.28
	0.2 - 0.05	0.36	0.05	-0.31	0.38	0.95	+0.57	1.52	1.56	+0.04
rA1	>4	0.12	0.16	+0.04	0.56	0.64	+0.08	1.25	1.19	-0.06
	4 - 2	0.08	0.08	0.00	0.57	0.85	+0.28	1.64	1.39	-0.25
	2 - 0.2	0.07	0.10	+0.03	0.57	0.57	0.00	1.17	1.05	-0.12
	0.2 - 0.05	0.37	0.30	-0.07	0.18	0.37	+0.19	1.14	0.76	-0.38
rA3	>4	0.11	0.16	+0.05	0.53	0.64	+0.11	1.11	0.87	-0.24
	4 - 2	0.08	0.02	-0.06	0.43	0.68	+0.25	1.00	0.73	-0.27
	2 - 0.2	0.05	0.05	0.00	0.29	0.36	+0.07	0.54	0.39	-0.15
	0.2 - 0.05	0.37	0.21	-0.16	0.27	0.90	+0.63	1.14	1.25	+0.11
gA3	>4	0.10	0.10	0.00	0.70	0.81	+0.11	1.44	1.25	-0.19
	4 - 2	0.09	0.08	-0.01	0.74	0.88	+0.14	1.47	1.25	-0.22
	2 - 0.2	0.08	0.11	+0.03	0.53	0.64	+0.11	1.16	0.93	-0.23
	0.2 - 0.05	0.34	0.14	-0.20	0.50	0.89	+0.39	1.44	1.15	-0.29

△ Difference between material with and without treatment at 100°C.

than trivalent cations[34].

In the situations that have more of P there were also larger amounts of organic carbon and Al, and less resistance of the soil organic matter to oxidation (Table 5 and 6). As the soil organic matter is negatively charged, it is able to complex significant amounts of phosphate when it is associated with cations[35]. The higher relationship of P with exchangeable-Al than with amorphous-Fe is because phosphate only reacts with ions at the sites of exchangeable cations[36]. The ability of soil organic matter to complex phosphate was reduced with soil cultivation.

5 CONCLUSIONS

The resistance of the organic matter to oxidation is related to the structure. It varies between classes of aggregates and soil depth, and it is also influenced by soil cultivation. The quantities of organic matter, Fe, Al, P and humic substances vary between classes of aggregates and are influenced by the soil cultivation. The macroaggregates have more fulvic acids and fewer humic acids than the microaggregates. The macro- and microaggregates differ in organic matter-Fe complexation, where the macroaggregates show a positive correlation and the microaggregates a negative correlation between amorphous-Fe and organic carbon. This behaviour is influenced by the kind of organic matter.

Humic substances, particularly the humin fraction of the soils from the Amazonian Basin should be better characterized, with the objective of understanding their low resistance to oxidation, which can explain part of the high rate of erosion when these soils are under agriculture use.

REFERENCES

1. F.P. Velasco and J.M. Lozano, An. Edaf. y Agrob., 1979, 37, 871.
2. P. Sánchez,'Suelos del Trópico: características e manejo', IICA, San José, Costa Rica, 1981, Chapter 5, p. 167.
3. J.V. Longo, Ms Thesis, Imprensa da Universidade de Viçosa, Viçosa, Minas Gerais, Brasil, 1982, 66p.
4. R. Arya, B.R. Gupta, and P.D. Bajpal, J. Ind. Soc. Soil Sci., 1984, 32, 26.
5. P. Duchaufour, Geoderma, 1976, 15, 31.
6. E. Barriuso, J.M. Portal, P. Faive, and F. Andreux, Pedologie, 1984, 34, 257.
7. E. Barriuso, J.M. Portal, and F. Andreux, Can. J. Soil Sci., 1987, 67, 647.
8. S. Karlsson, K. Kakansson, and B. Allard, J. Env. Sci., 1987, A22, 549.
9. J.A. Van Veen and E.A. Paul, Can. J. Soil Sci., 1981, 61, 185.
10. H.F. Birch, Plant Soil, 1958, 10, 9.
11. M. Chatterjee and K. Ghosh, J. Ind. Soc. Soil Sci., 1981, 29, 184.
12. D.C. Joshi, J. Ind. Soc. Soil Sci., 1981, 29, 25.
13. R.F. Harris, G. Chester, and O.N. Allen, Adv. Agron., 1966, 18, 107.
14. W.H. Metzger and J.C. Hide, Amer. Soc. Agron. J., 1938, 30, 833.
15. T.M. Adams, J. Agric. Sci. Camb., 1982, 98, 335.
16. E.T. Elliott, Soil Sci. Soc. Am. J., 1986, 50, 627.
17. N. Salomon, Soil Sci. Soc. Amer. Proc., 1962, 26, 51.
18. L.R. Weber, Soil Sci. Soc. Amer. Proc., 1965, 29, 39.

19. M.A. Tabatabai and J.J. Hanway, Soil Sci. Soc. Am. J., 1968, 32, 588.
20. G. Catroux and M. Schnitzer, Soil Sci. Soc. Am. J., 1987, 51, 1200.
21. R.L. Loureiro, A.A. Dias, and H, Magnago, 'Vegetação', In: 'Brasil: Departamento Nacional de Produção Mineral'. Projeto Radan Brasil, Rio de Janeiro, 1980, Folha Jurema, p. 325.
22. M.L. Jackson, 'Soil Chemical Analysis', Prentice-Hall Inc., Englewood Cliffs, 1958. p. 145.
23. M.M. Kononova, 'Soil Organic Matter', Pergamon Press, New York, 1966, p. 132.
24. J.A. McKeague, 'Manual on Soil Sampling and Methods of Analysis', Can. Soc. Soil Sci., Otawa, 2d. ed., 1987, p. 60.
25. B.V. Defelipo and A.C. Ribeiro, 'Análise Química do Solo', Conselho de Estensão da Universidade de Viçosa, Viçosa, Minas Gerais, Brasil, 1981, Boletim 29, 17p.
26. T.S. Tisdall and J.M. Oades, J. Soil Sci., 1982, 33, 141.
27. J.M. Oades, Plant Soil, 1984, 76, 319.
28. G.A. Orioli and N.R. Curvetto, Plant Soil, 1978, 50, 91.
29. F.E. Allinson, 'Soil Organic Matter and Its Role in Crop Production', Elsevier Scientific, New York, 1973, Chapter 5, p. 277.
30. M.S. Parra, Ms Thesis, Imprensa da Universidade de Viçosa, Viçosa, Minas Gerais, Brasil, 1986, 94p.
31. O.F. Saraiva, PhD Thesis, Imprensa da Universidade de Viçosa, Viçosa, Minas Gerais, Brasil, 1987, 175p.
32. F. Andreux and S.P. Becerra, Turrialba, 1975, 25, 191.
33. M. Schnitzer and K. Ghosh, Soil Sci., 1982, 134, 354.
34. H.W. Fassbender, 'Química de Suelos, com Enfase en Suelos de America Latina', IICA, San José, Costa Rica, 1984, Chapter 6, p. 168.
35. A. Wild, J. Soil Sci., 1950, 1, 221.
36. R.L. Parfitt, Adv. Agron., 1978, 30, 1.

The Effect of Long-continued Organic Manuring on Some Physical Properties of Soils

D.A. Rose

DEPARTMENT OF AGRICULTURAL AND ENVIRONMENTAL SCIENCE, THE UNIVERSITY, NEWCASTLE UPON TYNE NE1 7RU, UK

1 INTRODUCTION

Many investigators have observed the effects of added organic manures or wastes on physical properties of the soil. This work was summarized in a critical review which concluded that the main consequences of added organic matter were to decrease the bulk density of the soil and to increase the capacity of the soil to hold water[1]: regression equations were derived to predict the extent of such changes, if the change in the amount of organic carbon in the soil and the proportion of sand were known[1]. Other studies concluded that the addition of organic matter significantly increased the quantity of water in the soil available to plants[2], though the critical review suggested that the amount of available water was not greatly changed[1]. Such changes in the physical properties of soil following the addition and incorporation of organic matter are a consequence of concomitant changes in the structure of the soil.

Soil structure, however, is a labile property varying with time and altered by management, especially in agricultural soils. Measurements of soil physical properties *in situ* or on undisturbed cores in the laboratory refer to only one of many possible structural states and may be affected by the moisture status of the soil when the measurements were made or at the time of sampling. Most of the transient change in soil structure occurs in the macroporosity, which is affected by cultivations, whereas the microporosity of structural units (peds) or stable aggregates (crumbs) remains unchanged, or changes only slowly as a result of management practice. In particular, the structure of the pore space within stable aggregates is remarkably uniform; soil water characteristic curves (relating soil water content to soil matric potential) and hydraulic conductivity-water content relations are essentially unaffected by aggregate size[3], indicating the invariance of the microporosity within stable structural units.

Given this evidence of a stable microporosity, I used natural aggregates, packed in columns, as a model system to investigate in the laboratory the effects of long-continued applications of farmyard manure (FYM) on some physical properties of soils. I compared six

Table 1   Details of the soils used

Location	Crop	Plot	Annual FYM addition (tons/acre)	Year sampled	Cumulative FYM addition (t/ha)
Barnfield (Rothamsted)	Mangolds	80	-	1964	-
		20	14 since 1843	1964	4288
Broadbalk (Rothamsted)	Wheat	3	-	1964	-
		2B	14 since 1843	1964	4288
Hoos (Rothamsted)	Barley	6-2	-	1964	-
		7-2	14 since 1852	1964	3972
Saxmundham	Wheat	36	-	1965	-
		31	6 since 1899	1965	1009
Wellesbourne	Vegetables	22	-	1966	-
		21	27 since 1954	1966	870
Woburn (Lansome)	Vegetables	4	-	1965	-
		2	30 since 1942	1965	1808

pairs of soils, one of each pair with a long record of heavy dressings of FYM, the other without, from classical field experiments at Rothamsted (Barnfield, Broadbalk, Hoos), Saxmundham, Wellesbourne and Woburn (Lansome). Table 1 contains details of these soils.

The aggregated system

A porous system consisting of a bed of aggregates contains discrete groups of smaller pores within the aggregates, the micropores, separated by a more continuous system of larger pores between the aggregates, the macropores. Such a system with a bimodal pore-size distribution has a "double sigmoid" soil water characteristic curve[4], though Figure 1 shows only the upper sigmoid.

A bed of aggregates whose macropores drain readily to a uniform and well-defined water content, aggregate saturation θ_a, is a realistic model of a well-structured field soil draining readily after a thorough wetting to a well-defined field capacity. At field capacity, the micropores in the structural units provide a reservoir of water and nutrients for a growing crop, and the drained inter-aggregate pores offer the least complicated path between aggregate surface and soil surface for the diffusive exchange of the gases involved in soil respiration[5].

An aggregated system has a total porosity, ϵ_t, comprising the macroporosity, ϵ_v, between the aggregates and the microporosity, ϵ_a, within the aggregates. The aggregates themselves also have a porosity, ϵ_c, equal to the volume of the pores within the aggregates, ϵ_a, divided by the total volume of the aggregates, $1-\epsilon_v$. Hence[6],

$$\epsilon_t = \epsilon_v + \epsilon_a = \epsilon_v + \epsilon_c(1-\epsilon_v).$$

All porosities are ratios of volumes, $m^3\ m^{-3}$, whether the pore space is filled with air, with water, or with a mixture of both air and water. Note that $\theta_a = \epsilon_a$.

2 MATERIALS AND METHODS

Soils (Table 1) were sampled from the top 15 cm of the profile in late summer, air-dried in the laboratory, then sieved to produce stable aggregates (0.5-1 or 1-2 mm in diameter) for experiments. The particle-size distribution and pH (Table 2) were measured using the standard methods of the Soil Survey of England and Wales[7]. Organic-carbon content (Table 2) was estimated by the Walkley-Black method, using a factor of 1.3. Loss on ignition (Table 2) was determined by heating the soil to 800°C for 5 h.

The air-dry aggregates were packed uniformly into brass columns, 30 cm long and 38 mm in diameter, by slowly adding them while continuously vibrating the column until the soil stopped settling, taking care to avoid mechanical breakdown of the aggregates in the process. These columns, the model soil with minimum total porosity, defined the pore structure reported in this paper. The columns of aggregates were then used for flow measurements[4], to be discussed elsewhere.

The bulk density, μ_b, of these soils was found by dividing the oven-dry mass of the soil by the volume of the column. Soil particle densities, μ_s, were measured using a density bottle and water as the saturating liquid. The total porosity, $\epsilon_t = 1-(\mu_b/\mu_s)$. The microporosity, ϵ_a, is the volume of water in the soil at a matric potential of -5 kPa for 1-2 mm aggregates and -10 kPa for 0.5-1 mm aggregates[6]. The macroporosity, $\epsilon_v = \epsilon_t-\epsilon_a$. The porosity of the aggregates, $\epsilon_c = \epsilon_a/(1-\epsilon_v)$, and the proportion of macropores in the total porosity is ϵ_a/ϵ_t.

Soil water characteristic curves were measured on aggregates packed in perspex rings, 1 or 2 cm deep, to the same bulk density as the model soils in the brass flow columns. The techniques were standard[8]: suction plates were used for matric potentials ranging from 0 to -20 kPa, pressure plates from -25 kPa to -1.5 MPa and vacuum desiccators below -3 MPa. All water contents were measured gravimetrically as wetness w kg(H_2O) kg^{-1}(soil) and converted to volumetric water contents $\theta = \mu_b w$ m^3(H_2O) m^{-3}.

3 RESULTS AND DISCUSSION

Detailed results are given in Tables 2 and 3 and in Figure 1.

Table 2   Effect of FYM on soil composition.

Property	Unit	Barnfield (R) -	Barnfield (R) +	Broadbalk (R) -	Broadbalk (R) +
Coarse sand (600-2000 μm)	%	6	6	5	8
Medium sand (200-600 μm)	%	3	3	3	3
Fine sand (60-200 μm)	%	5	6	12	12
Silt (2-60 μm)	%	53	53	60	56
Clay (<2 μm)	%	33	32	20	21
Soil texture *	-	ZCL	ZCL	CL	CL
pH (1:2.5) in 0.01 M $CaCl_2$	-	7.5	7.0	7.3	6.8
pH (1:2.5) in water	-	7.7	7.3	7.5	7.0
Loss on ignition	%	5.6	8.1	4.8	7.9
Loss on pretreatment with H_2O_2	%	3.7	6.7	5.9	8.9
Organic carbon	%	0.69	2.23	0.87	3.12

- = no FYM applied;   + = FYM applied.
* CL = clay loam;   SCL = sandy clay loam;   SL = sandy loam;
  ZCL = silty clay loam.
  Textural classifications of the Soil Survey of England and Wales.

Effect of FYM on soil composition

Table 2 shows that long-continued applications of FYM increased the organic-C content of each soil, by absolute amounts ranging from 1.2% (Woburn) to 2.5% (Hoos), and by ratios ranging from 1.66 (Woburn) to 3.61 (Barnfield).   However, the composition of the mineral phase remained unchanged.

The FYM caused increased physical losses on ignition and chemical losses by oxidation.   The effect on pH was mixed: FYM decreased pH on three soils and increased pH on three soils.   The direction of change depended on whether the soil was initially more acid (pH < 6.1 in water, pH subsequently increasing with FYM) or less acid (pH > 7.2 in water, pH decreasing with FYM) than the applied FYM.

Hoos (R)		Saxmundham		Wellesbourne		Woburn	
-	+	-	+	-	+	-	+
4	6	3	4	7	7	19	11
3	3	19	20	42	40	23	19
10	10	27	28	20	20	37	42
59	57	21	21	17	18	9	14
24	24	30	27	14	15	12	14
ZCL	ZCL	CL	SCL	SL	SL	SL	SL
5.8	6.2	7.1	6.8	5.7	6.0	5.5	5.7
6.1	6.5	7.2	7.0	6.1	6.3	5.8	6.0
4.8	8.8	5.2	6.7	3.5	5.6	6.5	7.3
5.5	9.7	5.8	7.5	3.7	5.9	8.1	15.8
1.02	3.52	1.51	3.06	1.63	3.39	1.83	3.03

Effect of FYM on soil structure

Table 3 contains detailed results of the changes in soil structural properties following long-continued applications of FYM. All results reported are the mean of triplicates. No standard deviations are quoted because differences between replicates were usually in the third decimal place.

For each of the twelve properties listed in Table 3, the effect of FYM is to alter the property in the same direction for all pairs of soils studied. The main result of adding FYM is to increase the total porosity (row D) by greatly increasing the microporosity (row F) while decreasing the macroporosity (row E). As a result, both the porosity of the aggregates (row G) and the proportion of the total pore space that is within the aggregates (row H) is increased. Note that the values of ϵ_c are much higher than those found using a non-polar saturating liquid[6] (0.23 for Barnfield (-), 0.20 for Woburn (-) and 0.25 for Woburn (+)), which prevented soil swelling, and a different experimental technique[6].

Table 3 Effect of FYM on some physical properties of soils which relate to soil structure

	Barnfield (R)		Broadbalk (R)		Hoos (R)		Saxmundham	
	−	+	−	+	−	+	−	+
A	1.104	1.033	1.050	0.969	1.041	0.949	1.098	1.081
B	2.659	2.577	2.696	2.605	2.628	2.519	2.518	2.494
C	0.415	0.401	0.390	0.372	0.396	0.377	0.436	0.434
D	0.585	0.599	0.610	0.628	0.604	0.623	0.564	0.566
E	0.329	0.320	0.365	0.347	0.363	0.339	0.335	0.292
F	0.256	0.279	0.245	0.281	0.241	0.284	0.229	0.275
G	0.381	0.410	0.386	0.430	0.378	0.430	0.345	0.388
H	0.438	0.466	0.402	0.447	0.398	0.456	0.407	0.485
I	0.232	0.270	0.233	0.290	0.231	0.299	0.235	0.281
J	0.103	0.122	0.072	0.109	0.081	0.124	0.095	0.112
K	0.114	0.126	0.075	0.106	0.085	0.117	0.104	0.122
L	0.142	0.153	0.170	0.175	0.156	0.167	0.154	0.182

A = Bulk density — μ_b — Mg m^{-3}
B = Soil density — μ_s — Mg m^{-3}
C = Proportion of solid — $1-\epsilon_t$ — m^3 m^{-3}
D = Total porosity — ϵ_t — m^3 m^{-3}
E = Macroporosity — ϵ_v — m^3 m^{-3}
F = Microporosity — ϵ_a — m^3 m^{-3}
G = Aggregate porosity — ϵ_c — m^3 m^{-3}
H = Proportion of micropores — ϵ_c/ϵ_t — -
I = Field capacity* — w_{fc} — kg(H$_2$O) kg^{-1}(soil)
J = Wilting point — w_{wp} — kg(H$_2$O) kg^{-1}(soil)
K = Wilting point — θ_{wp} — m^3(H$_2$O) m^{-3}
L = Available water capacity — $\theta_{fc}-\theta_{wp}$ — m^3(H$_2$O) m^{-3}

Wellesbourne		Woburn		Effect of FYM	
-	+	-	+	on property	
1.064	1.032	1.037	1.006	-1.5% (Saxmundham) to -8.8% (Hoos)	A
2.639	2.585	2.596	2.537	-1.0% (Saxmundham) to -4.2% (Hoos)	B
0.403	0.399	0.399	0.396	-0.6% (Saxmundham) to -4.9% (Hoos)	C
0.596	0.601	0.601	0.604	+0.4% (Saxmundham) to +3.2% (Hoos)	D
0.437	0.400	0.425	0.393	-2.7% (Barnfield) to -13% (Saxmundham)	E
0.159	0.201	0.176	0.211	+9.1% (Barnfield) to +26% (Wellesbourne)	F
0.283	0.335	0.305	0.347	+7.7% (Barnfield) to + 18% (Wellesbourne)	G
0.267	0.335	0.407	0.485	+6.5% (Barnfield) to +25% (Wellesbourne)	H
0.150	0.195	0.169	0.209	+17% (Barnfield) to +30% (Wellesbourne)	I
0.058	0.085	0.059	0.083	+18% (Saxmundham) to + 52% (Broadbalk, Hoos)	J
0.062	0.087	0.061	0.084	+11% (Barnfield) to +40% (Broadbalk, Wellesbourne)	K
0.097	0.114	0.115	0.127	+3.2% (Broadbalk) to 18% (Saxmundham)	L

*The volumetric water content at field capacity, θ_{fc}, corresponding to a matric potential of -5 kPa, has the same value as the microporosity, (row F), except for the Saxmundham soils, whose values are θ_{fc} = 0.258 and 0.304 m^3(H$_2$O) m^{-3} for those receiving no FYM and FYM respectively.

Effect of FYM on soil water

Figure 1 shows the soil water characteristic curves (fitted by simple exponential equations) for the Hoos soils, typical of the behaviour of the other five pairs. At every matric potential, both the gravimetric wetness and the volumetric water content (not shown) are greater for the FYM soils, the ratio being a maximum near -1 MPa, with a value of 1.6 for gravimetric wetness. [The value of the ratio for volumetric water content is found by multiplying the ratio of the gravimetric wetnesses by the ratio of the bulk densities, i.e. by 0.949/1.041 = 0.912 for Hoos.]

The available water capacity (AWC, row L in Table 3) is estimated as the difference between the volumetric water contents at field capacity (approximated by a matric potential of -5 kPa) and at wilting point (approximated by a matric potential of -1.5 MPa). The addition of FYM increases the AWC in each of the six soils, confirming earlier results on undisturbed soil cores[2], though the increase for Broadbalk is relatively small. Note that an AWC of x m^3(H$_2$O) m^{-3} is equivalent to one of 1000x mm(H$_2$O) m^{-1}.

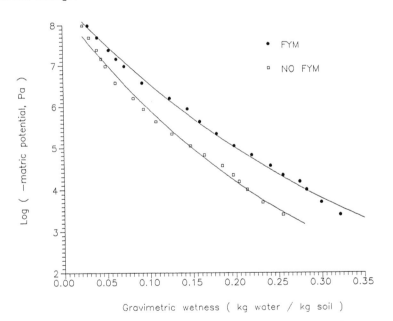

Figure 1 Soil water characteristic curves for Hoos soils

These results for AWC cannot be compared directly with the more complex behaviour analyzed in the critical review[1], for three reasons. First, the North American approximation of field capacity as a matric potential of -33 kPa was used. Second, the review surveyed a wider range of soil texture and showed, by multiple regression analysis, that increases in water-holding capacity at matric potentials of -33 kPa and -1.5 MPa depended not only on the increase in organic-C content but also on the proportion of sand in the soil, though the definition of sand differs between the USA and the Soil Survey of England and Wales. As a result, the increase in w_{fc} was greater than that in w_{wp} for fine-textured soils and *vice versa* in coarse-textured soils. Third, it is most unlikely that the data used in the regression equations referred to soils at the same stages of their temporal cycle of soil structure.

4 CONCLUSION

The results presented here confirm that FYM has beneficial effects on those physical properties of the soil that depend on the geometry of the pore space. In particular, FYM promotes an increase in the microporosity and the water available to plants. The agronomic benefits of these improved physical properties may be small, but will be increasingly important as crop yields approach more closely to their potential maxima.

ACKNOWLEDGEMENTS

I am grateful to my erstwhile colleagues, C.L. Bascomb and Margaret Chater of Rothamsted Experimental Station, for obtaining much of the data in Table 2.

REFERENCES

1. R. Khaleel, K.R. Reddy and M.R. Overcash, J.Environ.Qual., 1981, 10, 133.
2. P.J. Salter and J.B. Williams, J.Agric.Sci., 1969, 73, 155.
3. M. Amemiya, Soil Sci.Soc.Am.Proc., 1965, 29, 744.
4. D.A. Rose, J.Phys.D: Appl.Phys., 1968, 1, 1779.
5. J.A. Currie, Br.J.Appl.Phys., 1961, 12, 275.
6. J.A. Currie, J.Soil Sci., 1966, 17, 24.
7. B.W. Avery and C.L. Bascomb (eds), 'Soil Survey Laboratory Methods', Soil Survey of England and Wales, Harpenden, 1982, Chapter 3, p.14.
8. A. Klute (ed.), 'Methods of Soil Analysis', American Society of Agronomy, Madison, WI, 1986, Part 1, Chapter 26, p.635.

The Influence of Soil Compaction on Decomposition of Plant Residues and on Microbial Biomass

Ernst-August Kaiser, Gabriele Walenzik, and Otto Heinemeyer

INSTITUT FÜR BODENBIOLOGIE, BUNDESFORSCHUNGSANSTALT FÜR LANDWIRTSCHAFT, BUNDESALLE 50, D-3300 BRAUNSCHWEIG, GERMANY

1 INTRODUCTION

Soil compaction resulting from the passage of vehicles and implements is increasingly recognized as a problem in the agricultural crop production[1]. Compaction-induced deterioration of soil structure influences most physical, chemical and biological soil properties and processes[2]. The decrease of total soil porosity mainly results from a reduction in the volume of large pores (>50µm)[3,4], which are important for gaseous exchange[5]. Compaction-induced changes in soil aeration influence the abundance of soil microflora and their activity[6]. In compacted soil a three- to four-fold higher denitrification rate was found than in the control[7].

On the other hand microorganisms have beneficial effects on soil structure. An increase of aggregate stability due to microbial activity was observed [8-10].

The soil microbial biomass is also an important part of the soil organic matter, which influences the compactibility of arable soils[1], and is the main mediator of carbon turnover[11]. Close relationships between both were described as dependent on agricultural practices such as tillage[12], cropping systems[13], and residue incorporation[14,15].

So far there is a lack of information about the influence of soil compaction on soil microbiota and microbial processes. Therefore compaction-induced changes of soil microbial biomass, metabolic activity, and decomposition of plant residues were investigated to evaluate ecological consequences. The detection of changes requires great effort, since there is a great temporal and spatial variability of microbial soil properties due to the development of physical and chemical soil characteristics and plant growth.

2 MATERIALS AND METHODS

Field experiment

To evaluate effects of compaction on physical, chemical and biological soil properties and processes, a mechanical load was applied by controlled agricultural traffic in a field experiment on a loamy silt soil (pH 6.8, 10 mM $CaCl_2$). In order to simulate crop and soil management practices eight treatments (with three replicates) differing in timing (different field operations) and intensity (tractor weight)[16] of compactive effort were used. The cropping system was a common crop rotation of sugar beet, winter wheat and winter barley.

Microbial biomass carbon. Microbial biomass carbon (C_{mic}) was estimated by the SIR-method (substrate induced respiration)[17]. Soil cores (0-30 cm) were sampled monthly and stored at 4°C for subsequent analysis. The stored samples were sieved (2 mm) and adjusted to room temperature for 3 days. CO_2 production rate was estimated using a new developed set-up[18].

Basal respiration. The basal respiration [µg CO_2-C h^{-1} g^{-1} soil] was measured as the averaged CO_2-evolution of nonamended soil samples after an equilibration period of 10 hours [19].

Metabolic quotient. To calculate the metabolic quotient [qCO_2, mg CO_2-C h^{-1} g^{-1} C_{mic}][20], C_{mic} and basal respiration were estimated in paired samples (50g soil, dry weight) with 4 replicates.

Organic carbon. The soil organic carbon content (C_{org}) was estimated with three replicates by dry combustion (induced furnace, Leco) after removing inorganic C[21]. To remove carbonate-C, HCl (10%) was added dropwise. Subsequently the sample was dried at 70°C.

Bulk density. Soil bulk density [d_b, Mg m^{-3}] was determined in the 0-10, 10-20 and 20-30 cm soil depths by measuring the dry weight of 100 cm^3 undisturbed core samples (6 replicates).

Laboratory experiment

Topsoil (0-30 cm) from the experimental field [0.206 g (H_2O) g^{-1} (dry soil], 50 µg NO_3-N g^{-1} soil) was sieved (2mm), and mixed with ground ^{14}C-labelled wheat straw (1 mg straw g^{-1} soil), filled in tubes (4 x 5cm) and hydraulically compacted to different total pore space (Table 1). Six replicates were examined.

Table 1 Experimental soil compactive treatments (n=6).

COMPACTIVE TREATMENT	bulk density expected [Mg/m^3]	bulk density estimated [Mg/m^3]	percent water-filled pore space [#]
I	1.39	1.39 ±0.009	52.1
II	1.51	1.51 ±0.012	63.8
III	1.65	1.66 ±0.021	78.9

[#] The percentage of water-filled pore space (% WFPS) was calculated using the equation:[23]

$$\% \text{ WFPS} = \frac{\text{percent volumetric water content}}{\text{percent total pore space}}$$

Incubation procedure and analysis. The tubes were placed in jars, continually purged with humidified CO_2-free air and incubated at 25°C for 52 days. The aeration system[22] allowed daily sampling of CO_2 from the absorption medium (50 cm³ 0.5N NaOH). In using aliquots total CO_2-C was estimated by titration (1N HCl) and ^{14}C-CO_2 measured by liquid scintillation counting.

After 52 days microbial biomass carbon was estimated using the chloroform fumigation incubation (CFI) method[24], following the procedure described by MARTENS[25].

^{14}C-fixation in C_{mic} was calculated from the difference between ^{14}C-CO_2 release of the fumigated and the unfumigated sample.

Residual ^{14}C in the soil samples was measured with three replicates by combustion of subsamples in a Packard Sample Oxidizer 306.

Statistical analysis

Analysis of variance was used to determine variances attributed to effects of compactive treatments on the estimated soil properties for each month.

3 RESULTS AND DISCUSSION

Results of monthly soil microbial biomass (C_{mic}) and metabolic quotient (qCO_2) determinations are presented in Figure 1 and 2. Fluctuations due to season and crop-development can be recognized.

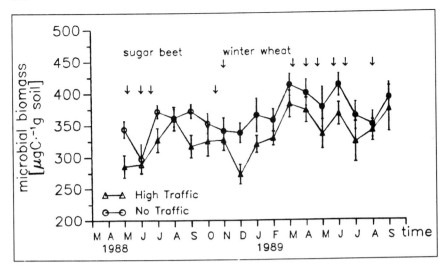

Figure 1 Microbial biomass contents in top soil (0-30 cm) of plots under different wheel-induced compactive treatments (arrows = time of wheel traffic, n=36, confidence level P= 5%)

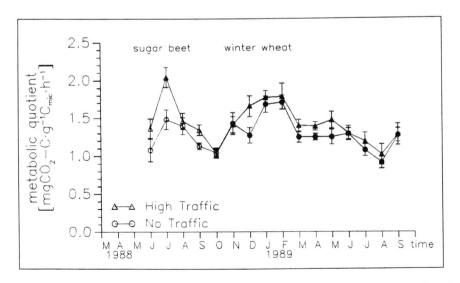

Figure 2 The metabolic quotient in top soil (0-30 cm) of plots under different wheel-induced compactive treatments (n=36, confidence level P= 5%)

To demonstrate effects of compaction the results from plots with extreme treatments are presented. Effects of the other six compactive treatments (not presented) were not detected.

In the "HIGH TRAFFIC"-plots, where the wheel-induced compaction efforts of a tramline were simulated, a lower microbial biomass, and a higher qCO_2 were estimated compared to the "NO TRAFFIC"-plots. The observed differences were consistent but in most cases not statistically significant. The most distinct difference was detected in May 88 and December 88, after the soil was ploughed (incorporation of plant residues) and the crops were sown.

Due to compactive treatment of the two extreme plots different soil bulk densities (d_b) developed. The compaction induced changes of the microbial environment are reflected by the values of % WFPS [23] (Table 2). When due to rain the values from November 88 increased to a level above 60 at December 88 in the "HIGH TRAFFIC"-plots, a decrease in microbial biomass carbon (C_{mic}) together with an increase in qCO_2 was found. In contrast on the "NO TRAFFIC"-plots, where the same rain caused a change in % WFPS from 40 to 54, no effect on microbial biomass could be detected, nor on qCO_2.

From the qCO_2-results we hypothesized that in compacted soil microorganisms consume (require) more energy for survival than in uncompacted soil.

Table 2 Development of % water-filled pore space (% WFPS), microbial biomass [C_{mic}, µg C g^{-1} soil) and the metabolic quotient [qCO_2, mg CO_2-C h^{-1} g$^{-1}C_{mic}$] influenced by different wheel-induced compactive treatments (n=36)

WHEEL-INDUCED COMPACTIVE TREATMENT	Date	% WFPS	C_{mic}	qCO_2
NO TRAFFIC [d_b= 1.23 Mg m^{-3}]	Nov.88	40.1	340	1.4
	Dec.88	54.4	340	1.4
HIGH TRAFFIC [d_b= 1.37 Mg m^{-3}]	Nov.88	51.3	335	1.4
	Dec.88	66.5	260	1.7

Table 3   Microbial biomass [C_{mic}, µg C g^{-1} soil], the metabolic quotient [qCO_2, mg CO_2-C h^{-1} g^{-1} C_{mic}], and basal respiration [µg CO_2-C h^{-1} g^{-1} soil] after 52 days of wheat straw decomposition (n=6, * P=5%)

	COMPACTIVE TREATMENT		
	I	II	III
d_b [Mg m^{-3}]	1.39	1.51	1.66
% WFPS	52.1	63.8	78.9
CFI-method C_{mic} initial = 450	460* ± 13	340 ± 33	320 ± 38
qCO_2	0.31* ± .08	0.49 ± .14	0.53 ± .07
basal respiration	35 ± 8	39 ± 10	40 ± 8

The results of the ^{14}C-straw decomposition laboratory experiment were similar to those from the field experiment. Microbial biomass was estimated after 52 days of incubation. After this time there were no limitations in the application of the CFI-method[25].

The results (Table 3) show a significant lower microbial biomass and a higher qCO_2 in highly compacted soil (III) compared to the lowest compaction level (I). The % WFPS (estimated at the beginning and the end of the experiment) was constant over time for each compaction level. The relationship between C_{mic}, qCO_2 and % WFPS found in the field experiment was supported.

Analysing the straw decomposition, which was estimated by the release of ^{14}C-CO_2, an increasing percentage of straw was mineralized due to a decreasing pore space (Table 4). The ^{14}C-residual data did not support this, which may be due to a lower accuracy of determination, since there were only 0.5 g ground soil used.
The differences in the percentage of straw decomposition probably result from a greater surface of contact between soil and plant material due to compaction. The ^{14}C-fixation in the microbial biomass resulting from mineralization did not vary with treatment.

Similar mineralization rates were found for whole wheat straw after 6 months in different soils, with 4.5

Table 4 Balance of wheat straw decomposition after 52 days (n=6, * P=5%)

	COMPACTIVE TREATMENT		
	I	II	III
^{14}C-release (1) [% of initial]	52.9* ± 2.6	55.1 ± 3.3	57.6 ± 5.2
^{14}C-residual [% of initial]	53.3 ± 4.7	53.4 ± 5.3	51.7 ± 3.9
CO_2-release [$\mu g CO_2$-C·g^{-1}] total (A)	1010 ± 170	860 ± 100	960 ± 90
straw (B)	210* ± 10	220 ± 10	230 ± 20
soil (A-B)	800 ± 150	640 ± 80	730 ± 70
^{14}C-fixation (2) in C_{mic} [% of initial]	5.4 ± 0.2	5.3 ± 0.2	5.4 ± 0.5
efficiency of decomposition (2/1)	0.101	0.095	0.093

to 8.2% incorporation of ^{14}C into the microbial biomass[26].

In calculating the quotient of ^{14}C-fixation in C_{mic} and ^{14}C-release, a lower efficiency of straw decomposition was detected in samples of high bulk density.

The effects observed in the field and laboratory experiments may be due to a lack of aeration caused by altered water-filled pore space. This results from an increasing anaerobic microbial activity at % WFPS greater than 60 [23]. HUYSMAN et al.[27] also indicated an anoxic metabolism in compacted soils due to a decreasing gas-diffusion status. This is supported by the observed differences in total CO_2-release between the extreme compactive treatments (I and III) in the laboratory experiment (Table 4).

Another indication is a higher basal respiration at treatment III, possibly supported from utilization of anaerobically formed metabolites.

Using ^{14}C-glucose PARSON and SMITH[28] reported a greater CO_2-evolution under aerobic than under anaerobic conditions. This also resulted in a lower efficiency of ^{14}C-glucose assimilation.

Table 5   Soil organic carbon depending on wheel-induced compactive treatment in top soil (0-30 cm) (n=27, * P=5%)

WHEEL-INDUCED COMPACTIVE TREATMENT	organic carbon [%]	
	Nov.88	Dec.88
NO TRAFFIC	1.05 ± 0.018*	1.06 ± 0.010*
HIGH TRAFFIC	0.98 ± 0.007	0.97 ± 0.021

As a consequence it can be assumed that compaction of soil will lead to decreased soil organic matter contents (C_{org}). Direct experimental evidence for this assumption was found by lower C_{org} values in plots of higher bulk density after 6 months of experimental treatment (Table 5). ALEXANDER (1986)[29] also calculated a negative correlation between bulk density and the soil organic matter.

The presented observations from the field experiment result from two dry seasons. Therefore only small compaction-induced deterioration of soil structure and short periods with a high % WFPS were observed. The estimated effects of compactive treatment will probably increase due to a higher soil moisture.

4   SUMMARY

In a field experiment controlled agricultural traffic was applied to investigate the influence of wheel-induced soil compaction. Soil microbial biomass and its metabolic quotient were determinated monthly. A compaction-induced decrease of microbial biomass content could be detected. The results of the field experiment were supported by a laboratory experiment, where the decomposition of ^{14}C-labelled straw under different compaction levels was estimated. A higher energy demand per unit biomass and a lower efficiency of plant residue decomposition was correlated to soil pore space reduction. The observed effects of compaction on soil microorganisms are probably due to a lack of soil aeration.

5 ACKNOWLEDGEMENTS

We thank S. Schintzel, A. Oehns-Rittgeroth and B. Volkmar for expert technical assistance and T. von Frenkell-Insam and A. Kaiser for language editing. For support with technical equipment we are grateful to F.-F. Gröblinghoff and K. Haider. This work was funded by the Federal Ministry of Research and Technology.

6 REFERENCES

1. B.D. Soane, Soil Tillage Res., 1990, 16, 179.

2. I. Hånkanson, W.B. Voorhees, and H. Riley, Soil Tillage Res., 1988, 11, 239.

3. B. Meyer, Mitteilgn. Dtsch. Bodenkundl. Gesellsch. 1982, 34, 149.

4. G. Drumbeck, und T. Harrach, Mitteilgn. Dtsch. Bodenkundl. Gesellsch., 1985, 43/I, 213.

5. J.A. Currie, J.Soil Sci., 1984, 35, 1.

6. M.M. Landina and I.L. Klevenskaya, Soviet Soil Sci., 1984, 16, 46.

7. L.R. Bakken, T. Børresen, and A. Njøs, J. Soil Sci., 1987, 38, 541-552.

8. J.M. Tisdall, and J.M. Oades, J.Soil Sci., 1982, 33, 141-163.

9. J.M. Lynch, and L.F. Elliott, Appl. Environ. Microbiol., 1983, 45, 1398-1401.

10. L.D. Metzger, D. Levanon, and U. Mingelgrin, Soil Sci. Soc. Am. J., 1987, 51, 346-351.

11. W.B. McGill, K.R. Cannon, J.A. Robertson, and F.A. Cook, Canadian Journal of Soil Science, 1986, 66, 1-19.

12. J.W. Doran, Biol. Fert. Soils, 1987, 5, 68-75.

13. T.-H. Anderson and K.H. Domsch, Soil Bio. Biochem., 1989, 21, 471-479.

14. T.-H. Anderson and K.H. Domsch, In Proceedings of Fourth International Symposium on Microbial Ecology (F. Megusar and M. Gantar, Eds.), pp 476-471. Slovene Society for Microbiology, Lublijana, 1986.

15. D.S. Powlson, P.C. Brooks, and B.T. Christensen, Soil Bio. Biochem., 1987, 19, 159-164.

16. G. Olfe und H. Schön, In: KTBL (eds.) Bodenverdichtung, Darmstadt, 1986, Schrift 308, 35-47.

17. J.P.E. Anderson, and K.H. Domsch, Soil Biol. Biochem., 1978, 10, 215-221.

18. O. Heinemeyer, H. Insam, E.- A. Kaiser, and G. Walenzik, Plant and Soil, 1989, 116, 191-195.

19. H. Insam, and K.H. Domsch, Micro. Ecol., 1988, 15, 177-188.

20. T.-H. Anderson, Ph.D. Thesis, University of Essex, 1988.

21. D.W. Nelson and L.E. Sommers, In: Methods of Soil Analysis. Part 2 (A.L. Page, R.H. Miller and D.R. Keeny, Eds.) pp. 539-594. American Society of Agronomy, Madison, 1982.

22. F.-F. Gröblinghoff, K. Haider, and Th. Beck, VDLUFA-Schriftenreihe, 1989, 28, 893-908.

23. D.M. Linn and J.W. Doran, Soil Sci, Soc. Am.J., 1984, 48, 1267-1272.

24. D.S. Jenkinson and D.S. Powlson, Soil Bio. Biochem., 1976, 8, 209-213.

25. R. Martens, Soil Biol.Biochem., 1985, 17, 57-63.

26. D.E. Stott, G. Kassim, W.M. Jarrel, J.P. Martin, and K. Haider, Plant and Soil, 1983, 70, 15-26.

27. F. Huysman, W.M.I.S. Gunasekara, Y. Avnimelch, and W. Verstraete, Bio. Fert. Soils, 1989, 8, 231-234.

28. L.L. Parson, and M.S. Smith, Soil Sci.Soc Am.J., 53, 1023-1085.

29. E.B. Alexander, Soil Sci. Soc. Am. J., 1980, 44, 689-692.

The Cotton Strip Assay: Field Applications and Global Comparisons

Gill Howson

INSTITUTE OF TERRESTRIAL ECOLOGY, MERLEWOOD RESEARCH STATION, GRANGE-OVER-SANDS, CUMBRIA LA11 6JU, UK

SUMMARY

Cotton strip assay can give an indication of changes in soil decomposition activity over a wide range of sites used for ecological studies by measuring the loss in tensile strength (TS) of cotton material buried in the soil. Some examples are presented together with a world wide comparison of its use.

1 BACKGROUND

Plant nutrition in most semi-natural ecosystems including forests and some forms of agriculture is largely dependent on nutrient cycling. In turn this is maintained by continued decomposition of organic matter on and within the soil. Decomposition is influenced by litter quality, temperature, moisture, nutrient availability and other soil properties. The decomposition of organic matter determines the rate of nutrient circulation; accumulated litter, for example on forest floors, indicates an imbalance between production and decomposition rates[1]. There are several ways of measuring the decomposition rate in the field.

1. Respiration of plant material, soil, or individual substrates is a direct measure of catabolic processes but on its own, however, it is difficult to interpret.

2. Weight loss from whole plant material contained in mesh bags over a period of time gives an integrated loss due to catabolism, leaching and comminution. However, for a variety of reasons and especially in global comparisons a single standard substrate may be more desirable. It also gives integrated loss due to catabolism but is not subject to the chemical and physical variability of plant litters.

3. Since plant litters have a high initial cellulose content, eg straw has 80%, Bracken petioles 75%, <u>Molinia caerulea</u> (L) 70% oven dry weight, (pers. comm. Merlewood Research Station Chemical Service) cellulose in various forms can be used as a standard substrate to remove the direct effect of substrate quality (an important source of variability) for intensive comparisons. Cotton tape was used in Australia for soil studies[2] and Borregaard cellulose board[3], unbleached calico[4] and strips of linen[5] have all been used to compare potential decay rates under field conditions.

To investigate the rate of breakdown of cellulose under different conditions and to quantify it so that meaningful comparisons could be made between different sites or different climatic regimes supporting different vegetation types a standard test was needed. Soil is a highly heterogeneous medium and experience has shown that meaningful results can only be obtained by a high degree of experimental replication. In studying a range of sites or plots any technique must therefore be cheap and simple so that large numbers of samples can be easily handled. It appeared that a commercial cloth testing system could be used in an ecological context and cotton was chosen because of its high cellulose content (93% alpha-cellulose). The degree of decomposition is assessed by the loss of tensile strength of the cotton. A standardized procedure has been developed[6,7] and the cotton strip assay is used in assessing the effects of natural variation or management practices on potential decomposition, where a comparative and integrated measurement rather than an instantaneous measurement is required. The cotton strip assay has been used extensively in many types of environment with a wide geographical spread to assess the potential for microbial cellulolytic decomposition within soils[8] and to compare activity in a wide range of sites. The assay has been criticized[9], but results have given useful insights in ecological studies where more intensive methods would not be possible. Just a few examples are discussed in this paper. It has also been modified for experimental studies on macro- and micro-scales[10]. Further technical aspects of the cotton strip assay are described in Latter et al[11]. It is not intended as a substitute for good research on organic matter turnover <u>per se</u>.

In normal field use strips (30 cm x 10 cm) are inserted vertically into the soil using a small straight spade. A standard cloth is supplied by The British Textile Technology Group, Manchester M20 8RX. Following burial for some weeks or months according to site conditions, the loss in tensile strength on retrieval gives an integrated index of activity over the periods of time in the field. Tests at intervals along the length of the strip, conveniently every 4 cm, show changes down the soil profile.

Two types of controls are used, fabric and field. Fabric controls are untreated cotton used to correct for any differences in batches of material. Field controls are placed in the soil profile and removed immediately to assess the effect of handling and field placing.

Cotton decay is measured by tensile strength loss (CTSL) which is the field control TS - final TS. Hill et al.[12,13] have described the change in tensile strength over time using a linearizing transformation.

Briefly,

$$\text{Cotton rottenness, CR} = \sqrt[3]{\frac{\text{field control TS} - \text{final TS}}{\text{final TS}}}$$

$$\text{The process rate CRR} = \frac{\text{CR} \times 365}{t}$$

and is the rate of rotting expressed as a yr^{-1} value, where t is time in days.

The linearizing transformation was tested by estimating CR from CTSL using data from a forestry field experiment where strips were buried then retrieved after differing periods of time (Figure 1).

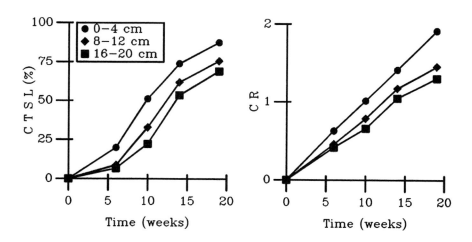

Figure 1. Loss in tensile strength in relation to time in soil in a field experiment at Gisburn Forest, Lancashire (Brown & Howson 1988). Combined data for 4 monoculture plots presented as unlinearized (CTSL %) and linearized CCR).

Tensile strength results below 5% CTSL or greater than 95% CTSL are of doubtful value and very high CTSL in particular give distorted values in the transformation procedure. Ideally CTSL around 50% provide maximum information[13].

2 EXAMPLES OF THE USE OF THE COTTON STRIP ASSAY

Effect of pure and mixed stands of forest species on soil

Alder (<u>Alnus glutinosa</u>), Scots pine (<u>Pinus sylvestris</u>), Sessile oak (<u>Quercus petraea</u>) and Norway spruce (<u>Picea abies</u>) were planted in 1955 at Gisburn Forest, Lancashire. 20 years later, standard chemical analysis developed for agricultural soils could not readily detect differences in the soils under the pure and mixed stands of the experimental plots.

The cotton strip assay showed clear between species effects and CTSL mirrored differences in the heights of a given tree species according to whether it was in a pure or mixed stand. Significant differences between tree treatments were shown in the mixtures for alder and spruce (Figure 2). The height growth was improved, apparently through modification of soil nutrient availability, which was confirmed by forest floor lysimeter studies[14] of mineralized N (NO_3+ NH_4) and by KCl extractable inorganic N estimates in the uppermost mineral soil (Table 1). Both measures of mineralized N also related well with cotton strip decomposition and tree height measurements[15].

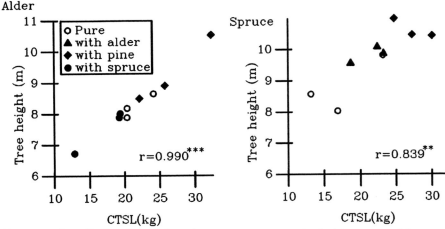

Figure 2. Relationship between tree heights at 26 years (1981) and tensile strength loss of cotton strips after 9 weeks (1980) in pure and mixed stands at Gisburn (means per plot). (** significant at P < 0.01; *** P < 0.001) (Brown 1988)

	NO$_3$-N g/m^2/yr	P g/m^2/yr
Pure Spruce	0.26	0.21
Spruce (with Oak)	0.3	0.04
Spruce (with Alder)	0.5	0.19
Spruce (with Pine)	0.9	0.38

Table 1. Mean gross fluxes of NO$_3$-N and P from forest floor lysimeters under Norway spruce canopies, pure and mixed at Gisburn (block 2 only, 1983/84).

Seasonal patterns of CTSL under the 4 pure tree species at Gisburn Forest showed that decomposer potential is greater under alder and Scots pine than under sessile oak and Norway spruce. Late summer to early autumn was the time of peak decomposition activity for all the tree species except oak, which was bimodal. (Figure 3)

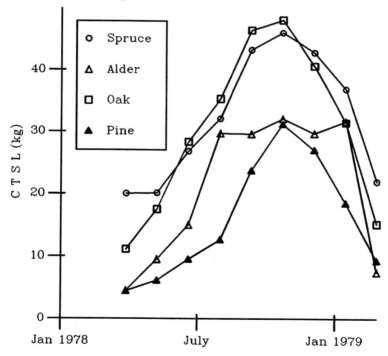

Figure 3. Changes with season in tensile strength loss of cotton strips (CTSL) following insertion under 4 tree species at Gisburn (1978-79). Nine overlapping burial periods, each of 12 weeks, means of all depths (ie whole strip averages) for 3 blocks combined points being placed on the retrieval date (Brown & Howson 1988)

The pattern of CTSL down the soil profile with a decline from 0-20 cm, varied with the seasons, oak and spruce showing a greater decline particularly in the high decomposition seasons, whereas under alder and pine the difference was not so marked[16]. (Figure 4)

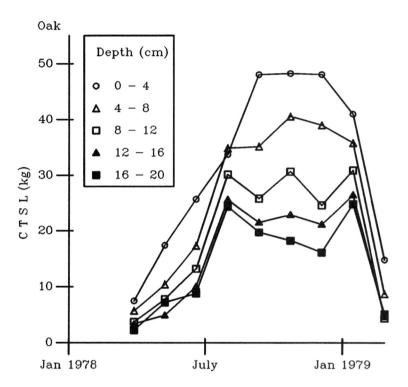

Figure 4. Changes with depth of the seasonal curves for tensile strength loss under oak. (Brown & Howson 1988)

Earthworm populations which are very different in the alder and pine compared to the oak and spruce may possibly account for the changes in CTSL down the soil profile with a higher number of earthworms (mainly Lumbricus rubellus) in the alder and pine (Table 2). The spruce and oak plots contained very low numbers of earthworms and this may have influenced decomposition rates with less organic matter being carried down the soil profile by them. However there is a need for conclusive evidence that the presence of earthworms stimulates cellulolytic activity in these soils in order to distinguish cause from effect.

	Mean Number per trap^{-1}	Mean Biomass per trap^{-1} (g)
Pine	16.06	9.19
Alder	13.88	9.33
Spruce	2.09	0.71
Oak	2.09*	0.71

Table 2. Mean numbers and biomass (preserved fresh weight) of earthworms per trap under 4 tree species. Two trapping periods and 2 blocks, combined, at Gisburn 1981. *Estimated by observation.

Demonstrating the effects of forestry management practices

Three intensities of line thinning treatments had been applied in a randomized block experiment in a Forestry Commission Sitka spruce (<u>Picea sitchensis</u>) plantation at Elwy Forest, north Wales. The 3 tree thinning intensities were, (i) high (ii) intermediate and (iii) unthinned, whereby 50, 30 and 0% respectively of the trees were removed along the planted rows to leave a row (or rows) of tree stumps between a row (or rows) of trees. Cotton strip assay was used in 4 consecutive 3-month periods and paired strips were placed between trees and between stumps. Litter fall was monitored and also litter accumulation on the forest floor. The CTSL showed the same basic pattern (Figure 5) of decline with depth seen at Gisburn forest and analysis of variance showed that CTSL was significantly lower in the unthinned plots during 2 of the 3 monthly periods (July - October ($P < 0.05$), January - April ($P < 0.01$)) than the combined thinned plots. The Jenny index $\underline{K}$, was used to estimate the forest-floor decomposition rates and was calculated from annual litterfall figures and accumulated litter in the L layer and were in the order: unthinned > intermediate > high-thinned which follows the decomposition of the cotton strips. The unthinned stands had a slower decomposition rate and hence a build-up of litter which is not just due to higher litterfall. Management practices resulted in changes in moisture status of the soil, light intensity and temperature, which all affect soil organisms and decomposition rates. The cotton strip assay was able to detect subtle seasonal differences between tree and stump positions in the thinned plots with a higher decomposition rate at the tree area and a lower rate at the stump area between January-July, while the reverse was true for the period July-October[17].

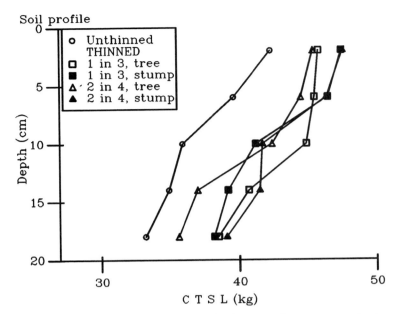

Figure 5. Tensile strength loss of cotton strips in thinning treatment plots. Change down profile for one block, autumn 1979. (Howson 1988)

Global applications

Delegates to a symposium[18] were asked to provide cotton strip assay data from various sites around the world (Figures 6 & 7) on tensile strength loss and environmental parameters and these were then analysed using a GENSTAT statistical package.[19]

Multiple regression analysis and analysis of variance suggested that potential evaporation (PEVAP), which reflects temperature and rainfall was the major factor influencing the decay of the cotton strips with a range from 0-11 mm/day. The equation:

$$CRR = 0.20 + 3.13 (PEVAP)$$

explained 77% of the variance and this may be useful in deducing how long to leave strips inserted at a new site. PEVAP is readily available from published tables[20] (Muller 1982) and Meentemeyer and Berg[21] have also emphasised the close correlation between plant litter decomposition and estimated actual evapotranspiration.

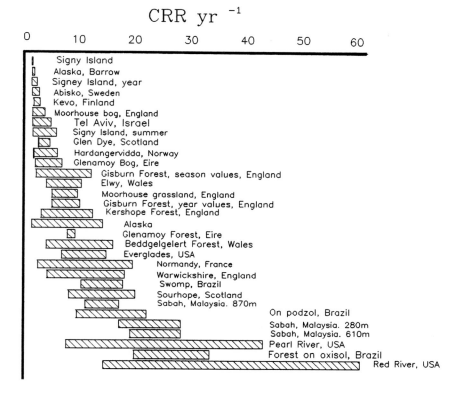

Figure 6. Ranges of reported CRR values for worldwide sites; showing high variability on the more active sites. Only a few values represent a full years data most values being influenced by the seasonal climate of the respective insertion periods. (Appendix 1, Cotton Strip Assay; an index of decomposition in soils).

3 CONCLUSIONS

The interaction of substrate quality, physico-chemical environment and decomposer organisms determines the rate of litter decay and using a standard material enables us to understand some of the factors involved in decomposition by removing the effect of one of these major variables ie the substrate quality.

It is possible using the cotton strip assay to demonstrate seasonal patterns in decomposition activity and the pattern of change with depth, in the soil, during the year. Extrapolation of annual estimates from a single seasonal measurement would be dangerous and could lead to a biased estimate of annual loss. Comparison of different site or management effects needs several seasonal measurements to enable an annual estimate of CTSL to be made. The cotton strip assay is sensitive to the microbial cellulolytic activity in a wide range of soils and has indicated that the complex interactions between environmental factors and decomposition processes, need further study.

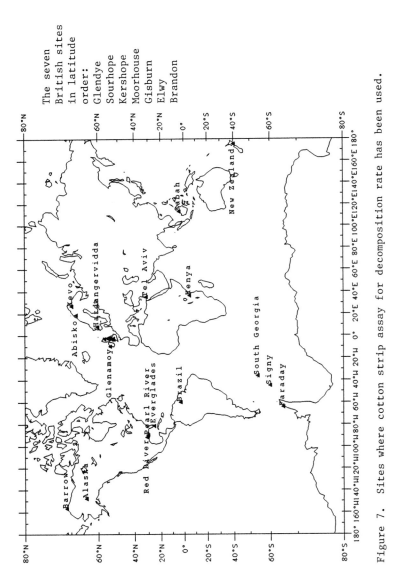

Figure 7. Sites where cotton strip assay for decomposition rate has been used.

REFERENCES

1. N. Malmer, Oikos (suppl.), 1969. 12, 79.
2. A.D. Rovira, Aust. Conf. Soil Sci., Adelaide, 1953. 1, 3.17, 1.
3. T. Rosswall, "Soil organisms and decomposition in tundra", edited by A.J. Holding, O.W. Heal, S.F. MacLean and P.W. Flanagan, Stockholm, Tundra Biome Steering Committee, 1974. p 325.
4. O.W. Heal, G. Howson, D.D. French and J.N.R. Jeffers, "Soil organisms and decomposition in tundra", edited by A.J. Holding, O.W. Heal, S.F. MacLean and P.W. Flanagan, Stockholm, Tundra Biome Steering Committee, 1974. p 341.
5. K. Kuzniar, Inst. Bad. Lesn. Rozpr. i Spraw, 1948. 50, 1.
6. P.M. Latter and G. Howson, Pedobiologia, 1977. 17, 145.
7. "Cotton strip assay an index of decomposition in soils", edited by A.F. Harrison, P.M. Latter and D.W.H. Walton, (ITE symposium no 24), Grange-over-Sands, Institute of Terrestrial Ecology, 1988. p 166.
8. P.M. Latter and A.F. Harrison, "Cotton strip assay an index of decomposition in soils", edited by A.F. Harrison, P.M. Latter and D.W.H. Walton, (ITE symposium no 24), Grange-over-Sands, Institute of Terrestrial Ecology, 1988. p 68.
9. P.J.A. Howard, "Cotton strip assay an index of decomposition in soils", edited by A.F. Harrison, P.M. Latter and D.W.H. Walton, (ITE symposium no 24), Grange-over-Sands, Institute of Terrestrial Ecology, 1988. p 34.
10. P. Holter, "Cotton strip assay an index of decomposition in soils", edited by A.F. Harrison, P.M. Latter and D.W.H. Walton (ITE symposium no 24), Grange-over-Sands, Institute of Terrestrial Ecology, 1988. p 72.
11. P.M. Latter, G. Bancroft and J. Gillespie, Int. Biodeterior. 1988. 24, 25.
12. M.O. Hill, P.M. Latter and G. Bancroft, Can. J. Soil Sci., 1985, 65, 609.
13. M.O. Hill, P.M. Latter and G. Bancroft, "Cotton strip assay an index of decomposition in soils", edited by A.F. Harrison, P.M. Latter and D.W. Walton (ITE symposium no 24), Grange-over-Sands, Institute of Terrestrial Ecology, 1988. p 21.
14. K. Chapman, PhD Thesis, University of Lancaster, 1986. A.H.F. Brown, Institute of Terrestrial Ecology Annual Report 1986/87, p 71.
15. A.H.F. Brown, "Cotton strip assay an index of decomposition in soils", edited by A.F. Harrison, P.M. Latter and D.W. Walton (ITE symposium no 24), Grange-over-Sands, Institute of Terrestrial Ecology, 1988. p 80.

16. A.H.F. Brown and G. Howson, "Cotton strip assay an index of decomposition in soils", edited by A.F. Harrison, P.M. Latter and D.W. Walton (ITE symposium no 24), Grange-over-Sands, Institute of Terrestrial Ecology, 1988. p 86.
17. G. Howson, "Cotton strip assay an index of decomposition in soils", edited by A.F. Harrison, P.M. Latter and D.W. Walton, (ITE symposium no 24), Grange-over-Sands, Institute of Terrestrial Ecology, 1988. p 94.
18. "Cotton strip asssay an index of decomposition in soils", edited by A.F. Harrison, P.M. Latter and D.W. Walton, (ITE symposium no 24), Grange-over-Sands, Institute of Terrestrial Ecology, 1988.
19. P. Ineson, P.J. Bacon and D.K. Lindley, "Cotton strip assay an index of decomposition in soils", edited by A.F. Harrison, P.M. Latter and D.W. Walton (ITE symposium no 24), Grange-over-Sands, Institute of Terrestrial Ecology, 1988. p 155.
20. M.J. Muller, Selected climate data for a global set of standard stations for vegetation science, The Hague, Junk. 1982.
21. V. Meentenmeyer, B. Berg, Scand. J. For. Res., 1986, 1, 167.

Section 4

Soil Organic Matter Turnover

Introductory Comments

by T.R.G. Gray, University of Essex

I was once told by a soil scientist that soil would be an excellent subject for study if it were not for the microorganisms that inhabit it and interfere with solution of the more central questions of soil physics and chemistry. This section of the present volume might have reinforced his view but to me it demonstrates the crucial role played by microorganisms in determining the nature of soil organic matter and transforming a body of gases, liquids and solids into a life-support system.

When largely insoluble and polymeric plant material is deposited on or in soil, it is already colonised by a microflora characteristic of the environment from which it came, e.g. the phylloplane, decaying wood or the root surface. Those members of this microflora adapted to the environmental conditions found in soil will survive and multiply but the majority will disappear gradually and be replaced by a succession of indigenous soil organisms. This transition will be accelerated by the movement and comminution of the plant material by the soil fauna which increases the surface area for colonisation and decay and transfers organic matter into mineral horizons. Pieces of organic matter act as foci for heterotrophic microbial activity and in particular for depolymerisation reactions which produce soluble and insoluble (often colloidal) intermediates. The soluble and colloidal materials may be leached through the mineral soil and deposited at sites far from their point of origin. Some of these important topics are examined by Lynch in his introductory paper.

The above sources of organic matter are not pure substances but complex and variable mixtures of different compounds which decay at different rates. Soluble and proteinaceous materials have short half-lives, cellulosic substances intermediate half-lives and lignified materials very long half-lives. Ball and Allen discuss the role of actinomycete enzymes in solubilising different sized particles of lignocellulose from wheat straw. Harkness, Harrison and Bacon remind us that sharp pulses of ^{14}C introduced into the atmosphere through the nuclear testing programme provide a method for studying the turnover times of soil organic matter and mass transfer in the carbon cycle.

During the course of breakdown, organic materials release highly reactive intermediate compounds which can condense with one another to produce acidic humic compounds. These are amongst the most long-lived substances and can have profound effects on the soil and aquatic environments into which they move. Humic compounds may form complexes with the inorganic fraction of the soil and have unexpected effects on microbial metabolism as Lynch (and others in Section 1 of this volume, e.g. Hayes, Malcolm and MacCarthy, etc) remind us.

The chemical nature of the plant material added to a soil can affect the chemistry of the humic materials formed but it is also possible that interference with the global environment will change the quantity and nature of organic matter added to the soil. In the context of agriculture today, the use of continuous monocultures instead of traditional crop rotations, as well as the increasing levels of greenhouse gases in the atmosphere, may have far reaching but as yet imperfectly understood consequences. Anderson and Gray highlight the differences in microbial metabolism which occur between agricultural cropping systems, stressing the increase in both qCO_2 and qD values in monoculture soils, whilst Kuikman, Lekkerkerk and van Veen show how increasing levels of carbon dioxide in the atmosphere may increase the input of easily decomposable organic matter, reduce the turnover of resistant organic fractions and thus increase the organic matter content of soil.

Also, modern agriculture, other than organic farming, applies little organic matter to arable land and so most organic inputs come largely from roots or from ploughed-in aerial crop debris, e.g. wheat straw (see Rule *et al* in Section 5). Since the rates of decomposition of organic matter are slow, it may take some time for soil to pass through a transitional state and reach a new equilibrium point. Depending upon the soil type, there may be less organic matter in the soil when the equilibrium is reached. Can rates of change be measured with any confidence? How long does it take to re-establish equilibrium? Does it matter? These are all questions addressed in this section of the book and Johnson and Prince show how monitoring of organic matter contents of fen silt soils with permanent grassland change after ploughing have shown initial dramatic falls in organic matter and a subsequent period of slow decline approaching but not reaching equilibrium after nearly 30 years.

As more and more land is taken out of agricultural production and used for other purposes such as forestry, we can also predict that further changes will occur in the underlying soils. These are explored by Whiteley who shows that in the first few years after planting with several species, a net release of soil organic nitrogen occurs. However, longer term changes may involve decreases in total carbon and nitrogen contents, accompanying a change in the distribution of these materials in the soil profile.

Sources and Fate of Soil Organic Matter

J.M. Lynch

MICROBIOLOGY & CROP PROTECTION DEPARTMENT, HORTICULTURAL RESEARCH INTERNATIONAL, LITTLEHAMPTON, WEST SUSSEX, BN17 6LP, UK

1 INTRODUCTION

The dominant components of soil organic matter are humic materials which are relatively inert biologically and have half-lives in excess of 1000 years.[1]

Plants and autotrophic micro-organisms are the primary producers in terrestrial ecosystems and thus all soil organic matter originates from them. Components of this organic matter provide binding and aggregating agents which are critical in providing soil its structure. Aside from any phylogenetic considerations, it can be assumed that autotrophic microbes such as algae, well-known for their capacity to produce polysaccharides as soil aggregating agents[2], were probably the initial colonizers of soil particles such as volcanic ash to generate a substrate suitable for the growth of the roots of higher plants.

The form of input of plant materials into the soil varies greatly between different ecosystems and crops. Table 1 estimates the input to the surface layer in a temperate arable soil.

The pattern of input down the soil profile becomes quite different with the inputs from living and dead roots making a relatively greater contribution than that from straw.

The implication of some of these inputs has been discussed elsewhere.[3,4,5] Here, instead, a summary of the salient features in relation to current opportunities and some of the myths and fallacies that appear to have been generated is given.

Table 1   Approximate annual input of substrates (primary productivity) to the surface 5 cm of an arable soil (with a total soil biomass of c. 400 kg C ha^{-1}) (from Lynch & Panting[6])

Source	Amount (kg/ha^{-1} year^{-1})
Root decomposition	400
Root exudation	240
Straw residues	2800
Autotrophic microbes	100
Total	3540

2 HUMIC ACIDS

The reason for such long half-lives of the soil humic fraction is due to their resistance to degradation by micro-organisms. In a recent study, enrichment cultures were attempted with natural humic substances and synthetic humic compounds (two melanoidins and one synthetic polymer) as the sole carbon source in liquid media.[7] The resistance to degradation did not seem to be related to sample origin or extraction procedure; fulvic acids were as refractory as humic acids. Further aeration did not appear to be the critical factor in biodegradation.

The refractory nature of humic acids is probably useful in their stabilizing influences on soil.[8] These can have a variety of effects on plants either as phytotoxins or in mediating nutrient uptake processes.[9] Relatively less attention has been paid to their effects on microbial processes but these can be beneficial[10] or harmful.[11] Recently the effects of humic acids on nitrification were studied.[12] The sources of the acids were samples of commercial peat and prepared mushroom wheat straw compost respectively. Both types of humic acid at concentrations in excess of 100 μg cm^{-2} inhibited the activities of nitrifying bacteria in a continuous-flow column. Lowering the pH of the system also inhibited nitrification. However in the presence of the compost humic acid, and to a lesser extent the commercial, humic acid at less than 100 μg cm^{-3}, the pH-inhibition of nitrification was relieved. The similarity between nitrification activity as a function of effluent pH in the presence of commercial humic acid indicates that humic acid might act at least in part as a buffer. This explanation did not apply however to the compost humic acid, because even at reduced effluent pH, nitrification still occurred at a comparatively high rate. Although the chemical bases of these effects are still unclear they bear on the question of the relative role of acid rains and humic acids on nitrification.

3 RHIZODEPOSITION

Rhizodeposition is a consequence of the plant's photosynthetic processes and is defined as the total amount of carbon coming from the roots.[13] The material is composed of exudates, secretions, lysates and gases (including ethylene and CO_2). A major but indeterminate proportion of this derives from microbial metabolism of the soluble and insoluble components of the rhizodeposition. The only satisfactory method of estimation of this pool is to grow plants exclusively on a source of $^{14}CO_2$, at constant specific activity, with roots contained in a sealed pot of non-sterile soil. The CO_2 loss can be estimated by periodic flushes of the root container with air and trapping the released $^{14}CO_2$ in sodium hydroxide or ethanolamine in methanol. For plants up to 3 weeks old, rhizodeposition can account for about 40% of the total dry matter production by the plant (Table 2).

Table 2 Percentage of net fixed carbon moved to roots and the percentage of net fixed carbon moved to the roots which is lost when shoots of plants are exposed continually to $^{14}CO_2$ (plants grown in soil). Conditions varied but were c. 18°C, 16 h day length, 0.03% CO_2 and none were sterilized.

Plant	Age (d)	% net photosynthate lost as			Ref.
		Respiration (A)	Rhizodeposition in soil (B)	Respiration and rhizodeposition in soil (A+B)	
Wheat	21	9.5	6.5	16.0	14
	21	16.9	11.8	28.7	15
	21	23.0	17.1	40.1	16
	28	17.2	4.8	22.0	17
Barley	21	9.0	9.9	18.9	14
	21	25.9	11.3	37.2	16
Maize	14	4.5	14.6	19.1	18
Tomato	14	8.6	30.1	38.7	19
Pea	28	23.3	12.8	36.1	19

There seems to be little difference between the pattern from monocotyledonous cereals, legumes and dicotyledonous plants. The process has been investigated using the $^{14}CO_2$ technique in laboratories in Australia, Germany, France and the Netherlands and generally there is a reasonable consistency between different laboratories. It is of course more difficult to investigate the carbon flow in perennial crops (trees and grasses) using continuous ^{14}C labelling; then the more equivocal method of pulse-labelling has to be used.

The *pros* and *cons* of such approaches have been discussed in depth by Whipps[20] but suffice to say for the purposes here that the amounts of carbon entering the soil per unit of photosynthate in annual and perennial systems is likely to be quite similar. Thus it is perplexing why so many plant and crop physiologists still continue to ignore this large component of plant's photosynthate which is the driving force for the microbial and faunal populations of the rhizosphere.

Since the studies in the 1920s and 1940s in the USA and Canada, there has been relatively little investigation on the microbial successions in the rhizosphere as a crop matures. Also, in terms of entry into the soil, the microbial population colonising senescing shoot tissues will govern the saprophytic decay of the tissue as it enters the soil. The C:N ratio of these tissues can be rather high because much of the N has been harvested (e.g. in grain). By contrast, root tissues usually have a lower C:N ratio more favourable for decomposition and therefore decay of roots is much more rapid. Microbial activity in the rhizosphere leads to the production of a variety of metabolites which can have a direct influence on plant growth (e.g. ionophores and growth regulators) or influence plant growth indirectly through a modification of soil structure or properties (e.g. polysaccharides).[3]

4 PLANT RESIDUE DECOMPOSITION

The decay of shoot tissues depends on their form. Green leaves decay more rapidly than stems. One of the most common forms of senescent shoots to enter the soil is straw which is lignocellulosic. The cellulosic fraction decays more rapidly than lignin (Table 3) and indeed lignin can present a structural barrier to the decay of the total cellulosic fraction. Because of the potential of harnessing lignocelluloses in plant residues and wood products, there has been major attention given to the anatomy, physiology, biochemistry, chemical and molecular biological aspects of lignocellulose biodegradation. For example, the American Chemical Society held a major symposium on this general topic area.[22] In terms of specifics, *Trichoderma* spp. are major agents of cellulolysis and the Royal Society of Chemistry recently published a volume on this topic.[23]

One of the topics of environmental interest to soil scientists and agronomists has been the question of straw burning and whether besides being a misuse of natural resources this has any influence on crop yield. Studies at Letcombe showed that in autumn on heavy clay soils which readily become anaerobic, direct-drilling could often allow a crop to be sown where there would

Table 3   Decomposition rate constants (k) for the major components of buried wheat straw measured September 1977-August 1978. (After Harper & Lynch[21])

Components	k (d^{-1})
Hemicellulose	0.0073
Cellulose	0.0074
Lignin	0.0005

not have been one otherwise where ploughing was necessary; spring cropping would have been the only viable alternative. However under such conditions the early stages of straw decomposition could yield phytotoxic concentrations of organic acids, notably acetic, propionic and butyric.[24] Initial observations on both the effect and interpretation were confirmed in the Pacific Northwest of the United States.[25] A problem of investigating the cause and effect in the field is that the acids were only formed in high concentrations on the surface of straw decomposing anaerobically and the acid did not diffuse very far from the straw surface;[26] the phytotoxic effect depended on the establishing seedling root coming into contact with the mat of decomposing straw, a likely occurrence. It was naïve therefore, as some have done, to try to measure organic acid concentrations in soil as such, especially in soils of good structure which are well-aerated.

It would also be naïve to suggest the organic acid toxicity as the unique adverse effect involved. In the Pacific Northwest, it was found that plant growth-inhibiting pseudomonads could be isolated from root cortices of wheat plants growing adversely in the presence of straw[27] and indeed recently a process has been described whereby this property can be harnessed as a microbial herbicide.[28]

Another possible interpretation of the 'straw effect' is the colonization of roots by *Pythium*.[29] In trying to outline the potentially beneficial effects of harnessing crop residues, a process was described where the cellulosic component of straw could be used as a carbohydrate source for an aerobic cellulolytic micro-organism (*Trichoderma harzianum*) which could channel carbohydrate energy to an anaerobic nitrogen-fixing bacterium (*Clostridium butyricum*). The association was protected by a polysaccharide-producing bacterium (*Enterobacter cloacae*) which not only removed excess oxygen from the consortium but also stabilized soil structure.[30] It seems likely that such associations occur in nature and indeed an association involving *Cellulomonas* as the cellulase-positive organism and

Azospirillum as the nitrogenase-positive organism was described subsequently.[31] However therein lies a problem in that our commercial sources of funding demanded that it would be necessary to introduce or elevate the natural process by inoculation and this would of course be difficult in soils where such consortia already seem to function. A more realistic approach which appeared at the time and is even more relevant today is to use consortia as inocula to degrade crop and wood wastes in piles, thereby generating peat substitutes for use in horticulture.[32]

5 CONCLUSIONS

In managing soil organic matter, it seems useful to develop predictive models of the dynamics of decomposition processes and product formation based on sound laboratory and field observation. This has been done satisfactorily for residue decomposition[33] and recently developments have been made on a model for the utilization of rhizodeposition products.[34]

Finally it is perhaps with some regret that soil science has been poorly-funded in recent years in comparison with the more glamorous fields such as biotechnology. However, it now appears to be coming 'full-circle' because in introducing new plant and microbial genotypes into soils, it is crucial that the biodiversity of organisms and their processes is monitored to avoid any detrimental effects on production and utilisation of soil organic matter.

REFERENCES

1. C.A. Campbell, E.A. Paul, D.A. Rennie and K.J. McCallum, Soil Sci., 1967, 104, 217.
2. B. Metting, 'Micro-algal Biotechnology', Ed. by M.A. Borowitzka and L.A. Borowitzka, Cambridge University Press, Cambridge, 1988, p. 288.
3. J.M. Lynch, Ed., 'The Rhizosphere', Wiley, Chichester, 1990.
4. J.M. Lynch, 'Soil Biotechnology', Blackwell Scientific, Oxford, 1983.
5. J.M. Lynch and M.D. Wood 'Russell's Soil Conditions and Plant Growth', 11th edition, Ed. by A. Wild, Longman, London, 1988, p. 526.
6. J.M. Lynch and L.M. Panting, Soil Biol. Biochem., 1980, 12, 29.
7. C.Y. Kontchou and R. Blondeau, Can. J. Soil Sci., 1990, 70, 51.
8. M.H.B. Hayes, 'Microbial Adhesion to Surfaces', Ed. by R.C.W. Berkeley, J.M. Lynch, J. Melling, P.R. Rutter and B. Vincent, Ellis Horwood, Chichester, 1980, p. 263.

9. D. Vaughan and R.E. Malcolm, Eds., 'Soil Organic Matter and Biological Activity', Martinus Nijhoff/Dr W. Junk, Dordecht, 1985.
10. S.A. Visser, Soil Biol. Biochem., 1985, 17, 457.
11. D.J. Hassett, M.S. Bisesi and R. Hartenstein, Soil Biol. Biochem., 1987, 19, 111.
12. M.J. Bazin, A. Rutili, A. Gaines and J.M. Lynch, FEMS Microbiol. Ecol., in press.
13. J.M. Whipps and J.M. Lynch, Proc. Phytochem. Soc. Eur., 1985, 26, 59.
14. D.A. Barber and J.K. Martin, New Phytol., 1976, 76, 69.
15. J.M. Whipps and J.M. Lynch, New Phytol., 1983, 95, 605.
16. J.M. Whipps, J. Exp. Bot., 1984, 35, 767.
17. R. Merckx, A. Den Hartog and J.A. van Veen, Soil Biol. Biochem., 1985, 17, 565.
18. J.M. Whipps, J. Exp. Bot., 1985, 36, 644.
19. J.M. Whipps, Pl. Soil, 1987, 103, 95.
20. J.M. Whipps, 'The Rhizosphere', Ed. by J.M. Lynch, Wiley, Chichester, 1990, p. 59.
21. S.H.T. Harper and J.M. Lynch, J. Soil Sci., 1981, 32, 627.
22. N.G. Lewis and M.G. Paice, 'Plant Cell Wall Polymers: Biogenesis and Biodegradation', American Chemical Society, Washington, D.C., 1989.
23. C.P. Kubicek, D.E. Eveleigh, H. Esterbauer, W. Steiner and E.M. Kubicek-Pranz, *Trichoderma reesei* Cellulases: Biochemistry, Genetics, Physiology and Applications', Royal Society of Chemistry, Cambridge, 1990.
24. J.M. Lynch, J. Appl. Bacteriol., 1977, 42, 81.
25. J.M. Wallace and L.F. Elliott, Soil Biol. Biochem. 1979, 11, 325.
26. J.M. Lynch, K.B. Gunn and L.M. Panting, Pl. Soil, 1980, 56, 93.
27. L.F. Elliott and J.M. Lynch, Soil Biol. Biochem., 1984, 16, 69.
28. L.F. Elliott and A.C. Kennedy, US Patent App. 6/6/1988 as 7-207592.
29. A.D. Rovira, L.F. Elliott and R.J. Cook, 'The Rhizosphere', Ed. by J.M. Lynch, Wiley, Chichester, 1990, p. 389.
30. J.M. Lynch and S.H.T. Harper, Phil. Trans. R. Soc. Lond., 1985, B310, 221.
31. D.M. Halsall, G.L. Turner and A.H. Gibson, Appl. Environ. Microbiol., 1985, 49, 423.
32. J.M. Lynch, J. Appl. Bacteriol. Symp. Supp., 1987, 71S.
33. E.B. Knapp, L.F. Elliott and G.S. Campbell, Soil Biol. Biochem., 1983, 15, 455.
34. J.M. Lynch and M.J. Bazin, Trans. XIV Congr. Int. Soil Sci. Soc., 1990, 3, 4.

The Potential of Bomb-^{14}C Measurements for Estimating Soil Organic Matter Turnover

D.D. Harkness[1], A.F. Harrison [2], and P.J. Bacon[2]

[1] NERC RADIOCARBON LABORATORY, SCOTTISH UNIVERSITIES RESEARCH AND REACTOR CENTRE, EAST KILBRIDE, GLASGOW G75 0QU, UK
[2] INSTITUTE OF TERRESTRIAL ECOLOGY, MERLEWOOD RESEARCH STATION, GRANGE-OVER-SANDS, CUMBRIA LA11 6JU, UK

1 INTRODUCTION

Recording the distribution of organic matter on and within a soil is relatively simple. Far less so is quantitative assessment of the characteristic turnover (or mean residence) times for the detrital plant-derived carbon. Isotopic tracing is an obvious option and particularly so given the natural occurrence of three carbon isotopes, (^{12}C, ^{13}C and ^{14}C).

The introduction of plant debris artificially spiked with radioactive carbon (^{14}C) into soil has been used successfully in controlled experiments[1-4]. However, the approach is restricted to the study of rapid mineralisation processes (< 5 years) by the time required for the radioactive label to reach biogeochemical equilibrium with the less-easily decomposed plant residues. Both the older components in contemporary soils and relict palaesols have been analysed by conventional radiocarbon dating[5-7]. In such instances, the application is limited to organic matter older than ca 200 years; this is due to a combination of two factors, i) the low concentration of naturally produced ^{14}C in living plants (13.5 dpm g^{-1}C) and ii) its relatively long (5568 year) radioactive half life.

An opportunity to bridge the crucial gap between rapid and slow turnover rates has been provided, inadvertently, by nuclear weapons test programmes carried out during the late 1950's and early 1960's. These explosions injected a sharp pulse of ^{14}C directly into the Earth's atmosphere. The subsequent and ongoing dispersal of this 'bomb-^{14}C' excess by biogeochemical processes has provided unique, but short-lived, conditions in which to trace and quantify mass transfer within the natural carbon cycle. In this post-bomb situation the resolution of turnover times is no longer constrained by the slow rate of radioactive decay but is limited by the steady decrease in atmospheric ^{14}C concentration as the bomb-produced excess ^{14}C is transferred to the World's oceans[8]. The resultant temporal variations of bomb-^{14}C in atmospheric carbon dioxide and contemporaneous plants are well documented[9-10]. The value of this transient ^{14}C enrichment in soil organic matter studies was first discussed in 1963[11] and since then several publications have described the use of bomb-^{14}C as a tracer for the input and turnover of organic matter in various soil plant systems[12-16]. However, due to the labour intensive and experimentally demanding nature of low-level ^{14}C measurement and conflicting priorities in environmental research, this tool has still to realise its full potential in the advancement of soil research.

In this paper we highlight the type of information that can be recovered and in context of two quite different studies viz.

i) The development and testing of a mathematical model to describe carbon transfer and nutrient cycling within an established mixed-deciduous woodland.
ii) A evaluation of the effects of afforestation on the organic status of an upland soil.

Detailed accounts of the respective sites, the sampling strategy and experimental procedures are published elsewhere[15-16]. Consequently in this discussion primary emphasis is placed on the research strategy and the conclusions reached.

2 MODELLING CARBON DYNAMICS

Objectives and testsite

This study theme is ongoing and with the ultimate aim of developing appropriate mathematical models that can be used to i) quantify carbon transport within natural soil systems and ii) predict the likely long-term impact of management on the physical characteristics and nutrient status of forest and agricultural soils.

In the first instance, we have considered a hypothetical soil in which the component carbon is assumed to be in steady state i.e., there is a fixed and constant balance in the accretion, transport and respiratory loss of organic matter. The site used to test and develop the model against experimentally measured data is Meathop Wood (Natl Grid Ref SD 435 795) which has a recorded history over the past 300 years. This mixed-deciduous type woodland was studied during the International Biological Programme. The soil is classified as acid brown earth with pH values ranging from 4.3 to 7.0 but predominantly < 5.3.

Model construction.

As a first stage in model development, the temporal

trends in ^{14}C enrichment anticipated for different carbon turnover rates were computed. In addition to the assumption that the ecosystem maintained steady state conditions this mathematical treatment also implied uniform mixing within a defined carbon reservoir and that photosynthetic fixation of carbon dioxide from the contemporary atmosphere provided the primary input source for organic carbon.

In the absence of anthropogenic disturbance to the atmospheric ^{14}C concentration i.e., before AD 1900, the ^{14}C activity of a uniformly mixed organic pool at year 't' can be expressed as

$$A_t = A_o e^{-\lambda T}$$

where A_o is the natural ^{14}C concentration in the atmosphere, T the mean age or turnover time of the carbon reservoir, and the ^{14}C radioactive decay constant (1.245 x 10^{-4} yr^{-1}). For the present century, a steady state reservoir response to progressive changes in atmospheric ^{14}C concentration can be described by the relationship

$$A_t = A_{(t-1)} e^{-k} + (1-e^k) A_i - A_{(t-1)} \lambda$$

where $A_{(t-1)}$ is the reservoir ^{14}C activity established for the previous year; A_i the input ^{14}C activity prevalent in atmospheric CO_2 during the preceding growth season; and k is an exchange rate constant which is by definition the reciprocal of mean residence time i.e., $k = 1/T$

Time series curves were generated for selected values of 'T' using the well documented changes in UK atmospheric ^{14}C concentrations since AD 1900 (Fig.1)[8-10].

Sampling Programme

To allow direct measurement of the progressive changes in bomb-^{14}C enrichment within the Meathop Wood ecosystem a series of samples was taken at intervals between 1972 and 1988. These were supplemented by bulk soil collections archived from 1961, 1969 and 1970. The overlying litter (Ol and Of) layers and the soil humus in 5cm increments to 15cm depth were monitored routinely. As the study progressed, other parts of the ecosystem were investigated to help clarify the emergent patterns of carbon transfer. Examples here include, soil humus to 35cm depth, partially decomposed plant debris within the soil horizons, branch litter and CO_2 in the air at various heights within and above the tree canopy.

All organic material intended for ^{14}C measurement was digested at 80°C in 0.5M HCl and then washed to neutral pH. Soil samples were then screened through a 1mm steel mesh to remove stones and all larger fragments of plant debris. It is important therefore to realise that 'soil humus' as discussed in terms of measured turnover rates is defined essentially by this physical separation

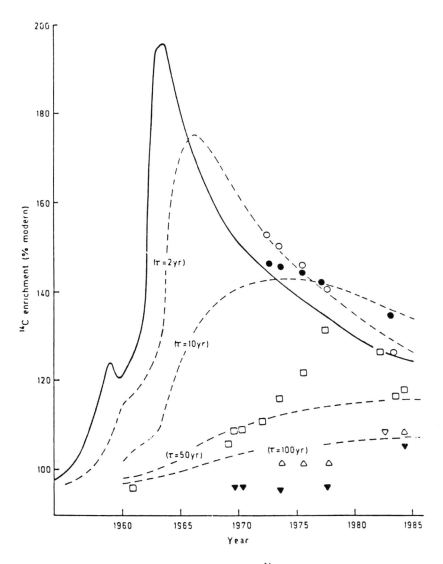

Figure 1. Comparison of trends in ^{14}C enrichment output by the first stages "steady state" model with values measured for samples from Meathop Wood. —— free atmosphere; --- model output for specified turnover period (τ); ○ leaf litter; ● fermentation layer; □ 0 to 5 cm soil humus; △ 5 to 10 cm soil humus; ▼ 10 to 15 cm soil humus. Redrawn from Harkness et al., 1986 (Ref. 15).

procedure.

^{14}C Measurement

Radiometric enrichments were determined by low level liquid scintillation counting of benzene synthesised from the sample carbon. In keeping with international convention[17] the ^{14}C enrichment values were calculated as % Modern and at the ± 1σ level for overall analytical confidence,

$$\% \text{ Modern} = \frac{A_s \times 10^2}{A_o}$$

where; As is the sample activity (cpm g^{-1} C) normalised for isotopic fractionation and adjusted, where necessary, for radioactive decay in the period between sample collection and measurement. A_o is 95% of the activity (cpm g^{-1} C) of benzene prepared from the international reference standard (N.B.S.) oxalic acid; batch SRM-4990) normalised for isotopic fractionation and adjusted for radioactive decay relative to AD 1950.

Comparison of modelled and measured ^{14}C enrichment values

A comprehensive and detailed listing of the measured ^{14}C enrichment, and related data, representing the past 30 years at Meathop Wood is published elsewhere[8,15]. Those values of particular relevance are summarised in Figure 1 against theoretical curves generated from the basic steady state model to represent specific turnover times. It is clear that the superficial leaf litter comprising the Ol and Of layers is readily characterised by turnover times of ca 2 and ca 8 years respectively. By contrast, the measured and modelled patterns of ^{14}C enrichment in the soil humus are at marked variance. The immediate implication is that the basic steady state model does not describe adequately the partitioning and/or transport of organic matter within the mineral soil. However, two features in the pattern of deviations are particularly instructive viz.

    a) For the 0 to 5 cm humus there is a delay of several years relative to the atmosphere before the onset of a rapid rise in bomb-^{14}C enrichment, and an equally sharp decline from a maximum value attained about 1980. Therefore, despite the relatively rapid turnover time for superficial leaf litter an additional biogeochemical delay has to be considered in modelling the transfer of atmospheric carbon to the soil humus via photosynthesis. This factor is particularly important where estimates of the transfer and turnover rates for organic matter are based on post-bomb ^{14}C values recorded over a short period or in a single year.

b) By 1985 the ^{14}C enrichment in the 0 to 5 cm soil humus had fallen well below the levels recorded for the post-bomb atmosphere and this can only be explained by the presence of a significant amount of pre-bomb carbon. Therefore, contrary to the model assumption, the humus in this and, by implication, the underlying soil horizons cannot be regarded as a uniformly mixed carbon pool.

Model Refinement

Improvement of the mathematical model therefore required, in the first instance, a sub-division of the total soil humus into discrete pre-bomb (slow cycling) and post-bomb (fast cycling) derived fractions. It was obvious from the temporal distribution pattern (Fig.1) that the maximum pulse of bomb-^{14}C had passed through the upper soil in less than 20 years. The mean turnover rate and relative proportion of the slow cycling component could then be evaluated from a reconstruction of the natural (pre-bomb) age/depth gradient. This was achieved by extrapolation to the soil surface from ^{14}C measurements of soil humus in the 35 to 15 cm depth range collected in autumn 1977. At that time there was no evidence of bomb-^{14}C having penetrated below 10 cm (Fig.1). This analysis (Table 1) shows an effectively constant turnover rate of ca 350 years for the slow cycling carbon in the topmost 15 cm and with soil carbon ages gradually reaching ca 600 years towards the base of the rooting zone.

Table 1 Calculated distribution and age of 'slow cycling' carbon in Meathop Wood soil humus.

Depth Horizon (cm)	Relative Contribution to Total Soil Carbon	Conventional ^{14}C Age(yrs)
0 to 5	0.37	370
5 to 10	0.75	330
10 to 15	0.95	350
15 to 25	0.98	450
25 to 35	1.00	590

Division of the total soil humus into appropriately sized 'fast' and 'slow' cycling components results in a much closer model approximation of the distribution of ^{14}C activities measured since 1961. However, the time lag in peak amplitude relative to the atmospheric ^{14}C record is not fully resolved. In particular, the fast cycling component, which comprises ca 65% of the humus in the top 5 cm of soil, still seems to demand a step-wise delay of several years for the input availability of photosynthesised carbon but which is then rapidly incorporated into the humus carbon pool. A predominant input from leaf litter and/or root exudates seems unlikely since that pathway would require a more immediate and over-damped response from the bomb-^{14}C pulse. Indeed the available evidence points to the fact that much of the organic debris deposited on and within the Meathop soil is decomposed and respired without ever becoming incorporated into the soil humus. Root fragments and other partly decomposed plant debris within the soil in spring 1984 clearly provide a potential source of both pre- and post-bomb carbon (Table 2); likewise decomposing branch litter which may be incorporated into the soil by faunal mixing and/or eluviation. Either input mechanism could account for the initial delay in the transfer of bomb-^{14}C to the soil humus.

Table 2 Measured ^{14}C concentrations in litter, branch debris, soil humus and soil debris during spring 1984.

Sample		^{14}C Enrichment ^{o}/o Modern±1.0
Litter (Ol)		126.0
Litter (Of)		134.5
Branch debris	(0.2 to 0.5cm diam)	126.2
Branch debris	(0.5 to 1.0cm diam)	131.7
Branch debris	(1.0 to 2.0cm diam)	152.2
Branch debris	(2.0 to 5.0cm diam)	117.9
Soil humus	(0 to 5cm depth)	117.0
Soil humus	(5 to 10cm depth)	109.0
Soil humus	(10 to 15cm depth)	105.0
Soil debris	(0 to 5cm depth)	126.5
Soil debris	(5 to 10cm depth)	113.5
Soil debris	(10 to 15cm depth)	110.0

In essence, the isotopic data recorded so far in the Meathop study point to the fact that the ultimate degree of refinement attainable in modelling carbon transfer within plant/soil systems must be constrained by the physical and biogeochemical complexities of the organic matter held within the ecosystem. Foremost is the fact that direct input of atmospheric carbon to the soil humus by photosynthetic fixation cannot be assumed; an initial delay of several years may have to be recognised for temperate woodlands. It is also important in certain contexts to differentiate between organic matter that is susceptible and resistant to microbial decomposition, i.e. the definition of a uniformly mixed carbon reservoir is important and especially so where it is necessary to resolve bomb-^{14}C tracer against an appreciable background of pre-bomb organic residues.

It must be emphasised that the ongoing Meathop study represents an attempt at comprehensive and highly descriptive modelling. In many contexts a far less ambitious use of natural ^{14}C measurement can produce useful information that is otherwise unobtainable.

Nutrient cycling

An initial incentive in developing the Meathop

organic matter turnover model was an option for quantitative evaluation of nutrient cycling within the ecosystem[18]. It is interesting then to note a close agreement between the values obtained for N and P mineralisation rates calculated on the basis of modelled carbon turnover times with independently determined values. Dividing the total N and P contents of the surface litter and soil layers within the rooting zone by the mean residence time derived for their respective organic carbon inventories produced mineralisation rates totalling 165 kg ha^{-1} yr^{-1} for N and 11.3 kg ha^{-1} yr^{-1} for P. These values compare with 170 kg ha^{-1} N and 11.4 kg ha^{-1} P previously estimated as the annual demand from the woodland vegetation.

3 AFFORESTATION OF UPLAND SOIL

Objective and investigation sites.

The study summarized in this context formed part of a wider investigation of the effects of various tree species on typical upland soil[19].

It has been estimated that as a consequence of the projected expansion of commercial forestry till the end of this century as much as 12% of the total UK land area will be under managed tree production. Much of this investment is likely to involve the planting of exotic conifers, mainly grown in monoculture, on upland tracts previously given to rough pasture.

The study was based on a series of Sitka spruce stands and adjacent rough grassland within the Bowland Forest (Natl Grid Ref SD 750585). Individual tree stands had been planted on hill grazing land at ca 280m altitude and at known dates between 1953 and 1968. Therefore, when sampled in autumn 1984 the individual plots provided a collective frame for 31 years of progressive change in soil character and relative to the original pasture.

Sampling programme and analytical procedures

Field collection was by replicate coring; at least 6 cores were taken at random and then bulked for each site. The mineral soil representing the top 5cm of the profile was separated and screened through a 4mm steel mesh.

Previous ^{14}C measurements on the grassland and other tree plots[20] had confirmed the presence of organic components with 'fast' and 'slow' mineralisation rates in the soil humus. Therefore, a chemical separation procedure (Fig.2) was used to resolve the imprint of bomb-^{14}C over the residue of much older more stable carbon. The separation was carried through quantitatively to allow determination of the wt% carbon content of the three defined organic fractions and in the

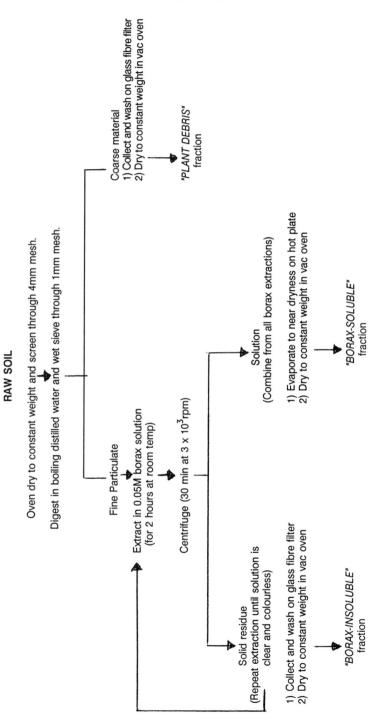

Figure 2. Soil separation procedure.

total soil.

^{14}C enrichment values were determined and calculated as already described. Where necessary the ^{14}C value for total carbon was calculated as,

$$\% \text{ Mod}_{(d)} f_{(d)} + \% \text{ Mod}_{(s)} f_{(s)} + \% \text{ Mod}_{(i)} f_{(i)}$$

where subscripts d, s and i denote respectively the plant debris (>1mm), borax-soluble and borax-insoluble fractions and f is the corresponding proportional contribution to the total carbon pool.

The ^{14}C enrichment pattern

The most significant feature of the ^{14}C pattern (Table 3) is that during 30 years of tree growth the organic matter held in the top 5cm of soil shows very little influence from the bomb-^{14}C peak in atmospheric CO_2.

Table 3   Carbon content ^{14}C enrichment under Sitka spruce stands (0.to 5cm soil collected November 1984).

Year Planted	Plant Debris (>1mm)		Amorphous Soil (<1mm)			
	wt%	^{14}C% Mod	Borax-soluble		Borax-insoluble	
			wt%C	^{14}C% Mod	wt%C	^{14}C% Mod
Control	2.9	109	3.7	108	12.8	101
1953	1.9	107	4.2	106	5.7	97
1956	0.6	105	3.8	100	7.2	90
1965	1.2	117	2.2	112	8.1	101
1967	2.0	123	2.0	110	7.9	101
1968	1.1	121	2.8	115	8.5	101

Despite an accumulation of relatively fresh (5 ± 2 year old) organic litter virtually none of this material had entered the soil even as finely divided plant debris. Paradoxically where a small amount of bomb-^{14}C labelled carbon exists, in the plant debris and 'borax-soluble' humus, this is greater for the younger plots. This trend indicates a residue from original vegetation (grass) mixed into the soil during initial ploughing of the site for planting. Again this reinforces the observation that little or no organic matter is being introduced into the soil humus from the trees.

Changes in the total carbon inventory.

Although the ^{14}C measurements show that there is very little input of fresh organic matter, tree growth is clearly causing other significant changes in the total carbon pool (Fig.3).

During the first 10 to 15 years after tree planting there is a marked loss of soil carbon but thereafter it appears to reach a state of mass balance. This effect may well be a consequence of site preparation rather than

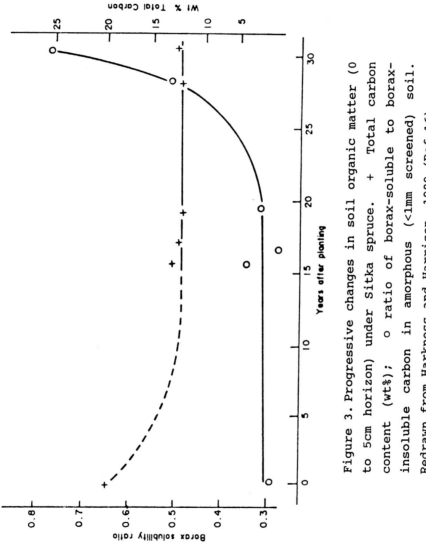

Figure 3. Progressive changes in soil organic matter (0 to 5cm horizon) under Sitka spruce. + Total carbon content (wt%); o ratio of borax-soluble to borax-insoluble carbon in amorphous (<1mm screened) soil. Redrawn from Harkness and Harrison, 1989 (Ref 16).

a direct influence from the trees. Ploughing and site drainage would aerate the topsoil giving the opportunity for enhanced microbal degradation and a consequent net loss of organic matter relative to the original peaty gley.

After ca 20 years of tree growth and although the total carbon content is effectively constant (with apparently no significant transfer between the atmosphere/vegetation and the soil humus) there is a marked increase in the ratio of borax-soluble to borax-insoluble humus. This change in organic composition can only result from the influence of the trees and would seem to imply a breakdown of the more stable organic matter of the original grassland with carbon turnover rates in excess of 1500 years.

4 FUTURE POTENTIAL

Ongoing dispersal of bomb-^{14}C through the biogeochemical carbon cycle provides a unique opportunity for quantitative evaluation of the pathways and transfer rates of organic carbon within and through soils. However, the ability to resolve these naturally induced variations will decrease progressively over the next several years as the man made tracer tends towards a uniform distribution.

A major disincentive to applying this approach in the characterisation of soil organic matter has been the labour-intensive and experimentally demanding nature of low level ^{14}C measurement. In particular, the need to synthesise upwards of 2g benzene from the raw sample carbon followed by an extended period (several days per sample) of radiometric counting in highly specialised detection systems[21]. Recent developments in the measurement of natural ^{14}C concentrations by accelerator mass spectrometry (AMS)[22] go a considerable way in alleviating these practical constraints. For example, by this latter method, 1 mg of sample carbon in the form of CO_2 can be analysed in approximately one hour.

If the bomb-^{14}C transient is to be exploited to best advantage there is a priority need, certainly within the next decade, to set up appropriate field collection programmes and in collaboration with experimental facilities that can provide ^{14}C analyses by the radiometric and/or accelerator (AMS) methods.

REFERENCES

1. D.S. Jenkinson,' Experimental Pedology', Butterworth, London, 1965.
2. L.H. Sorenson, Soil Science, 1967, 104, 234.
3. D.S. Jenkinson, Soil Science, 1971, 111, 64.
4. J.F. Dormar, Canadian Journal of Soil Science, 1975, 55, 473.
5. C.A. Campbell, E.A. Paul, D.A. Rennie and K.J. McCullum, Soil Science, 1967, 104, 217.
6. H.W. Scharpenseel and H. Schiffmann, Zeitschrift für Pfanzenernährung und Bodenkunde, 1977, 140, 159.
7. B. Guillet, 'Constituents and Properties of Soils', Academic Press, London, 1982.
8. A. Walton, M. Ergin and D.D. Harkness, Journal of Geophysical Research, 1907, 75, 3089.
9. M.S. Baxter and A. Walton, Proc. Roy. Soc., London, 1971, 321, 105.
10. R. Nydal, K. Louseth and F.H. Skogseth, Radiocarbon, 1980, 22, 626.
11. D.S. Jenkinson, 'The use of Isotopes in soil organic matter studies', Pergamon Press, Volkenrode, 1963.
12. S.J. Ladyman and D.D. Harkness, Radiocarbon, 1980, 22, 885.
13. J.D. Stout and K.M. Goh, Radiocarbon, 1980, 22, 892.
14. B.J. O'Brien, Soil Biology and Biochemistry, 1984, 16, 115.
15. D.D. Harkness, A.F. Harrison and P.J. Bacon, Radiocarbon, 1986, 28, 328.
16. D.D. Harkness and A.F. Harrison, Radiocarbon, 1989, 31, 637.
17. M. Stuvier and H.A. Polach, Radiocarbon, 1971, 19, 355.
18. A.F. Harrison, D.D. Harkness and P.J. Bacon, 'Nutrient Cycling in Terrestrial Ecosystems', Elsevier Applied Science, London and New York, 1990.
19. J.M. Ogden, PhD Thesis, University of Stirling, 1986.
20. S.J. Ladyman, PhD Thesis, University of Strathclyde, 1982.
21. D.D. Harkness and H.W. Wilson, Proc. 8th Internat. Radiocarbon Conf., 1972, Roy. Soc. New Zealand, I, B102.
22. A.E. Litherland, Phil, Trans. Roy. Soc. London, 1987, A323, 5.

The Influence of Soil Organic Carbon on Microbial Growth and Survival

Traute-Heidi Anderson[1] and T.R.G. Gray[2]

[1] INSTITUT FÜR BODENBIOLOGIE, BUNDESFORSCHUNGSANSTALT FÜR LANDWIRTSCHAFT, BRAUNSCHWEIG, GERMANY
[2] DEPARTMENT OF BIOLOGY, UNIVERSITY OF ESSEX, COLCHESTER CO4 3SQ, UK

1 INTRODUCTION

Recently, a major effort has been spent in research on the three carbon compartments: <u>soil organic carbon</u>, <u>organic litter</u> and <u>soil microbial biomass</u> (Figure 1) with respect to <u>organic matter dynamics</u>, <u>decomposition processes</u> and <u>microbial interactions</u>[1]. The study of the intricate inter-relationship between these carbon sources has high priority in microbial ecological research. It is important to determine carbon flux rates entering and leaving the microbial biomass pool, as well as determining the extent to which soil carbon specifies the kind of microbial community developing.

It is known that specific plant communities are controlled by environmental factors and so it is reasonable to assume that soil microbial communities are influenced in a similar manner. Distinct microbial successions on different organic substrates exemplify this. The recognition of autochthonous and zymogenous organisms[2] and sugar fungi and lignin fungi[3], although simplistic, is still relevant today.

At the community level we might ask whether external factors control different microbial communities in a steady state to such an extent that they can actually be differentiated. The determination of community potentials, i.e. specific rate constants of activity, could be a way of answering this.

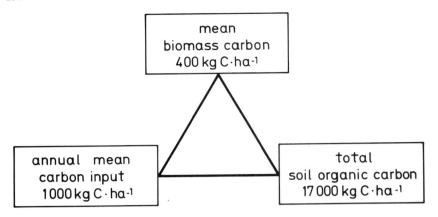

Figure 1. Partitioning of carbon inventories. The values are representative for a permanent monoculture plot without organic matter input. They correspond to 1.13% of total soil organic carbon and ca. 270 $\mu g\ g^{-1}$ microbial carbon, assuming the weight of 1 ha of soil to be 1.5×10^6 kg (to a depth of 12 cm).

2 MATERIALS AND METHODS

Soils

Sampling of agricultural plots is described in detail in Anderson and Domsch[4]. Forest soils were sampled at the end of each month throughout the year. The litter and the O_f layers were removed and the O_h layer scraped off with a small shovel and retained. The underlying A_h horizon was sampled up to a depth of 5 cm. At each sampling time, 4 to 5 samples were collected and either measured separately (<u>Fagus</u> soils) or bulked (<u>Pinus</u> soils).

Temperature experiments

Soils were sieved and adjusted to a water potential of c. -300 kPa and stored at 15°C in a temperature-controlled incubator for 18 months. The soil containers were sealed with parafilm to allow gas exchange but prevent water loss. Water loss was insignificant for the first 8 months but amounted to 4% after 18 months. In order not to disturb the soil samples, water loss was not compensated for by addition of fresh water.

For the 35°C incubation experiment, the incubation chamber was maintained at about 90% relative humidity. Percent water loss did not exceed 0.4% during the incubation time of 10 weeks.

Microbial biomass carbon estimations

Microbial biomass carbon (C_{mic}) estimations were made according to the substrate induced respiration method[5]. 25g and 100g dry weight samples were used for forest and agricultural soils respectively.

Calculation of quotients

For the calculation of the qCO_2 (unit CO_2-C unit^{-1} biomass-C h^{-1}) of the basal respiration rate (not glucose-activated respiration), the hourly mean CO_2 output over 10 h was taken after soils had reached a relatively constant CO_2 production rate after 20 h of incubation at 22°C.

The initial microbial biomass and the total microbial-C loss after a set time were determined and the qD (<u>sensu</u>[6]) calculated using the following equation

$$qD = [(C_{mic})_{t1} - (C_{mic})_{t2} / (C_{mic})_{t1}] / t_2 - t_1$$

Soil carbon determinations

Soil carbon (C_{org}) was estimated after removing inorganic carbon[7]. Carbonate carbon was removed with 10% hydrochloric acid and the samples subsequently dried on a sandbath at 70°C. The remaining organic carbon was determined by dry combustion (Leco).

3 RESULTS

The $C_{mic}:C_{org}$ ratio

Odum[8] related bioenergetics to ecosystem development in terrestrial ecosystems. Thus, soil ecosystems should develop towards a state of equilibrium if the environment remains constant (excepting regular small natural fluctuations) over a long period of time. Microbial communities should become more highly organised and efficient with time. When equilibrium is attained, input to any particular C compartment should equal output. A state of equilibrium would be characterised by relatively constant levels of microbial biomass on a year by year basis and by the constancy of the $C_{mic}:C_{org}$ ratio. In order to test this hypothesis, $C_{mic}:C_{org}$ ratios of soils in near steady states in more than 130 agricultural plots in the temperate climatic zone of Central Europe were investigated[4,9]. This showed that the microbial biomass was linearly correlated with the soil organic carbon content of plots over a range of 0.5 to 3.0%. However, when comparing $C_{mic}:C_{org}$ relationships in both permanent monocultures (M) and continuous crop rotations (CR) plots (without manure amendments), it was shown that the regression lines derived from CR plot data had a steeper slope. This suggested that CR plots had a higher microbial-C biomass level per unit soil carbon (Figure 2); 2.9% of the total C_{org} was in the microbial biomass as compared with 2.3% for monoculture plots. These differences were highly significant.

Insam et al.[10] extended these studies and investigated the $C_{mic}:C_{org}$ ratio of plots in different climatic regions. As would be expected, they found that the prevailing climate affects the $C_{mic}:C_{org}$ ratio. They also found higher microbial biomass-C contents per unit soil C in continuous crop rotations. The much wider spectrum of $C_{mic}:C_{org}$ ratios recorded in the earlier literature ranging from 0.25% to over 7.0%[11-15] are probably due to variations in sampling time, vegetational cover, management and analysis methods. Evidently, the pool of available carbon is temporarily higher in organically amended plots. Thus, in plots with green manure treatment, the microbial biomass-C increased to 4% of total soil carbon (Figure 2). However, a level of 4% microbial biomass-C is short-lived, occurring immediately after amendment and soon falling back to a level characteristic of the plot[4].

The explanation of higher microbial-C levels per unit soil carbon in long-term continuous crop rotation plots is different. It is assumed that microbial communities in CR plots with higher yield coefficients (Y) have evolved due to the more heterogeneous organic matter input. In such cases, more C from the annual organic matter input would be converted into microbial C and so raise the $C_{mic}:C_{org}$ ratio. We report here studies in which CR and M plots were used to determine whether <u>quantitative</u> and <u>qualitative</u> differences in physiological constants could be determined between simple and complex soil microbial communities.

Assessment of microbial activity; the use of physiological quotients

A number of permanent monoculture (M) and continuous rotation (CR) plots were studied and their qCO_2 (mg CO_2 per mg microbial-C per h) and qD (mg microbial carbon loss per mg initial microbial-C per h) values recorded[6]. The mean (qCO_2) was almost twice as high in M compared with CR plots [Table 1a].

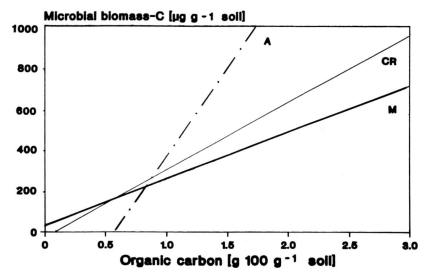

Figure 2. Correlation between microbial carbon and organic carbon of soils with continuous monoculture (M), continuous crop rotation (CR) and monoculture plots amended with green manure (A); M, r = 0.93 (n = 34); CR, r = 0.97 (n = 57); A, n = 13.

Table 1a. Comparison between the metabolic quotients qCO_2 and qD of microbial biomass from long-term M and CR plots at 22°C.

Type of management	qCO_2 (mg CO_2/mg C_{mic}/h)			qD^a (mg C-loss/mg C_{mic}-init./h)		
Monoculture	10.97 x 10^{-4} SE 0.68	"	n = 17	2.12 x 10^{-4} SE 0.11 x	"	n = 12
Crop rotation	6.45 x 10^{-4} SE 0.45	"	n = 19	1.41 x 10^{-4} SE 0.07	"	n = 14

a - based on total C-loss after 5 wk incubation; SE, standard error, significantly different at P = 0.001

Since the qCO_2 values could neither be correlated with C_{org} nor pH of the plots, it was assumed that specific community respiration rates differ in the two soil systems. In addition, qD values were also higher in M plots, indicating a less efficient utilisation of the available energy sources. Since sampling time and soil handling were identical for all the samples studied, these qualitative differences can only be associated with differences in the physiological community potentials between the two systems. Estimations of the CO_2 evolution rate per g soil or C-loss rate per g soil did not show these differences [Table 1b]. The usefulness of quotients like qCO_2 and qD is further illustrated with additional experiments on microbial-C-loss. Table 2 shows total microbial-C loss in 7 soils after 18 months of incubation at 15°C. If, in a more conventional manner, the total C-loss is related to the C_{org} contents of the samples, the two parameters are not correlated. However, when qD values are determined a clear relationship with C_{org} is seen (Figure 3). Soils with a high organic matter content lost less microbial-C per unit biomass per unit time than soils with low C_{org} contents.

The biological link between soil organic matter and microbial biomass efficiency is clearly demonstrated by the fact that microbial biomass-C loss can be accelerated by incubation at high temperatures. Figure 4 shows the qD of 3 M versus 3 CR plots incubated at 35°C. The qD is greater in soils with the lowest C_{org} content and in addition in soils of monoculture origin.

Factors beside carbon influencing microbial activity

During the comparison of M and CR plots which encompassed soils differing in type, texture, organic matter content and pH, an interesting phenomenon was observed frequently. Soils differed in their ability to assimilate glucose. Two types of respiration curves were found. Figure 5a illustrates the respiration curve of two soils (I and II) which had received the same amount of glucose (3000µg g^{-1}) at the onset of the experiment. The soils were from the same sampling site, both were luvisols, and the pH, C_{org} and initial microbial biomass sizes were comparable (Figure 5a). Moreover, the plots differed in soil management. Soil I had had a permanent sugar beet crop (17y) with NPK amendments, while soil II had been a black fallow for 33y with no amendment. Although glucose-C was in excess, the microbial response (CO_2 evolution rate increase) was lower in the latter. The possibility that nutrients other than C might be limiting in soil II was tested by adding either nitrogen or phosphorus or a combination of both in a ratio of 4N and 1P unit per 10 C units. The results are illustrated in Figure 5b. At

Table 1b. Comparison between respiration rate and death rate of microbial biomasses from long-term M and CR plots.

	Monoculture			Crop rotation	
Soil[a] code	Microbial-C (mg/100g)	Resp. rate (ml CO_2/h)	Soil code	Microbial-C (mg/100g)	Resp. rate (ml CO_2/h)
2/1	45.16	0.092	1/1	21.99	0.008
7/1	5.18	0.016	1/2	30.19	0.002
14/1	21.99	0.047	1/3	21.59	0.028
15/1	22.19	0.034	4/1	11.18	0.015
16/2	43.02	0.149	4/3	17.99	0.024
19/6	23.19	0.047	4/4	15.18	0.026
19/7	24.59	0.062	12/1	72.14	0.059
20/1	31.60	0.080	16/4-2	15.98	0.016
23/2	19.36	0.039	17/1	12.78	0.016
23/5-2	22.39	0.035	17/2	12.78	0.011
23/7	10.78	0.026	17/3	18.58	0.024
23/12	14.18	0.023	17/4	15.38	0.015
23/20	8.59	0.017	20/2	34.40	0.065
24/6	21.19	0.040	20/3	43.81	0.076
24/7	19.19	0.056	20/4	50.42	0.073
24/9	18.11	0.029	22/3	67.63	0.084
24/10	18.30	0.032	22/4	76.44	0.122
			22/7	85.45	0.144
			22/8	83.05	0.149

Soil code	Initial microbial-C (mg/100g)	Death rate (mg/100g/5wk)	Soil code	Initial microbial-C (mg/100g)	Death rate (mg/100g/5wk)
2/1	47.48	10.27	4/1	15.08	1.98
16/2	44.61	8.10	4/2	16.94	2.30
19/2	32.00	5.61	4/4	19.27	2.89
19/3	31.76	4.16	12/1	75.96	5.73
19/6	24.13	3.74	16/4-2	16.21	1.83
23/2	23.59	3.60	16/5-2	30.05	3.73
23/5-2	29.99	2.50	17/1	13.78	1.60
23/7	11.88	1.50	17/3	20.52	2.94
23/20	14.78	2.10	20/3	45.61	6.01
24/5	26.28	4.13	20/4	35.07	3.07
24/7	25.99	4.40	22/3	74.31	8.18
24/10	29.03	6.20	22/4	68.52	5.93
			22/7	86.32	12.88
			22/8	83.05	9.80

a - soil codes correspond with those in Anderson and Domsch[4]

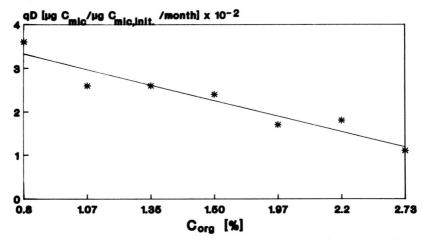

Figure 3. Correlation between microbial biomass-C loss per unit initial biomass and unit time and the organic C content of 7 agricultural soils. Values are based on total microbial-C loss after 18 months of incubation at 15°C ($r = -0.96$).

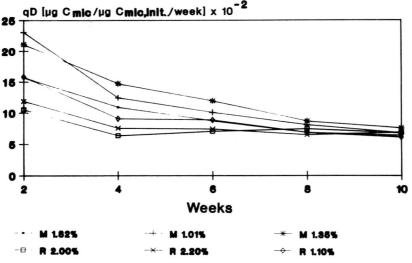

Figure 4. The effect of high temperature on the survival of soil microbial biomass when incubated at 35°C over a period of 10 weeks. Soils differed in management and % C_{org} content. M, monoculture; R, crop rotation. Differences between M and CR soils are only significant at $p = 0.01$ to 0.05 up to the fourth week of incubation.

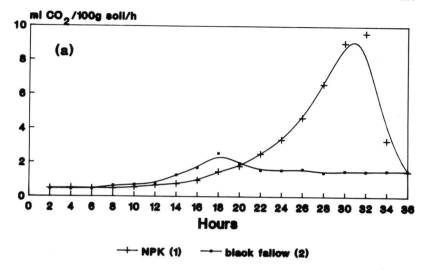

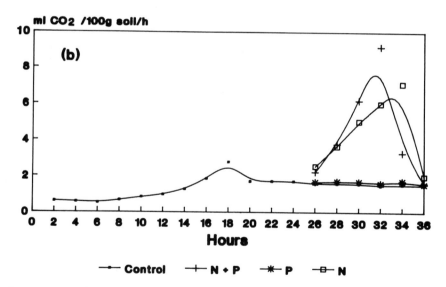

<u>Figure 5a, b</u>. Respiration rate of microbial biomasses from plots under different management after amendment with 3000 µg g^{-1} glucose. (a) Plot under sugar-beet with NPK treatment (1) and black fallow with no crop and no amendment (2). (b) Increase in the respiration rate of the black fallow after amendment with N or N+P in a ratio of 4N:1P:10C.

the time when respiration was depressed in soil II (after the 20th hour), amendment with available N (NH_4NO_3) initiated further glucose consumption, as evidenced by a dramatic increase in the CO_2 output rate, which was even more pronounced with N + P additions. P (Na_2HPO_4) additions alone had no affect. It is clear that the specific nutrient status of a soil controls the utilisation of available C and therewith the specific level of microbial biomass which can develop.

Observation of this effect is not new[16], but the more sensitive method of detection certainly is. By using increasing concentrations of glucose, it is possible to induce in every soil, one or other of the two types of respiration curve observed for soils I and II (Figure 5a). Figure 6 demonstrates experiments done with a chernozem soil with added glucose concentrations in the range of 500 to 6000 $\mu g\ g^{-1}$. The <u>natural</u> status of available nutrients in this soil was sufficient to support complete glucose assimilation up to 1000 $\mu g\ g^{-1}$ causing the respiration rate to decline following the respiration peak. However, higher glucose concentrations (3000 $\mu g\ g^{-1}$) initiated a steady state CO_2-evolution rate (the 'shoulder') following the respiration peak observed for soil II which lacked available nutrients from the start (Figures 5a and 5b). By adding a N + P cocktail at this stage, further uptake of carbon was induced, leading to a large increase in the respiration rate (Figure 6). Such indications of nutrient deficiencies were first observed in a beech forest soil (rendzina) where glucose consumption was accelerated by the addition of P (Domsch and Winkelmann, unpublished).

Differences in nutrient availability could also be the explanation for the following observation in two forest stands. By studying microbial biomass development in <u>Fagus</u> and <u>Pinus</u> stands on a monthly basis over a period of one year, a large increase of the $C_{mic}:C_{org}$ ratio from 0.8 to 1.6% was observed in the <u>Fagus</u> soil over the summer months (Figure 7). This did not occur in the <u>Pinus</u> soil. Here a relatively constant $C_{mic}:C_{org}$ ratio (with a mean around 0.7%) was observed throughout the year. These differences were found in both the O_h and A_h horizons. The increases of microbial-C per unit C_{org} of the <u>Fagus</u> soil were not related to differences in temperature or water regimes, nor to the pH or C_{org} contents which were similar in both soils [Table 3]. Furthermore, both stands were close to one another (18 miles apart). However, quality and frequency of organic matter input differ in deciduous and coniferous woodlands. The increase of the $C_{mic}:C_{org}$ ratio in the beech soil might be attributed to a higher amount of carbon becoming available from the previous autumnal litter input, or a higher amount of

The Influence of Soil Organic Carbon on Microbial Growth and Survival 263

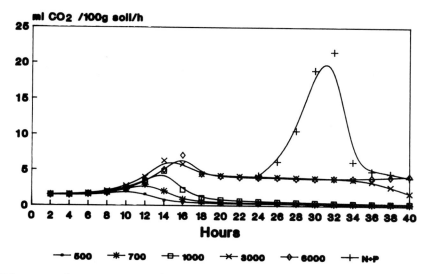

Figure 6. Comparison of respiration rates after increasing amendment of glucose ($\mu g\ g^{-1}$) of a chernozem soil and the effect of additions of N and P after glucose saturation at 3000 - 6000 $\mu g\ g^{-1}$ glucose.

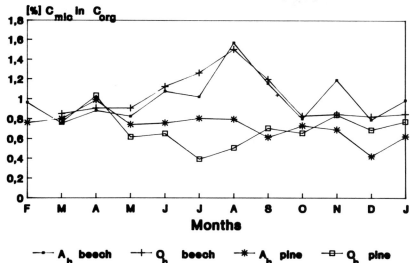

Figure 7. Comparison of the $C_{mic}:C_{org}$ ratio of a beech and a pine stand during the course of one year.

Table 2. Total microbial-C loss after 18 months of incubation at 15°C.

Soil code	C_{org} (%)	Initial microbial-C (mg/100g soil)	Microbial C-loss (mg/100g soil)
S-I	0.80	25.59	16.81
155	1.07	29.59	13.87
153	1.35	26.39	12.41
148	1.50	54.42	23.89
151	1.97	47.61	14.71
Jerx	2.20	73.64	23.89
138	2.73	81.65	15.48

Coefficient of variation maximal 5%

Table 3. Soil properties and conditions of a beech wood and a pine stand[a].

Soil	Microbial biomass (mg/100g)	C_{org} (%)	N_t (%)	P (µg/g)	K (µg/g)	pH (KCl)	Soil temp (°C)
Beech							
O_h	152.57	15.32	0.879	77.55	240	3.19	
SD	30.80						
							8.32
A_h	29.90	3.02	0.161	76.50	114	3.23	
SD	7.30						
Pine							
O_h	127.20	19.47	0.953	40.35	203	2.80	
SD	24.60						
							8.72
A_h	14.91	2.15	0.094	19.35	89	3.18	
SD	3.03						

a - values show annual mean; SD, standard deviation

utilisable N or P contained in the litter, or to a combination of both. A similar case was reported[17] for an agricultural plot where C_{mic} increased relative to C_{org} during the summer months. Subsequent analyses of N (total), P (lactate extract) and K contents gave no clue to the cause of this temporary increase in the $C_{mic}:C_{org}$ ratio of the beech stand. Present analytical methods may not be sensitive enough to detect changes in the <u>availability of nutrients</u> which can be detected by the microorganisms.

4 CONCLUSIONS

The data given here demonstrate the usefulness of quotients and ratios for elucidating effects of environmental changes on microbial communities. Only when metabolic rates are related to microbial biomass size instead of a weight or volume of soil, is direct comparison of true microbial activities in different soils possible. Besides obtaining a further quantitative measure for describing community activity, ecological hypotheses can now be investigated experimentally under controlled laboratory conditions.

5 ACKNOWLEDGMENTS

We wish to thank Professor K.H. Domsch for constructive discussion of this work and for reading the manuscript. Our thanks are extended to Mr K. Steffens for technical assistance. This research was in part funded by the German Research Foundation (DFG).

6 REFERENCES

1. M.J. Mitchell and J.P. Nakas, 'Microfloral and Faunal Interactions in Natural and Agro-Ecosystems', Nijhoff and Junk, Dordrecht, 1986.

2. S. Winogradsky, <u>C.r, Acad Sci, Paris</u>, 1924, <u>178</u>, 1236.

3. S.D. Garrett, <u>New Phytol.</u>, 1951, <u>50</u>, 149.

4. T.-H. Anderson and K.H. Domsch, <u>Soil Biol. Biochem.</u>, 1989, <u>21</u>, 471.

5. J.P.E. Anderson and K.H. Domsch, <u>Soil Biol. Biochem.</u>, 1978, <u>10</u>, 215.

6. T.-H. Anderson and K.H. Domsch <u>Soil Biol. Biochem.</u>, 1990, <u>22</u>, 251.

7. D.W. Nelson and L.E. Sommers, 'Methods of Soil Analysis' Ed. by A.L. Page <u>et al.</u> , American

Society of Agronomy, Madison, 1982, Vol. 2, p. 539.

8. E.P. Odum, Science, New York, 1969, 164, 262.

9. T.-H. Anderson and K.H. Domsch, 'Perspectives in Microbial Ecology' Ed. by F. Megusar and M. Gantar. Slovene Society for Microbiology, Ljubljana, 1986. p. 467.

10. H. Insam, D. Parkinson, and K.H. Domsch, Soil Biol. Biochem., 1989, 21, 211.

11. A. Ayanaba, S.B. Tuckwell and D.S. Jenkinson, Soil Biol. Biochem., 1976, 8, 519.

12. D.S. Jenkinson and D.S. Powlson, Soil Biol. Biochem., 1976, 8, 209.

13. J.P.E. Anderson and K.H. Domsch, Soil Sci., 1980, 130, 211.

14. T.M.C.M. Adams and R.J. Laughlin, J. Agric. Soil Sci., 1981, 97, 319.

15. D.S. Jenkinson and J.N. Ladd, 'Soil Biochemistry' Ed. by E.A. Paul and J.M. Ladd, Decker, New York, 1981, Vol. 5, p. 415.

16. R.J. Swaby, Australian Conference on Soil Science, Adelaide, 1953, 1, 3.76.2

17. J.M. Lynch and L.M. Panting, Soil Biol. Biochem., 1980, 12, 29.

Carbon Dynamics of a Soil Planted with Wheat under an Elevated Atmospheric CO_2 Concentration

P.J. Kuikman, L.J.A. Lekkerkerk, and J.A. Van Veen
INSTITUTE FOR SOIL FERTILITY RESEARCH, PO BOX 48, 6700 AA WAGENINGEN, THE NETHERLANDS

1 INTRODUCTION

Organic matter in soils is one of the vital constituents of the soil ecosystem. It is the sink and the source of plant nutrients, the substrate for microbial life in soil, it provides a large adsorbing surface for chemicals and it largely determines the porosity of a soil for air and water transport. The soil organic fraction consists of a myriad of compounds, single or in various complexes. The organic matter content of soils varies between nearly 0% (sand) and 100% (peat soils) and its size and composition changes with time. The dynamics of soil organic matter are the result of changes in its composition and content through biological and chemical processes.

Above ground primary production is the major input of organic matter for most soils. Photosynthetically fixed carbon is transferred to the soil from above ground plant parts via litter fall and from roots via root litter production and rhizodeposition. Manuring, using mostly plant residues, alone or in combination with animal waste, is another input of organic matter to arable soils and pasture. Output of organic matter from soil is mostly due to biological decomposition and mineralization.

The input of carbon via roots of growing plants is estimated to be between approximately 900 and 3000 kg $C.ha^{-1}.yr^{-1}$ for both arable soils and forests.[1] The proportion of photosynthetically fixed carbon that is released from the roots into the soil either as respired CO_2 or as organic compounds varies from 10 to 40%.[1] The quantity and the quality of the material released from roots into the soil depend on several factors, such as plant species, developmental stage of the plant and soil environmental factors (temperature, soil moisture and soil nutrient status).[2] Root derived carbon compounds such as exudates, mucilages and detritus are transformed by microbes which utilize them for biosynthesis and energy production. A large proportion of the root derived material is easily decomposed by microbes and as a consequence the place of input of this root-derived material, i.e. the rhizosphere, is the place with the highest microbial activity in soil. The utilization of this root-derived material by soil microbes is controlled by other soil biota and also by various soil related factors such as nutrient status, texture, structure and environmental factors such as moisture and temperature.[2-5]

2 EFFECTS OF ELEVATED CO_2 CONCENTRATION

An elevated atmospheric CO_2-concentration could directly or indirectly influence soil organic matter dynamics. Direct influences are unlikely, because concentrations of CO_2 of up to several thousands of ppm are commonly found in soil.

Indirect effects of an elevated CO_2-concentration such as the projected rise in temperature, as well as changes in soil moisture due to climatic changes, will affect plant growth, translocation of carbon into the soil/root system (quantitatively and qualitatively) and the turnover of organic matter in soil. It is reasonable to expect that a rise in temperature, particularly in temperate zones, will lead to an enhanced microbial decomposition of soil organic matter. Increased turnover of soil organic matter at elevated temperature might also result in an enhanced mineralization of nutrients, tied up in the organic matter fractions, and thus a higher availability of these essential nutrients to plants and consequently higher primary production. Similar feed-back mechanisms might also complicate the overall effect due to changes in soil moisture content and in other factors that are caused by global climatic changes. Even though these effects might be substantial, here we will only focus on the influence of elevated CO_2-levels on primary production and on the resulting changes in soil organic matter.

Atmospheric CO_2 and plant production

With regard to the effects of elevated CO_2-concentrations in the atmosphere, most research has so far been directed towards the effects on above-ground plant processes, such as photosynthesis, transpiration and biomass accumulation. A doubling of the atmospheric CO_2-concentration from 350 to 700 ppm will cause a higher assimilation rate, resulting in an average increase in yield of 41% for C3 crops, whereas the effects on C4 plants are expected to be less.[6] Curtis et al.[7] suggested that below ground rhizomes provided adequate sinks for the increased assimilation by plants because they did not observe any increased carbon accumulation in shoots. This is supported by Whipps[8] who found that increases in CO_2 enhanced the relative growth rate of maize roots more than the relative growth rate of the shoots. The percentage of root-translocated carbon that was released from the roots into the soil was not significantly affected by different CO_2-concentrations. Since this is one of very few studies on effects of elevated CO_2-concentrations on soil organic matter input processes, at present only speculations can be made on the effect of elevated atmospheric CO_2-concentrations on changes in the size of the carbon flow from plant roots into the soil. A point to consider is, that besides the direct effects of increased primary production on carbon flux in the plant-soil system, an increased plant biomass production also results in an enhanced demand for nutrients by crops. This could induce a more intensive competition for nutrients between plants and soil microorganisms at elevated CO_2-levels. Increased competition for nutrients between plants and soil biota at elevated CO_2-concentrations will also have feed-back effects on the functioning of terrestrial ecosystems.

Atmospheric CO_2 and soil processes

The effects of elevated atmospheric CO_2 on below ground processes, via primary production and carbon input into the soil, i.e. microbial turnover of soil organic matter, have received much less attention. One could hypothesize, that stimulation of the production of plant biomass by an elevated CO_2-concentration in the atmosphere would lead to a corresponding increase in soil organic matter content, caused by the increase in input, and decomposition rates remaining the same. However, there are a number of reasons for questioning a linear proportionality between plant productivity and soil organic matter level. Organic material that enters the soil either as root-derived material or as litter will be processed by microorganisms. The turnover of this organic matter through the microbial biomass will ultimately control both decomposition and retention of soil organic matter and thus the level of soil organic matter. Several factors such as the quality of the organic material, soil nutrient status, temperature and moisture affect the microbial utilization of this root-derived organic material.[2-5] These factors could well change if above ground primary production increases as a result of elevated atmospheric CO_2-concentrations. Thus, it is of great importance to know if qualitative and quantitative changes in carbon inputs to the soil will alter the rate of turnover of soil organic matter fractions. For instance, if increased nutrient demand by the plants cannot be met by the soil, the C/N ratio of plant tissues and consequently of plant litter might increase.[9] This in turn might influence the decomposibility of plant-derived materials, which is known for example, to be influenced by their nitrogen content.[10]

In order to get a better understanding of some of the key processes and interactions of root-derived materials and microbial activity at elevated atmosphere CO_2-levels, we recently started a project in which wheat plants are grown in an atmosphere with different levels of CO_2, labelled with ^{14}C. In the present paper some preliminary results of this study will be discussed, with emphasis on the pattern of translocation within the plant/soil system, soil/root-respiration and turnover of soil organic matter.

3 EXPERIMENTAL DESIGN

Seeds of spring wheat (*Triticum aestivum* L. cv. *Ralle*) were germinated in glass tubes for 10 days, and the seedlings then were transferred to columns, each containing 2.7 kg of loamy sand with a soil moisture content of 14% (v/w). Ten days later the columns were sealed at the shoot-root interface, using a silicone rubber (Dow Corning, Q3-3481). This was necessary to stop direct CO_2 exchange between soil and atmosphere, and to allow the ^{14}C to be partitioned between roots and shoots.
The columns were placed in two identical ESPAS (Experimental Soil Plant Atmosphere System) growth chambers and plants were grown at 350 and 700 ppm CO_2 in a $^{14}C-CO_2$-containing atmosphere. Each chamber contained 14 soil columns with wheat plants. The climatic conditions inside both chambers were the same and controlled by computer. At the start of the experiment the day length was 12 hours. After 2 weeks the light period was extended to 14 hours; after 5 weeks to 16 hours. Light intensity was between 20000 and 25000 lux. The relative humidity of the air was about 80%. The shoot- and root-compartment were separated by a base plate, which made it possible to adjust the temperature of the 2 compartments independently. The temperature of the shoot-compartment was $18°C$ during day time and $14°C$ during night time. For the root-compartment the temperature was $16°C$ and $14°C$ respectively. Water was given to the soil columns daily by connected irrigation tubes.

The soil, a loamy sand, was sieved (<2 mm) before starting the experiment. The water content was adjusted to 14% and the soil was mixed with nutrients. A fertilizer mixture was added, containing 30 mg N (NH_4NO_3), 70 mg P and 149 mg K (K_2HPO_4 and KH_2PO_4), all per kg moist soil, plus trace elements (Cu, B Mo, Mn, Zn and Fe). An extra dressing of nitrogen (60 mg N per kg soil) was added to the planted pots in the irrigation water after both 4 and 6 weeks of growth in the chambers. Plant growth is assumed not to be limited by nutrients or water in this experiment.

Three plants per treatment were harvested at 3 different dates, 22 (T1), 35 (T2) and 49 (T3) days after starting the experiment. At harvest, plant dry weight, the release of carbon components into the soil and the amount of ^{14}C and C-total within shoots, roots, soil/root-respiration and soil were measured. During plant growth, soil/root-respiration was measured 3 times a week for each column. The columns were flushed with CO_2-free air twice a day. The CO_2 in the soil atmosphere was collected in a 0.5 M NaOH-solution. Three times a week the carbon (total C and ^{14}C) in the NaOH-solution was determined by titration with 0.5 M HCl, after precipitation with $BaCl_2$. ^{14}C-CO_2-measurements were done by scintillation counting. Data were statistically analysed by analysis of variance.

4 CARBON PARTITIONING BETWEEN PLANT AND SOIL

Plant mass production

The dry weight of shoots of plants grown at 700 ppm CO_2 was up to 50% higher ($P<0.05$) at T2 and T3 compared to that of plants grown at 350 ppm CO_2 (Table 1). The relative distribution of ^{14}C between shoot, root, soil/-root respiration and soil did not differ significantly between plants grown at 700 ppm CO_2 or 350 ppm CO_2. This suggests that the distribution pattern of carbon within the plant, including the translocation to below-ground parts, is not changed by elevated CO_2-levels in the atmosphere. These results are therefore different from earlier reports,[7] in which root biomass increased more than shoot biomass. Our latest results, however, indicate that the increase in plant biomass production depends on the availability of nutrients in soil. At low soil mineral nitrogen contents, the relative increase in plant biomass at 700 ppm CO_2 compared to that at 350 ppm CO_2 was much smaller than at high soil mineral nitrogen contents.

Table 1   Dry mass (mg per column, 2370 g dry soil) and shoot:root ratio (S/R) of wheat grown at 350 and 700 ppm CO_2.

	T_1 (22d)		T_2 (35d)		T_3 (49d)	
	350	700	350	700	350	700
Shoot	1540	1940	5030	7690*	14020	20960**
Root	570	800	2420	3500*	4150	4870
Total	2110	2740	7450	11190**	18170	25830**
S/R ratio	2.7	2.5	2.1	2.2	3.3	4.3**

*: $P<0.05$; **: $P<0.01$

CO_2 respiration from roots and soil

Part of the CO_2 collected from the soil columns during the experiment came from the decomposition of native soil organic matter, present in the sieved soil at the start of the experiment. From the specific activity of the plant material and the specific activity of the respired CO_2, the contribution of this unlabelled material was calculated (Table 2). Up to T_1, unlabelled soil organic matter was decomposed at the same rate at both CO_2-levels, and contributed 25-40% of the total soil/root-respiration. At T_3 the turnover of unlabelled organic matter was responsible for 27% and 10% of the total amount of respired CO_2 for 350 and 700 ppm CO_2, respectively. It is striking, that the total respiration of unlabelled CO_2 from the soil/root compartment was less enhanced by an increasing CO_2 level than was the respiration of plant derived ^{14}C-material. At the end of the incubation, 53% more total-CO_2 and 74% more ^{14}C had been respired from soils exposed to 700 ppm CO_2 than soils exposed to 350 ppm CO_2, respectively.

Table 2 Cumulative evolution of labelled $^{14}CO_2$ and total $^{14+12}CO_2$ (mg per column 2370 g dry soil) from soil plus roots, by planted soils after 22, 35 and 49 days of exposure to either 350 or 700 ppm CO_2.

	T1(22d)		T2(35d)		T3(49d)	
	350	700	350	700	350	700
^{14}C-CO_2	157	288**	633	1139**	847	1544**
Total CO_2	261	399	843	1282*	1090	1707**

* : $P<0.05$; ** : $P<0.01$

5 CARBON BUDGET

A net soil carbon balance (Table 3) can be drawn up for the soil, by comparing the carbon coming from the roots plus that remaining in the soil as ^{14}C-labelled material (as determined by digestion of the soil) with the output of carbon by microbial decomposition of native soil organic matter ($^{12}CO_2$-respiration) (Table 3). At T_3 there was a net decrease of soil organic matter when plants were exposed to 350 ppm CO_2 (-45 mg C per column), and a net increase in soil organic matter when plants were grown at 700 ppm CO_2 (+84 mg C per column). Moreover, the root biomass at 700 ppm was larger than the biomass at 350 ppm CO_2. Since only the first part of the growing season, before flowering, is considered here, these short term effects of elevated CO_2 on soil organic matter dynamics should not be extrapolated to long-term carbon fixation in arable soils. It is known, from pulse labelling with ^{14}C-CO_2 of wheat plants in the field, that the pattern of carbon allocation changes drastically during the growing season. In plants older than about 8-10 weeks, 80-90% of the photosynthetically fixed carbon remains aboveground which is much more than in young plants (Swinnen, personal communication). Despite these uncertainties, the increased primary production will enhance the input of carbon into the soil, both during the growth of plants, as well as through root turnover and root litter production. The observed increase in soil organic matter might be enhanced by the reduced decomposition rate of native, more recalcitrant soil organic matter. This should be further investigated, as changes in soil organic matter could play a quantitatively important role in the global carbon budget. Investigations will particularly focus on the possible feed-back mechanisms, such as those influenced by the nutrient status of the soil and by temperature, on carbon dynamics in soil/plant ecosystems.

Table 3   Carbon budget for a soil planted with wheat and exposed to either 350 or 700 ppm CO_2. Input, output, net change and root biomass after 49 days are given as mg C per column, 2.37 kg dry soil

	350 ppm	700 ppm
T_1		
Input ^{14}C residue	25	61
Output $^{12}CO_2$ from SOM	104	111
Net change (after 49 days)	− 79	− 50
Root biomass	143	257
T_2		
Input ^{14}C residue	99	174
Output $^{12}CO_2$ from SOM	210	143
Net change (after 49 days)	− 111	+ 31
Root biomass	840	1403
T_3		
Input ^{14}C residue	198	247
Output $^{12}CO_2$ from SOM	243	163
Net change (after 49 days)	− 45	+ 84
Root biomass	1606	2054

6 CONCLUSIONS

The input of readily decomposable root-derived material to the soil was higher when plants were exposed to 700 ppm CO_2 than to 350 ppm CO_2. Because microorganisms preferred this material as their energy source the turnover of the more resistent native soil organic matter was reduced at 700 ppm CO_2. The non-soluble ^{14}C-residue in the soil was also increased at the highest CO_2-level. At 700 ppm CO_2, a stimulating effect on the organic matter level in soil was observed.

7 REFERENCES

1. J.A. Van Veen, R. Merckx, S.C. Van de Geijn, Plant Soil, 1989, 115, 179.
2. R. Merckx, A. Dijkstra, A. Den Hartog and J.A. Van Veen, Biol.Fertil.Soils, 1987, 5, 126.
3. R. Merckx, J.H. Van Ginkel, J. Sinnaeve and A. Cremers, Plant Soil, 1986, 96, 85.
4. R. Merckx, A. Den Hartog and J.A. Van Veen, Soil Biol.Biochem., 1985, 17, 565.
5. J.A. Van Veen and P.J. Kuikman, Biogeochem., 1990, in press.
6. J.D. Cure, Agric.Forest Meteor., 1986, 38, 127.
7. P.S. Curtis, B.G. Drake and D.F. Whigham, Oecologia, 1989, 78, 297.
8. J.M. Whipps, J.Exp.Bot., 1985, 36, 644.
9. P. Vitousek, Am. Nat., 1982, 119, 553.
10. J.M. Melillo, J.D. Aber, A.E. Linkins, A. Ricca, B. Fry and K.J. Nadelhoffer, Plant Soil, 1989, 115, 53.

Solubilisation of Wheat Straw by Actinomycete Enzymes

A.S. Ball and M. Allen

DEPARTMENT OF BIOLOGY, UNIVERSITY OF ESSEX, COLCHESTER CO4 3SQ, UK

1 INTRODUCTION

Agricultural residues such as grass lignocelluloses represent large renewable resources for which bioconversion to useful products is one of a number of alternative strategies currently under investigation. Wheat straw is a widely available substrate and its disposal represents an environmental problem. It is essentially lignocellulose-cellulose microfibrils embedded in a matrix of lignocarbohydrate comprising polyphenolic lignin covalently bound to hemicellulose. The bioconversion of this renewable resource is therefore of immense ecological and also biotechnological importance.

The biodegradation of lignocellulose is limited by the interactive nature of its structure, as well as the physical and chemical properties of the individual polymers. Lignin is the most recalcitrant and structurally variable of the polymers and its degradation represents the rate-limiting step in lignocellulose bioconversion. In order to improve the rate of bioconversion, it is first necessary to understand the processes involved in the degradation of native lignocellulose.

Actinomycetes represent the only prokaryotic group in which activity against all three major components of lignocellulose has been identified.[1] Actinomycetes are aerobic, Gram-positive bacteria, widely distributed in soils and composts. They are generally found growing in the hyphal form and are therefore well adapted to penetrate lignocellulosic material. Plant biomass is probably the most important natural growth substrate for these bacteria and it is not surprising that a range of biodegradative enzymes and activities has been evolved.

The principal product resulting from the solubilisation of lignocellulose by actinomycetes is a soluble, high molecular weight lignocarbohydrate complex (APPL).[2] Solubilisation can be extensive; up to 40% of lignin radiolabelled lignocellulose was solubilised by <u>Thermomonospora mesophila</u>.[3] However, rates of lignin mineralisation to CO_2 are low (<10%).[3] The ecological importance of lignocarbohydrate solubilisation is suggested by reports which link lignin degradation with humus formation in both soils and sediments.[4]

Although lignocellulose solubilisation by actinomycetes is well established, the processes involved are poorly understood. The enzymes responsible for solubilisation are largely unknown, but their extracellular nature has been demonstrated.[5] Enzymes involved in the direct attack on the lignin polymer such as peroxidases have been implicated, along with cellulases and hemicellulases, although other enzymes may be involved in the disruption of lignin-carbohydrate cross links[6], and oxidation reactions leading to solubilisation.[6,7] Support for this view comes from work indicating that the ability to attack lignin compounds is not correlated with lignocellulose solubilisation.[8]

This paper examines the solubilisation process for a number of actinomycetes using enzymological studies, and quantitative and qualitative analysis of the solubilised lignocarbohydrate complex. The effect of mechanical pretreatment of wheat straw on rates of solubilisation is also examined.

2 MATERIALS AND METHODS

Strains and growth conditions

Actinomycete strains were maintained as spore suspensions and hyphal fragments in 20% (v/v) glycerol at -70°C and routinely cultured on L-agar plates.[9] Distilled water suspensions of sporulating growth were used to inoculate shake flasks containing basal salts medium supplemented with pretreated wheat straw 0.2% (w/v);[10] ball-milled, blended or chopped (average particle size 50 µm, 2 mm and 3 cm respectively). Cultures were incubated for up to 12 days at 200 rpm and at 30°C, 37°C or 50°C as appropriate. Intracellular protein was used as a measure of growth and was determined by boiling harvested pellets for 20 minutes with 1M NaOH prior to protein estimation by the Lowry method.[11] Extracellular protein was also estimated by the method of Lowry.

Recovery of solubilised lignocarbohydrate

Cultures grown on wheat straw were centrifuged (10,000 g for 10 minutes at 4°C) and the supernatant acidified to pH 1-3 with HCl. The acid-precipitated product (APPL)[2] was removed by centrifugation (as above), washed twice and finally resuspended in distilled water. Dry weight measurement of APPL (expressed as mg produced g^{-1} straw) and protein determinations (Lowry method) on APPL fractions dissolved in 0.05 M NaOH (expressed as mg protein g^{-1} straw) were used to monitor production.

Enzyme assays

Culture filtrates were assayed for the presence of active enzymes. Release of reducing sugar from oat spelt xylan (Sigma Chemical Co.) and carboxymethyl-cellulose (CMC; low viscosity, BDH Ltd) were performed using a microtitre plate assay system described previously.[10] Arabinofuranosidase and acetyl esterase were assayed using the p-nitro-phenyl linked substrate described previously.[10] Peroxidase activity was recorded by measuring the formation of the dopachrome pigment (OD_{470}) from L-DOPA (Sigma Chemical Co.).[7]

HPLC analysis of APPL fractions

Analyses of APPL extracted from straw degraded by four **Streptomycete** strains were determined using an Ultrasphere ODS - 5µ (Altex USA) HPLC column eluted with acetonitrile (10-50%) at a flow rate of 0.4 ml min^{-1}, monitored at 276 nm. Standard substrates (0.02 mM in acetonitrile) included p-coumaric acid, syringic acid, ferulic acid, gallic acid and veratric acid (all available from Sigma).

3 RESULTS

Solubilisation of pretreated lignocellulose by actinomycetes

The solubilisation of lignocarbohydrate by six actinomycete strains during growth on chopped, blended and ball-milled wheat straw was examined. Although lignocarbohydrate solubilisation was found to vary between strains, maximum solubilisation rates were obtained by actinomycetes grown on ball-milled straw, while lowest levels of solubilisation were associated with chopped straw (30.8 and 0.3 mg APPL per mg protein respectively - Table 1). Maximum solubilising activity on all three pretreated straw samples was achieved using **S. viridosporus**, although there was an 8-fold difference between rates of solubilisation on chopped and ball-milled straw by this strain. Similar profiles were observed for **Streptomycete** strains EC1 and EC22 and **Amycolata autotrophica**. The results may suggest increased solubilisation due to a greater surface area available for enzyme attack. However, **S. badius** and **Thermomonospora mesophila** exhibited different profiles.

TABLE 1

Production of APPL by actinomycetes grown on pretreated wheat straw

Taxon and strain[1]	APPL (mg dry weight mg^{-1} intracellular protein) from		
	Chopped Straw	Blended Straw	Milled Straw
Streptomyces viridosporus T7A	3.8	12.2	30.8
" badius 252	0.8	0.9	1.6
" sp. EC1	0.6	1.2	3.4
" sp. EC22 (T)[2]	3.3	10.0	19.0
Thermomonospora mesophila DSM43048	1.6	2.7	2.6
Amycolata autotrophica DSM43099	0.3	0.3	1.7

[1] For sources of strains and incubation conditions see References 6, 8 and 10
[2] (T) = thermophile
Control experiments indicated no APPL produced in the absence of organisms

Values are a result of three determinations with all standard deviations between 0 and 15%.

Solubilisation of Wheat Straw by Actinomycete Enzymes

Solubilisation rates varied little between pretreatments with only a 2-fold difference in activities for each strain (Table 1).

Enzyme activities in culture supernatants from lignocarbohydrate solubilising Streptomycetes

Recent reports suggest that a correlation may exist between the production of APPL and the levels of various extracellular enzyme activities.[6] The production of a number of extracellular enzymes associated with lignocellulose degradation was examined for four Streptomycete strains (Table 2). Once again S. viridosporus exhibited maximal solubilising activity. Enzyme profiles were similar for all four strains, although differences in specific activities were observed. However single enzyme activity could not be correlated with rates of solubilisation.

HPLC analysis of the solubilised product

Degradation of the lignin component of the solubilised product by the four actinomycete strains was examined in more detail by HPLC analysis. The elution profiles of the four strains exhibited considerable differences (Figure 1). However some similarities in elution profiles were obtained when comparing APPL from the thermophilic Streptomycete EC22 and S. badius (Figures 1b and 1d). Interestingly, the elution profile from S. badius, the strain in which maximum APPL production occurred, was different to those of the other strains (Figure 1c).

A range of standards were used in an effort to identify the products obtained. However no standard could be calibrated to within $\pm$ 5% of the retention times of the sample profiles, making identification impossible.

4 DISCUSSION

The ability of actinomycetes to solubilise straw is again confirmed. Lignocarbohydrate solubilisation occurred during growth of actinomycetes on all three mechanically pretreated straw samples. Maximum levels of solubilisation occurred during growth on ball-milled straw by S. viridosporus. Generally as particle size increased, there was a subsequent decrease in yield of approximately 6-8 fold, suggesting that mechanical pretreatment of ligno-cellulose substrate results in a greater degree of enzymic activity due to an increase in surface area of the substrate.[12] However, the rates of solubilisation by S. badius and T. mesophila varied little between substrates. These two strains have been shown to possess extracellular enzymes capable of

FIGURE 1

Elution profiles of APPL from ball-milled straw produced by
a) <u>Streptomyces</u> <u>sp</u> EC1, b) <u>Streptomyces</u> <u>sp</u> EC22, c) <u>S. viridosporus</u>
and d) <u>S. badius</u>. Profiles were obtained by reverse phase
chromatography (mobile phase, acetonitrite/water (10-50% v/v)).

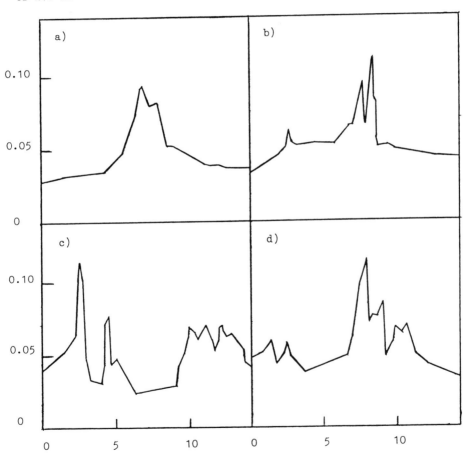

OD 276 nm

Time (mins)

TABLE 2

Extracellular enzyme activities of supernatants from Streptomycete cultures grown on ball-milled straw

STRAIN	Specific enzyme activities (IU mg^{-1} protein)					APPL (mg dry weight g^{-1} straw)
	Xylanase	Endoglucanase	Acetylesterase	Arabinofuranosidase	Peroxidase	
Streptomyces sp EC22	0.6	0.1	55	26	0.7	40
Streptomyces sp EC1	0.5	0.1	30	22	0.8	50
S. badius 252	0.4	0.2	71	48	0.9	50
S. viridosporus T7A	1.4	0.1	35	29	0.2	65

Values are a result of three determinations with all standard deviations between 0 and 15%

cleaving β-ether links, the most common bond type present in native lignins.[13] In eukaryotic-lignin degrading systems, cleavage of this bond is associated with a lignin peroxidase enzyme.[14] The effect of lignin degradation may be analogous to mechanical pretreatment in that it will open out the substructure of lignocellulose,[10] thereby increasing the surface area available for enzyme activity. It may be that for S. badius and T. mesophila, strains possessing lignin peroxidase activity, lignocarbohydrate solubilisation was not limited by surface area with the result that mechanical pretreatment had little effect on solubilisation rates.

The solubilised product resulting from lignocellulose degradation by actinomycete enzymes is a high molecular weight complex of lignin, carbohydrates and proteins. Although little is known about the enzymology, a number of enzyme activities including peroxidases, xylanases, and other hemicellulolytic enzymes and cellulases have been implicated in the process.[6,7,8] Examination of five enzyme activities involved in lignocellulose degradation revealed that all were present in culture filtrates, although no correlation between any single enzyme activity and lignocellulose could be observed.

Further evidence illustrating the complexity of lignocellulose solubilisation by actinomycetes comes from HPLC analysis of APPL from different Streptomycetes. Significant differences were detected in the elution profiles of each strain although no products could be identified. This provides further evidence for the diverse nature of APPL from different Streptomycetes.[15]

We conclude that mechanical pretreatment may be used to improve rates of lignocellulose solubilisation. In association with recombinant DNA techniques involving the cloning of important solubilising enzymes such as lignin peroxidase, solubilisation rates may be greatly enhanced. Future work will also be involved in studying further the ecological and biotechnological significance of lignin solubilisation by actinomycetes. APPL may be an intermediate in the formation of humic substances in soils and sediments.[4,16]

REFERENCES

1. A. J. McCarthy, FEMS Microbiol. Rev., 1987, 46, 145.
2. D. L. Crawford, A. L. Pometto III and R. L. Crawford, Appl. Environ. Microbiol., 1983, 45, 898.
3. A. J. McCarthy and P. Broda, J. Gen. Microbiol., 1984, 130, 2905.

4. E. Fustec, E. Chauvet and G. Gas, Appl. Environ. Microbiol., 1989, 55, 922.
5. J. C. Mason, M. Richards, W. Zimmermann and P. Broda, Appl. Microbiol. Biotechnol., 1988, 28, 276.
6. A. S. Ball, B. Godden, P. Helvenstein, M. J. Penninckx and A. J. McCarthy, Appl. Environ. Microbiol., 1990, In Press.
7. M. Ramachandra, D. L. Crawford and A. L. Pometto III, Appl. Environ. Microbiol., 1987, 53, 2754.
8. A. S. Ball, W. B. Betts and A. J. McCarthy, Appl. Environ. Microbiol., 1989, 55, 159.
9. D. A. Hopwood, M. J. Bibb, K. F. Chater, T. Kieser, C. J. Bruton, H. M. Kieser, D. J. Lydiate, C. P. Smith, J. M. Ward and H. Schrempf, 'Genetic Manipulation of Streptomycetes,' a Laboratory Manual. 1985, John Innes Foundation.
10. A. S. Ball and A. J. McCarthy, J. Gen. Microbiol., 1988, 134, 2139.
11. O. M. Lowry, N. J. Roseborough, A. L. Farr and R. J. Randall, J. Biol. Chem., 1951, 193, 265.
12. A. J. McCarthy, E. Pearce and P. Broda, Appl. Microbiol. Biotechnol., 1985, 21, 238.
13. B. Godden, A. S. Ball, P. Helvenstein, A. J. McCarthy and M. J. Penninckx, 1990. In preparation.
14. M. Tien and T. Kent-Kirk, Proc. Natl. Acad. Sci. USA, 1984, 81, 2280.
15. J. R. Borgmeyer and D. L. Crawford, Appl. Environ. Microbiol., 1985, 49, 273.
16. N. Senesi, T. M. Miano and J. P. Martin, Biol. Fertil. Soils, 1987, 5, 120.

Dynamics of Organic Carbon and Nitrogen Accumulation and Distribution in Soils Following Farm Woodland Planting

G.M. Whiteley

DEPARTMENT OF PURE AND APPLIED BIOLOGY, THE UNIVERSITY OF LEEDS, LEEDS LS2 9JT, UK

1 INTRODUCTION

The likelihood of a large increase in the area of farm woodland planting over a relatively short period of time has heightened the need for research into both the management of farm woodlands for efficient timber production and their impact on the environment. Within the context of north European agriculture, financial incentives for the development of alternative uses for surplus farmland, landscape improvement and planning restrictions in environmentally sensitive areas of various kinds are all tending to encourage this increase without particular reference to any soil type or former crop production system. However, in a wider context, the establishment of some form of woodland ground cover in degraded, eroded or infertile soils has always been a common practice.

A long-term field experiment to study the productivity and environmental impact of three tree species planted as farm woodlands in formerly long-term arable soil was established at the University of Leeds Field Station in 1988[1]. The site includes plots of each species planted on a two meter square grid and wider spaced rows in an experimental agroforestry arrangement. The latter will not be considered here because the changes, particularly during the early part of the maturation cycle, will be dominated by the crop and soil management of the intervening strips.

Re-afforestation of arable land is likely to have a large influence on the organic carbon and nitrogen equilibria resulting in a changing pattern of net gains and losses of carbon and nitrogen from the soil organic matter pool and the plant and microbial biomass fraction. The magnitude and duration of the net gains or losses of stored soil nitrogen is of particular interest to this study in considering the nitrogen nutrition of the trees and the effects of the change in land use on the leakage of nitrate in drainage water[2,3].

The site is on a brown earth soil of the Tickenham soil series[1], a well drained sandy clay loam topsoil and clay loam subsoil of about one meter depth which lies directly on an exposed magnesian limestone aquifer[1]. This aquifer is of smaller extent but shares some of the current problems associated with the effects of crop fertilization and the ploughing in of old pastures that have given rise to increased concentration of nitrates in the groundwater of larger sandstone and chalk aquifers and in the lowland farming areas of England[4].

The changes in net mineralisation or immobilisation rates during the maturation cycle of the trees and the magnitude and time taken to reach any new equilibria in stored organic nitrogen will determine when there might be periods of nitrogen shortage or excess of importance to growth or nitrate leakage. Losses during this period could have an effect on the amount of stored nitrogen which could be critical on some sites. A major problem with this type of research is that controlled field experiments accumulate data only very slowly. Using an alternative approach of collecting survey data from areas of known land use history has the major limitation of not being able to make treatment comparisons from replicated blocks. However, use of both surveys and past experimental records will provide some information about the likely behaviour of a specified soil series for comparison with results of the new experiments, as they accrue and any subsequent models derived from them. A problem with the use of past records or old established experiments is that often the preferred comparisons are constrained either through the original design of the experiment or the initial measurements taken[5].

All the organic carbon and nitrogen figures quoted here are for samples collected from the Tickenham soil series. Allowance has been made for variations in bulk density when comparing profile distributions and total quantities[6].

2 EVIDENCE FROM FIELD SAMPLING

A simplified scheme to organise information collected from various sources for this particular soil series would need to take account of both the changing equilibria of total stored carbon and nitrogen in the soil profile and any changes in their distribution with depth.

Short-term changes following new farm woodland planting.

After only the third year of the new farm woodland experiment it would be unlikely to be able to distinguish any changes from crude total analyses of carbon and nitrogen contents from field sampling. However, direct measurement of the annual fluxes of nitrate in the drainage water during this period[7] can be used to estimate the magnitude and direction of the changes in total soil

Table 1   Net annual mineralisation of carbon and nitrogen following new woodland planting on arable land.

Tree species	Year after planting	Range of total nitrate N loss (kg ha^{-1})	Carbon loss (%) (0-30 cm)
Ash only	1	18-24	0.004-0.005
Ash, sycamore and cherry	3	14-27	0.003-0.006

carbon and nitrogen (Table 1). The estimates of total nitrate N losses in Table 1 were calculated by multiplying the measured concentrations in soil water samples collected regularly in lysimeters during the 1989-1990 season by estimates of drainage volume from the soil water balance equation. From four replicate plots of each tree species in the third year after planting, it was not possible to distinguish statistically between species and so the range of nitrate N loss in Table 1 refers to the lowest and highest plot measurement irrespective of species. The figures for carbon loss simply indicate the equivalent annual magnitude of the loss of organic carbon from a 0-30 cm former ploughed horizon that would contain the same quantity of nitrogen as the annual flux from the soil profile, assuming no change in the carbon to nitrogen ratio.

Changes occurring during the first maturation cycle for trees planted on formerly arable land.

One source of information on longer-term changes is a comparison made between one field with long-term arable use and a neighbouring small broadleaf farm woodland planted at one side of the field in about 1950[1]. The distributions of organic carbon and total nitrogen in representative bulked samples taken from each of the areas is shown in Table 2. The changes over this period show the expected development of a litter layer and a shallower A horizon in the woodland soil than the depth of ploughing in the arable soil. Assuming a net change in the new woodland soil and the maintenance of the initial equilibria in the arable soil, the total net accumulation of soil nitrogen is only in the region of 200 kg ha^{-1}, or 10 kg N ha^{-1} per year over the 40 year period. This is lower than the expected rate of atmospheric deposition[8], thus indicating a relatively neutral role for nitrogen storage or release from the organic matter pool.

Table 2 Organic carbon and total nitrogen distributions under continuous arable cultivation and a neighbouring woodland

	Depth (cm)	Organic carbon (%)	Total Nitrogen (kg N ha^{-1})
Arable soil	0-30	1.74	3,706
	30-86	0.74	3,527
	Total in profile		7,233
Mature broadleaf woodland	0-12	3.02	2,328
	12-34	1.45	2,298
	34-86	0.66	2,897
	Total in profile		7,523

Apparent long-term equilibria in areas of known land use history.

Comparison between single areas as outlined above can serve only as illustrative examples because of the limmitations the size of area represented and the inability to distinguish the effects of land use from other aspects of the spatial variability in soil properties. To obtain more representative information about the long-term equilibria for this series under different land uses, a survey was conducted over a larger area in the surrounding district. From this representative areas of the soil series were identified with a single uniform land use extending in area to several hectares. The areas were large enough for aspects of soil variability associated with slope position to be in all the land use treatments. The results shown in Table 3 for carbon and nitrogen distribution and total nitrogen are the means of five bulked samples each collected from 20 soil cores from a 160 m^3 sampling grid. The pasture is a parkland managed for low intensity grazing. The mature beech woodland shows extensive regeneration and, in common with the conifer woodland, has an understorey dominated by bracken and brambles. The conifer woodland was planted more than 100 years ago with the original corsican pine stand subsequently replanted with a 40 year old stand of larch.

It is of interest to note that there are apparent differences in both the distributions of organic carbon and nitrogen and differences in total nitrogen between the four soil profiles. Despite the presence of a litter layer and a well structured and organic stained A_1 in both the woodland soil profiles, the total carbon and nitrogen contents are lower than the long-term arable profile.

Table 3 Comparisons of soil survey data for different land uses on the Tickenham soil series. (SE)

Land use	Depth (cm)	Organic carbon (%)	Total nitrogen (%)	Total nitrogen in the profile (kg ha^{-1})
Arable (>50 years)	0-30	2.07 (.16)	.19 (.024)	
	30-50	0.66 (.08)	.06 (.015)	10,297
Beech woodland (>100 years)	0-2	6.44 (1.0)	.43 (.028)	
	2-7	2.80 (.29)	.20 (.056)	
	7-25	1.64 (.04)	.15 (.048)	
	25-50	0.45 (.21)	.04 (.016)	8,383
Conifer woodland (>100 years)	0-3	16.0 (3.4)	1.1 (.260)	
	3-13	2.58 (.75)	.21 (.048)	
	13-28	1.19 (.16)	.10 (.022)	
	28-50	0.60 (.08)	.05 (.007)	7,344
Permanent pasture (>50 years)	0-12	2.75 (.24)	.27 (.079)	
	12-25	2.55 (.40)	.23 (.063)	
	25-50	1.42 (.22)	.13 (.034)	14,510

The chemical composition and mode of incorporation and decomposition of organic matter is likely to vary greatly between these four land use systems[9]. For example, mean carbon to nitrogen ratios ranged from between 10 and 11 to 1 throughout the pasture and arable profiles compared with ratios as high as 15 to 1 in the upper layers of the woodland profiles. The ratio in the litter layers here were very variable, as has been found elsewhere[10], with individual samples having a carbon to nitrogen ratio as high as 30 to 1.

Effects of soil disturbance.

Disturbance by regular ploughing is one of the main differences that would be affecting the environment for organic matter accumulation and mineralisation between the arable soil and the other three land use types in Table 3. The separation of this as a factor can be assessed from measurements taken from prolonged cultivation experiments on the same soil series. The longest of these has some records for organic carbon and nitrogen distribution extending over 17 years of continuous winter wheat, during which time the straw and stubble was burned annually. Table 4 is calculated from data presented by Chaney et al.[11], and shows some tendency for organic matter contents to be retained in the top 10 cm of the profile when annual disturbance is removed. The corresponding total nitrogen distributions were more variable between replicates but showed the same trends of distribution and total content.

Table 4 Effects of soil disturbance by annual ploughing on the distribution of organic carbon after continuous winter wheat.

Depth (cm)	After 3 years*	After 10 years		
		Ploughed	Direct drilled	(SE)
0-5		1.08	1.37	.05
5-10		1.11	1.24	.03
10-15	1.32	1.08	1.11	.03
15-20		1.08	1.11	.03
Total in top 20 cm (tonnes ha^{-1})	40.4	31.7	35.7	

* Mean for both cultivation treatments

It is of equal interest to note the large decline in total organic mater in the topsoil between years three and ten on the study. Only a mean value is given for year three because there were no detectable differences between treatments and depths. The difference between treatment may partly reflect more of the organic matter being distributed below the 20 cm depth of measurement in the ploughed soil. In studies on other soil series, prolonged direct drilling has been found to affect the distribution but not the total content of organic carbon or nitrogen when compared with annual ploughing[12]. Nevertheless, the most conservative estimate of the loss, at between 4 and 5 T ha^{-1} of organic carbon, equivalent to 400 to 500 kg N ha^{-1} is very high when compared with the long-term changes suggested in Table 3 for major changes of land use. It is however within the order of magnitude of annual leaching losses for cereal crops quoted elswhere[2] and recorded in a nitrogen recovery experiment on spring barley on the same soil series[13].

3 IMPLICATIONS FOR FARM WOODLAND PLANTING

The direct evidence collected for the first years after tree planting suggests a net release of soil organic nitrogen irrespective of tree species which is available to be leached. The evidence from Tables 2 and 3, for the expected longer-term changes on this soil series, does not show a substantial net accumulation of organic matter and nitrogen as has more often been reported when a woodland vegetative cover is established on soils that have been degraded or eroded through inappropriate management in the former arable land use.

Availability of soil nitrogen

For a farm woodland that was not fertilised after planting, the release of nitrogen from soil organic matter, including the residual decomposition of recent crop residues on this soil series, can maintain a supply of available nitrogen in the soil profile. When considering alternative planting arrangements for farm woodland trees on this soil series, in particular trees interspersed with retained pasture, the soil management of the site would become critical and account would need to be taken of the reduced mineralisation of nitrogen in addition to the direct effects of competition for nitrate already in the soil solution.

Aggregate stability and erosion protection.

On this soil series the land use comparison in Tables 2-4 have shown that the effects of land use on distribution are greater than on total quantities of carbon and nitrogen in the soil profile. This gives the mistaken impression to the casual observer of a larger organic matter pool in woodland soil than the long-term arable soil. The evidence suggests that the new farm woodland profiles will move relatively rapidly in the direction of an accumulation in a shallow surface horizon. This is attributable to both the pattern of deposition and associated biological activity in the soil and partly through the effects of reduced soil disturbance. The latter is apparent after 10 years direct drilling.

Increased surface protection from erosion would be an important consequence of this redistribution in soils at risk from erosion. For this soil, the topsoil is of a low stability[10]. For all the treatments in Tables 2 and 4 aggregate stability was measured using both wet sieving and turbidity methods[14]. It was possible to obtain an approximate linear correlation between stability index and the organic matter content of samples from different depths and treatment using both methods.

Methods will be needed to organise the increasing quantities of field data on how changes in land use can influence the long-term equilibrium of accumulation and loss of nitrogen and carbon from the pool of organic matter. To this end the traditional concept of a soil series (or groups of series) may be useful for predicting the behaviour of the organic matter pool as a source or sink for nitrogen in the environment. This could be done, for example, for groups of series associated with major sandstone or chalk aquifers as part of an assessment of the environmental impact of major land use changes into and out of woodland or grassland. Over a prolonged time scale it seems possible that the changing equilibria could be described in the form of simple models relating to quantities and distributions of organic matter constituents[15]. These models could then be refined and tested.

REFERENCES

1. G. M. Whiteley, Belowground Ecology, 1990, 1, part 3, 10.
2. O. Strebel, W. H. M. Duynisveld and J. Bottcher, Agriculture, Ecosystems and Environment, 1989, 26, 189.
3. O. Strebel and J. Bottcher, Agricultural Water Management, 1989, 15, 265.
4. S. S. D. Foster, A. C. Cripps and A. Smith-Carington, Philos. Trans. R. Soc. London, 1982, 296, 477.
5. M. F. Billett, F. Parker-Jervis, E. A. Fitzpatrick and M. S. Cresser, J. Soil Sci., 1987, 41, 133.
6. C. L. Bascomb, in, 'Soil Survey Technical Monograph No. 6', B. W. Avery and C. L. Bascomb, Harpenden, 1974, 29.
7. K. M. Rushton and G. M. Whiteley, in, 'Soils and Environment in Northern England', Proceedings of the North of England Soils Discussion Group, University of York, September 1990.
8. K. W. T. Goulding, Soil Use and Management, 1990, 6, 61.
9. R. G. Joergensen and B. Meyer, J. Soil Sci., 1990, 41, 17.
10. R. G. Joergensen and B. Meyer, J. Soil Sci., 1990, 41, 279.
11. K. Chaney, D. R. Hodgson and M. A. Braim, J. agric. Sci., Camb., 1985, 104,
12. D. S. Powlson and D. S. Jenkinson, in, 'Rothamsted Experimental Station Report for 1979', Part 1, June 1980, 233.
13. D. R. Hodgson, G. M. Whiteley and A. E. Bradnam, J. agric. Sci., Camb., 1989, 112, 265.
14. J. T. Douglas and M. J. Goss, Soil Tillage Res., 1982, 2, 155.
15. D. S. Jenkinson and L. C. Parry, Soil Biol. Biochem., 1989, 21, 535.

Changes in Organic Matter in Fen Silt Soils

P.A. Johnson and J.M. Prince

SOIL SCIENCE, ADAS, KIRTON, BOSTON, LINCS. PE20 1EJ, UK

1. INTRODUCTION

The role of organic matter in soil, in nutrient cycling and soil structure, has been extensively studied and reported elsewhere. For example Russell[1] includes discussion on the role of organic matter in several chapters of his book "Soil Conditions and Plant Growth". The report "Modern Farming and the Soil"[2] published in 1970 expressed concern at what were believed to be falling levels of organic matter in arable soils in the United Kingdom. A fall in organic matter content in a sandy loam over 97 years from 2.5% to 1.2% in an all arable rotation has been reported by Johnston[3], in contrast at Rothamsted on a silty clay loam the fall has been from 1.8% to 1.6% over 124 years.

In the early and mid 1960s, 14 sites were selected on the fertile siltlands of the Holland district of Lincolnshire where organic matter levels could be monitored. Monitoring continued, on an annual basis at first but latterly less frequently, until 1989.

Although curves have been fitted to the data the main aim of the paper is to make the data available to other workers.

Soils and Previous History

Thirteen of the sites selected were located on typical siltland soils of the Romney and Blacktoft series[4] and one on an organic soil of the Downholland series. Two of the Romney sites had been ploughed out of permanent grass in the previous year, as had the Downholland site. The date of ploughing out from grass of the other sites was unknown.

Sampling Methods and Analysis

At each site two points were selected and the samples were taken within a 5 m radius of these points. Results presented are a mean of the analyses, carried out by standard ADAS methods[5], of the two samplings points. It is believed that laboratory error could be $\pm 0.1\%$ and that sampling error may be twice the laboratory error.

Cowfield High

Organic matter % = $3.34 - 0.16 \log^n$ years
$r = 0.209$ ns

Holbeach High

Organic matter % = $1/[0.22 + (0.002 \text{ years})]$
$r = 0.72 \; p \; 0.02$

Cowfield Low

Organic matter % = $3.37 - 0.13 \log^n$ years
$r = 0.83 \; p \; 0.02$

Leverton Low

Organic matter % = $2.9 - 0.09 \log^n$ years
$r = 0.63 \; p \; 0.05$

The Cowfield sites, which have had similar cropping except for year 1, have had three 2-year leys in their sampling period, Leverton Low has had 11 brassicae crops leaving large organic residues, whereas Holbeach High has had a predominantly arable rotation. Fitting a $\log^n$ curve to the Holbeach High data reduced the level of significance of the curve.

Group 3: Moderate Rate of Fall. These three sites include the Leverton High site (fig 7) which has an identical known history to the Leverton Low site which has shown only a small fall in organic matter level. However the Leverton High site started with almost twice as much organic matter.

Leverton High

Organic matter = $5.43 - 0.08$ years
$r = 0.94 \; p \; 0.001$

Spalding High (fig 8) showed a similar rate of fall but from a slightly lower initial level. The history of this site shows two 2-year leys, which have apparently not affected the fall in organic matter.

Spalding High

Organic matter = $1/(0.20 + 0.006 \text{ years})$
$r = 0.74 \; p \; 0.05$

The third site in the group is Moulton (fig 9). This is the Downholland series site and the scatter of recorded points is very wide. The site was ploughed out of grass in 1967 and first sampled in 1969. Samples taken at 15-25cm depth at this site show similar but related variations in organic matter content which make it difficult to suggest that inclusions of subsoil material might explain some of the lower figures.

Moulton

Organic matter % = 15.47 - 0.14 years

Group 4: No change or slight rise. Both sites at Baldwykes (fig 10 and 11), the low sites at Spalding (fig 12) and Holbeach (fig 13) and Lutton (fig 14) either remained stable or showed very small rises in organic matter content.

Baldwykes Low

Organic matter % = 2.45 + 0.015 years
$r = 0.38$ ns

Baldwykes High

Organic matter % = 3.15 + 0.04 years
$r = 0.24$ ns

Spalding Low

Organic matter % = 2.61 + 0.07 $\log^n$ years
$r = 0.43$ ns

Holbeach Low

Organic matter = 2.57 + 0.0002 years
$r = 0.01$ ns

Lutton

Organic matter % = 2.14 + 0.03 $\log^n$ years
$r = 0.17$ ns

It is not known when any of these sites were ploughed out of grass. Baldwykes has had a ley arable rotation with 2-year leys every six years, Spalding had two 2-year leys as mentioned previously, Holbeach has been arable and Lutton has had 7 years of bulbs during its 26 year history. Thus there is nothing in common between the sites except their unknown period of cultivation after being ploughed out of grass.

2. CONCLUSIONS

Following the ploughing out of permanent grass on the silty soils of the Holland district of Lincolnshire there is a dramatic fall in organic matter content. After a varying number of years the rate of fall declines and eventually stabilises. The length of time before stable levels of 2.5-3.0% are reached will vary, possibly depending on the original organic matter content. It would therefore appear that, as suggested by Jenkinson[6], organic matter with a low half life is rapidly oxidised leaving that of longer half life until loss is equalled or possibly at some sites exceeded by inputs of fresh organic matter from crop residues. The cropping of the land during the period of fall did not appear to have any noticeable effect on the speed of decline.

REFERENCES

1. E W Russell, 1988 Soil Conditions and Plant Growth. Longman, England.
2. MAFF, 1970 Modern Farming and the Soil. HMSO, London.
3. A E Johnston 1978, Organic Matter and Soil Structure, National Agricultural Centre, Stoneleigh: Conference Paper - Soil Structure and Drainage.
4. Soils and Their Use in Eastern England 1984, Bulletin No. 13, Soil Survey of England and Wales, Harpenden.
5. Analysis of Agricultural Materials, Technical Bulletin 27, MAFF, London
6. Jenkinson D 1981, "The Fate of Plant and Animal Remains in Soil" in Chemistry of Soil Processes Ed. Greenland D Y and Hayes M H B, John Willey & Sons, London.

Section 5

Fertility and Soil Organic Matter

Introductory Comments

by A.E. Johnston, Rothamsted Experimental Station

Soil fertility is controlled by complex interactions between biological, chemical and physical properties. One attempt to unravel and quantify the contribution of soil organic matter or humus to all three is presented here. Field based assessments of the effects of humus on soil fertility are costly because they must be long-term, humus content changing so slowly in temperate climates. The humus content of silty clay loams at Rothamsted is still increasing, albeit slowly now, where farmyard manure (FYM, 35 t/ha) has been applied each year for more than 100 years but has been maintained by annual applications of fertilizers to arable crops. For more than 100 years yields on fertilizer- and FYM-treated soils were the same, provided fertilizer applications were adjusted to the needs of the crop, even though a three-fold difference in humus content had developed over this period. However in the last 10 to 20 years yields, especially of spring sown crops, have been larger on FYM-treated soils. The most probable reason for this change is new cultivars with a high yield potential which can be protected with weedkillers and pesticides. To achieve the increased potential requires optimum soil physical, chemical and biological conditions.
    For any one soil it is difficult to determine a critical humus content below which yields will be adversely affected. To extrapolate between soils of different texture is impossible because soil texture has such a large effect on the equilibrium humus level for the same farming system. Such levels are always lower on soils of lighter texture and many benefits from enhanced humus levels have been recorded on such soils.
    The ban on in-field straw burning will have a significant effect on agricultural practices. Effects of straw incorporation on soil humus and mineral nitrogen content, crop yields and the efficiency of some soil acting herbicides are discussed in two papers. Effects on humus content and yield have been small on MAFF Experimental Husbandry Farms but straw additions may eventually keep the humus level above

the critical value for that soil as on the sandy loam at Woburn. After more than 100 years of continuous arable cropping soil humus had declined to 1.2%; this was increased to 1.6% by incorporating 7.5 t/ha straw for each of six years. This extra humus consistently increased yields of arable crops over a wide range of amounts of fertilizer nitrogen.

Disposal of organic wastes to soil is only a problem if the wastes contain phytotoxic contaminants. In their absence organic wastes as diverse as a compost derived from organic town refuse and the effluent from a meat processing plant had no major deleterious effects in the short term. However, continued disposal of large volumes of effluent depends on maintaining adequate drainage.

Whilst adding organic matter to soils maintains microbial activity and humus improves its physical properties, recent research has addressed the benefits and problems arising from the mineralization of organically bound nitrogen. It can also be inferred that there is a pressing need to model these processes to predict rates and times of mineralization (see Section 4). The profitability of farming will be improved by using fertilizer nitrogen only to supplement the supply from the soil up to the optimum for the field's yield potential. A soil analysis for mineral nitrogen status will be required until soil nitrogen turnover can be predicted but the best time for sampling soils with different amounts and types of organic matter has yet to be decided. Even soils given different amounts of fertilizer nitrogen for many years, and hence different quantities of organic material returned each year, can contain different amounts of mineral nitrogen even though amounts of total nitrogen do not differ appreciably.

Unfortunately soil conditions for mineralization are often optimum in autumn when crop demand is small. Excess nitrate remaining in soil at this time may be at risk to loss by leaching when soils are at field capacity and there is through drainage.

Current interest in organic farming systems is catered for here only by a brief discussion of possible contributions from humus and some references; no factual evidence is presented. Although sizeable benefits from extra humus are given in other contributions where yield potential was protected by the judicious use of agrochemicals, these inputs are not allowed in organic farming systems. Some real comparisons would help discussion. The crucial factor is not whether organic farming systems produce more wholesome food, or are more financially viable, but whether such systems can produce enough food to feed a still increasing world population.

Soil Fertility and Soil Organic Matter

A.E. Johnston

LAWES TRUST SENIOR FELLOW, SOILS AND CROP SCIENCES DIVISION, AFRC INSTITUTE OF ARABLE CROPS RESEARCH, ROTHAMSTED EXPERIMENTAL STATION, HARPENDEN, HERTS. AL5 2JQ, UK

1 INTRODUCTION

In 1843 Liebig noted: "The fertility of every soil is generally supposed by vegetable physiologists to depend on .... humus. This substance, [is] believed to be the principle nutriment of plants, and to be extracted by them from the soil ....". Lawes and Gilbert quickly showed that the second part of this statement was not true. But the nebulous idea expressed in the first part has continued to be a topic of discussion. In 1977 E.W. Russell wrote: "It has long been suspected, ever since farmers started to think seriously about raising the fertility of their soils from the very low levels that characterised medieval agriculture, that there was a close relation between the level of organic matter, or humus, in the soil and its fertility. In consequence good farmers have always had as one of their goals of good management the raising of the humus content of their soils. But present day economic factors have forced many farmers to adopt practices which cause the humus content of their soils to fall to levels which they believe are lowering fertility. Thus the major problem facing the agricultural research community is to quantify the effects of the soil organic matter on the complex of properties subsumed under the phrase soil fertility, so that it can help farmers develop systems which will minimise any harmful effects this lowering brings about."

The paper from which this quotation is taken was published some seven years after the Strutt Report[1] which was commissioned following the two very poor harvest years of 1968 and 1969. Amongst many recommendations the Report suggested that arable soils should contain not less than 3% organic matter (1.75% organic carbon). This suggestion aroused much discussion and Russell[2] discussed some aspects of the role of humus in soil fertility. The Report also questioned the move away from mixed arable crop rotations to the more frequent growth of cereals. It is interesting now to conjecture why this concern was expressed because there is some evidence[3] that rotations with a greater proportion of cereals tend to maintain higher levels of soil humus compared to rotations with many root crops.

The quotation above from Russell succinctly summarizes three facets of a topic which has long been with the agricultural research and advisory services: is humus important in soil fertility?; over what time scales and with what practices do humus contents change?; can the various factors which contribute to the organic matter effect be separated and quantified? This paper aims to progress the discussion using results from experiments on the silty clay loam, 20-25% clay and 50% silt, developed in drift over Clay with flints at Rothamsted and the sandy loam with only 10% clay, developed in drift over Lower Greensand at Woburn. Average annual rainfall is about 700 mm at Rothamsted and 625 mm at Woburn.

2 THE ROLE OF SOIL ORGANIC MATTER

Soil organic matter contributes to soil fertility in a number of ways. It increases cation exchange capacity, especially in light textured soils,[3] and water holding capacity, especially that of available water.[4] During its oxidation N, P and S are released and some may become plant available at times in the growing season and positions in the soil profile which are difficult to mimic by the application of fertilizers. Humus helps stabilize soil structure, especially in soils of poor structural stability. Partially decomposed organic residues can prevent soil particles from coalescing and so maintain an open structure.

There are few suitable techniques for studying effects of humus on soil structure, especially those which could be applied in the laboratory and related directly to effects in the field. Results of some laboratory tests on soils with different levels of humus have been reviewed.[5] A map of the isodynes (lines of equal soil resistance) for Broadbalk mainly distinguished soils of different texture but within each textural class the smallest values were always on FYM-treated plots.[6] At that time these soils had about twice as much humus as fertilizer treated soils.

Whilst it may be difficult to quantify separately effects of extra soil organic matter on each of the above attributes this should not deter studies on any overall benefits which might accrue from maintaining or increasing levels of organic matter in soil. This

is especially so where increasing polarization of
farming systems has led to long periods in arable
cropping with the attendant risk that humus levels
will have been declining in these soils. At present
the one sure way to progress is to look for any
benefits to yield, for the crop integrates the various
effects of the physical environment in which it lives.

3 AMOUNTS OF ORGANIC MATTER IN SOIL

The amount of humus in soil depends on: i) the input
of organic material and its rate of oxidation; ii) the
rate at which existing soil organic matter decomposes;
iii) soil texture; iv) climate. Factors i and ii
depend on the farming system practised. All four
factors interact so that humus levels change towards
equilibrium values. In temperate climates these
changes tend to be slow and farmers could not be
expected readily to appreciate that such changes are
taking place.

In 1949 two Ley-arable experiments were started at
Rothamsted, one on a site long in arable cropping
where the soil had about 1.7% C; the other soil, with
about 3.1% C, had long been in permanent grassland.[7]
The yields of three arable "test" crops which followed
various 3-year leys ("treatment" crops) were measured
and compared with the yields of the same three arable
crops grown in an all-arable rotation. The crops for
this 6-year arable rotation were chosen to minimize
the build up of soil-borne pathogens. Some plots of
permanent grassland were retained on the grassland
site and, on the old arable soil, some were sown to
grass, which remained unploughed. In 1937 a similar
experiment, with a 3-year treatment phase but 2-year
test phase, had been started at Woburn but
unfortunately no continuous grass treatment was
included.[7]

Changes in organic carbon at Rothamsted (Fig 1A)
can be summarized as follows: on the soil ploughed out
from grass the humus level declined steadily but after
36 years it was still not as small as that in the old
arable soil retained in arable cropping. The humus
content of the old grassland soil increased a little
as a result of improved sward management but in the
old arable soil put down to permanent grass % C was
still well below that in the old grassland soil even
after 36 years. However, the changes in % C with time
under permanent grass in this experiment fit well a
relationship between accumulation of humus and time
for other Rothamsted soils (Fig 2). In Figure 2 the
200 and 300 year values are from the Park Grass
experiment which was sited on a field which had been
in grass for at least 200 years when the experiment

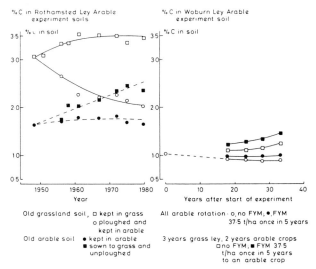

Figure 1 Change in % carbon with time in a silty clay loam at Rothamsted and a sandy loam at Woburn under contrasted farming systems.

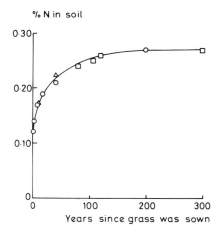

Figure 2 Relationship between % nitrogen in unploughed grassland soil and year since grass was sown on a silty clay loam, Rothamsted.

started in 1856. The "300 year" value was determined in 1959, 100 years after the start of the experiment. The relationship shows that it takes about 100 years for the equilibrium humus content of this silty clay loam to change from that characteristic of an old arable soil to that of a grassland soil and about 25 years to reach the halfway stage.

In both the Rothamsted Ley-arable experiments the effect of the 3-year leys on total soil carbon was surprisingly small.[3] Compared to the all-arable soils the average increase in % C in the 24th and 27th year after the experiment started was only about 10% for grass-clover and all-grass with N leys; the 3-year lucerne ley did not increase % C.

Under all treatments the humus content of the sandy loam at Woburn (Fig 1B) was less than the smallest amount in Rothamsted soil (Fig 1A) highlighting the important role of soil texture; increasing clay content stabilising soil organic matter.[8] The site chosen in 1937 for the Woburn experiment had been in a four-course arable rotation for most of the time since 1876.[7] Although the soil had only just over 1% C at the start, soil humus continued to decline during the next 30 years under continuous arable cropping. However, there has been no catastrophic fall in yields. In fact yields have tended to increase provided fertilizer applications, especially nitrogen, were well chosen and cropping sequences were such as to minimize the build up of soil-borne pathogens. Other examples of changes in soil organic matter with time have been given by Johnston.[3] In all cases the important observation is the slowness with which the total organic matter content of soil changes under temperate conditions.

Because soil organic matter changes only slowly with time under British climatic conditions it is not surprising that data from the ADAS Representative Soil Sampling Scheme show little change in organic matter levels within each farming system during the 10 years 1974-83.[9] The scheme aims to sample each year a number of fields representative of different farming systems. Nearly 70% of arable soils have between 2 and 5% organic matter and only about 10% have less than 2%. However, it is not known what, if any, limitations there are to yield on such soils and whether any part of such limitation could be attributed to the lack of humus.

Soil humus content can be increased, however, when large amounts of organic material are available. Silty clay loams in the Rothamsted Classical experiments which have received 35 t ha^{-1} FYM each year since their start in the 1840s and 1850s now have about 3.5% C compared to 1% C initially[10,11]. However, recent calculations[12] suggest that much organic matter was also lost from these FYM-treated plots by mineralization. Each year throughout the Broadbalk experiment between 120 and 135 kg N ha^{-1} of the 255 kg applied (225 kg in FYM plus 30 kg from other sources) cannot be accounted for either as N removed in the harvest crop or as an increase in soil organic N. Also between 1942-67 annual applications of 37.5 t ha^{-1} FYM to the sandy loam at Woburn increased organic carbon from 0.87 to 1.64% C; doubling the FYM application gave an increase to 2.26% C. However, once FYM was no longer applied soil humus content began to decline rapidly. In the first five years, 1968-72, soils which had received the single and double rate of FYM lost 0.17 and 0.41% C respectively. The average annual rate of decline was faster than the average annual rate of increase in the earlier years.[13]

4 YIELDS OF CROPS

One of the primary aims of the field experiments started by Lawes and Gilbert at Rothamsted in the 1840s and 1850s was to determine whether the major plant nutrients in FYM were equally important for crop growth and whether all crops responded similarly. Initial results quickly showed that small quantities of N P and K applied in the correct proportions as inorganic salts gave yields equal to those given by 35 t ha^{-1} FYM (Table 1). As these experiments continued with treatments applied cumulatively to each plot so humus content changed. It increased appreciably

Table 1 Yields, t ha^{-1}, of winter wheat and spring barley, grain at 85% dry matter, and roots of mangolds and sugar beet at Rothamsted

			Yield with	
Experiment	Crop	Period	FYM*	NPK fertilizers*
Broadbalk	Winter wheat	1852-56	2.41	2.52
		1902-11	2.62	2.76
		1955-64	2.97	2.85
		1970-75	5.80	5.47
Hoosfield	Spring barley	1856-61	2.85	2.91
		1902-11	2.96	2.52
		1952-61	3.51	2.50
		1964-67	4.60	3.36
		1964-67	5.00	5.00†
Barnfield	Mangolds	1876-94	42.2	46.0
		1941-59	22.3	36.2
	Sugar beet	1946-59	15.6	20.1

* FYM 35 t ha^{-1} containing on average 225 kg N; N fertilizer per ha: wheat, 144 kg; barley, 48 kg, except † 96 kg; root crops, 96 kg in FYM-treated soils. With fertilizer applications there was a small increase initially followed by a long period when % C remained constant.[10,11] In the 1960s there was about 2.5 times more humus in the FYM-compared to the fertilizer-treated soils but yields were still much the same with both treatments (Table 1). Yields of spring barley after the early 1900s were less with fertilizers than with FYM because the N applied, 48 kg ha^{-1}, as fertilizer was too little. When 96 kg N ha^{-1} was given in 1964-67 yields increased to those given by FYM; see also Fig 3 for yields in 1970-79. These results clearly indicated little benefit from extra organic matter in soil, a conclusion supported by results in the 1950s and early 1960s from the Ley-arable experiments at Rothamsted and Woburn. The ADAS Ley-fertility experiments on a wider range of soil types did show some small benefits to yields of wheat and barley on silty loams and loamy sands, but not on a silty clay loam.[14] Part of this benefit was ascribed to the greater drought tolerance of crops grown following leys.

Table 2 Optimum yields, t ha^{-1}, of potatoes, winter wheat and spring barley and N, kg ha^{-1}, needed to achieve it (in parenthesis) Rothamsted Ley-arable experiment

	Old grassland site*		Old arable site*	
	after clover ley	all-arable rotation	after clover ley	all-arable rotation
Potatoes 1968-70	52.0 (150)	51.7 (225)	44.2 (150)	43.9 (225)
Wheat 1969-71	7.48 (100)	7.63 (150)	7.40 (100)	7.33 (150)
Barley 1970-72	6.51 (50)	6.30 (126)	6.33 (88)	6.21 (126)

* In 1970 the all-arable rotation soils contained 2.1 and 1.6% C on the old grassland and old arable sites respectively

In the mid 1960s and early 1970s results from the Ley-arable experiments at Rothamsted and Woburn began to show some benefits for extra organic matter for some crops. Tables 2 and 3 show optimum yields of a range of arable crops and the N application needed to achieve them (in kg ha^{-1} in parenthesis) in the all-arable rotation or following a 3-year ley. In all cases less N was needed to achieve optimum yield after the ley. At Rothamsted the difference in soil humus in all-arable rotation soils between the two sites in 1970, 2.1 and 1.6% C, respectively, probably explains the difference in the optimum yields of potatoes, 8 t ha^{-1}, although it had little effect on best yields of either winter wheat or spring barley.

Table 3 Optimum yields, t ha^{-1}, of spring barley grain and sugar from sugar beet and N, kg ha^{-1}, needed to achieve it (in parenthesis) Woburn Ley-arable experiment

	After a ley*	All-arable rotation
Barley 1968-70	3.80 (50)	4.12 (100)
Sugar 1965-67	8.95 (176)	8.33 (220)

* Clover ley preceded barley, lucerne preceded sugar beet

The Classical experiment on Barnfield started in 1843 and P K Mg and FYM treatments have remained unchanged. In 1968 four amounts of N were tested on potatoes, sugar beet, spring barley and spring wheat each grown three times in rotation between 1968 and 1973. Irrespective of the amount of N applied yields of both root crops and spring barley, but not spring wheat, were larger on soils with extra organic matter from applications of FYM since 1843 (Table 4). The best yield of barley was achieved with a much smaller amount of N on the FYM-treated soil than on fertilizer-treated soils.

Table 4 Yields of potatoes, sugar beet, spring barley and spring wheat in 1968-73 on soils treated with fertilizers or FYM since 1843

Crop	Manuring	Fertilizer N applied			
		0	1	2	3
Potatoes	FYM	24.2	38.4	44.0	44.0
tubers	PK	11.6	21.5	29.9	36.2
Sugar beet	FYM	27.4	43.5	48.6	49.6
roots	PK	15.8	27.0	39.0	45.6
Barley	FYM	4.18	5.40	5.16	5.08
grain	PK	1.85	3.74	4.83	4.92
Wheat	FYM	2.44	3.73	3.92	3.79
grain	PK	1.46	2.97	3.53	4.12

N rates, kg ha^{-1}, 0, 1, 2, 3 to root crops 0, 72, 144, 216
to cereals 0, 48, 96, 144

These benefits from extra organic matter began to be measured as crops with a larger yield potential were introduced and agrochemicals became available to protect this yield potential through the control of weeds, pests and diseases. This is well illustrated by yields of three cultivars of spring barley grown in three successive periods on the Hoosfield Continuous

Soil Fertility and Soil Organic Matter

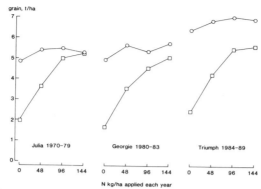

Figure 3 Yields of three cultivars of spring barley grown on soils continuously manured with PK fertilizers (□) or farmyard manure (○) since 1852, Hoosfield Barley Experiment, Rothamsted.

Barley experiment at Rothamsted (Fig 3). In each period four amounts of N were tested on soils with and without FYM since 1852. Although there was a large difference in soil carbon, 3.38 and 0.96% C with and without FYM, respectively, in 1975,[11] amounts of both readily soluble P and K were large and unlikely to restrict yields (137 and 140 mg kg^{-1} bicarbonate-soluble P and 827 and 329 mg kg^{-1} exchangeable K in soils with and without FYM, respectively). Best yields of cv Julia on both soils were not significantly different but cv Georgie and then Triumph yielded more on soils with more organic matter. For Triumph the yield difference was 1.2 t ha^{-1} grain.

New experiments designed and started since the mid-1960s have aimed to compare only the effects of different amounts of organic matter in soil by maintaining the readily soluble P, K and Mg constant across all treatments or by testing one or more of these nutrients in addition to organic matter.[3]

In one such experiment grass was grown on half the plots for 12 years, 1958-69, so that in 1969 the soil had 1.4% C, soils on the other plots had only 0.9% C. For the last five years of the treatment period different levels of bicarbonate-soluble P were established, in 1969 they ranged from 3 to 70 mg P kg^{-1}. Yields of potatoes, sugar beet and spring barley grown during 1970-72 were always larger within each range of soluble-P values on soils with more organic matter, especially at the lower soluble-P levels (Table 5). The benefit of extra matter in this experiment could well have been due to an improvement in soil structure. This is because when the soil from all 48 plots was sampled, air dried, ground to pass a

2 mm sieve and cropped in the glasshouse with ryegrass, the relationship between yield and soluble-P was the same for all soils.

Table 5 Yields, t ha^{-1}, of potatoes, sugar from sugar beet and spring barley on soils with different amounts of readily soluble P and at two levels of soil organic matter

Crop	% C in soil	P soluble in 0.5 M NaHCO$_3$, mg kg^{-1}				
		0-9	9-15	15-25	25-45	45-70
Barley	0.9	-	2.71	3.29	4.21	4.61
grain	1.4	3.18	4.78	5.20	5.00	5.46
Potatoes	0.9	-	31.5	36.7	39.3	44.0
tubers	1.4	26.6	43.0	45.3	46.4	47.8
Sugar	0.9	-	5.03	5.91	6.66	6.80
	1.4	2.45	5.92	7.06	6.73	6.85

The most complex recent experiment, on the sandy loam at Woburn, had an initial treatment phase which lasted six or seven years.[15] Treatments used to increase soil humus were FYM (250 t ha^{-1} total fresh weight), straw (45 t ha^{-1} total dry weight) and grass-clover and all-grass with N leys. For comparison two plots received inorganic fertilizers only. The inputs and offtakes of P, K and Mg were measured on all plots and on one fertilizer plot inputs were adjusted to match those on FYM plots whilst on the other fertilizer plot and the ley plots, the inputs were adjusted to match those on the straw plot. At the end of the treatment phase bicarbonate-soluble P and exchangeable K and Mg were very similar for the appropriate comparisons.[15] There were four test crops and on each eight amounts of N were tested. The test crops and range of N rates, kg ha^{-1}, were: potatoes, 0-350 kg; winter wheat, 0-175 kg; sugar beet, 0-280 kg; spring barley, 0-175 kg. Yields of all four crops, averaged over the four lowest and four highest rates of N, were always larger on soils with more organic matter (Table 6). Although there were large differences in readily soluble P, K and Mg between plots getting fertilizers only this had little effect on yield and the average is given in Table 6. Although very large amounts of N were tested, yields on low organic matter soils could not be increased to those on soils with more humus. This suggests that a factor other than N was involved and on this light textured soil it could have been a larger amount of available water.

Table 6 Yields, t ha^{-1}, of potatoes, winter wheat, sugar from sugar beet and spring barley on a sandy loam with different levels of organic matter

Crop	N rate	Preceding treatments and resulting % C in soil			
		Fertilizers 0.69	Ley 0.92	Straw 0.92	FYM 1.04
Potatoes tubers	Low* High	31.0 43.6	43.8 49.7	35.4 47.4	38.3 50.8
Wheat grain	Low High	3.17 5.23	5.14 5.66	3.60 6.10	3.85 5.73
Sugar	Low High	1.99 2.90	2.83 3.25	2.40 3.08	2.75 3.38
Barley grain	Low High	2.63 3.65	3.56 4.12	3.45 4.14	3.32 3.91

* Mean of four lowest and four highest rates of N tested on each crop (see text for range)

It is of great interest that there were benefits from ploughing in straw for all four crops especially now that there will be a ban on field-burning of straw in England and Wales after 1992. This result appears to conflict with others from straw disposal experiments, which included ploughing in straw, made on MAFF Experimental Husbandry Farms in the 1950s and 1960s.[16] The soil at Woburn with minimum organic matter returned contained much less organic matter, 0.69% C, than any comparably treated soil on the Husbandry Farms, range 1.17 to 2.10% C. This is the probable reason for the benefits measured at Woburn.[17]

5 SEPARATING THE N EFFECT OF ORGANIC MATTER FROM OTHER EFFECTS

Trying to distinguish and separate the various factors through which any effects of organic matter are achieved is difficult. Recently we have attempted to separate the effect of N from other possible, but not defined, factors, based on a scheme described previously,[18] using yields for winter wheat grown on Broadbalk at Rothamsted. Response curves were fitted to the yield data from fertilizer and FYM-treated soils from 1979-84. The form of the curve chosen was exponential plus linear which gave in each case an optimum yield and its associated N application. Curves of constant shape were fitted to the data for each of the six years. Each year the shape was determined only by the response to the five levels of fertilizer N on PK-treated plots. This curve was then shifted until it fitted the yields on the two FYM-treated soils, with and without fertilizer N, the horizontal and vertical shifts being determined separately. For both fertilizer and FYM treatments, i.e. low and high organic matter soils, the maximum yield and associated N dressing could be determined. For the six years the range of maximum grain yields on fertilizer plots was 6.46 to 7.48 t ha^{-1} at N dressings ranging from 252 to 270 kg ha^{-1}. For FYM plots maximum yields ranged from 7.22 to 9.21 t ha^{-1} at N dressings ranging from 155 to 223 kg ha^{-1}. Maximum grain yields overlapped slightly, nitrogen dressings not at all, and in each year FYM-treated plots gave larger maximum yields at lower N levels. The difference between years can be regarded as seasonal effects, differences between treatments within years as FYM effects.

Both sets of six response curves could be brought into coincidence, by suitable vertical and horizontal shifts bringing the maxima of the curves into coincidence. For fertilizer- and FYM-treated plots average maximum grain yields were 6.94 and 8.33 t ha^{-1} respectively at associated N dressings of 262 and 193 kg ha^{-1} respectively. The overall fit accounted for 98.9% of the total variation.

The FYM and fertilizer curves could also be brought into coincidence by horizontal and vertical shifts (Fig 4). The average horizontal shift, -69.2 kg N ha^{-1}, range -39 to -109, represents the N equivalent of the FYM-treated soils. The average vertical shift of 1.39 t ha^{-1}, range 0.1 to 2.3, represents a unique benefit other than that due to N applied as one dressing in spring.

6 DECLINE OF SOIL ORGANIC MATTER DERIVED FROM DIFFERENT SOURCES

Organic materials added to soils are acted on by the soil microbial biomass and soil organic matter consists of a stabilized residue of the added material and the products of microbial activity. Organic matter in the Market Garden experiment soils at Woburn was increased by applications of sewage sludge and sludge compost during 1942-61 and FYM and FYM compost during 1942-67. All four manures were applied at either 37.5 or 75 t ha^{-1} fresh material and treatments were replicated in two blocks.[13] Once application of the manures ceased, soil organic matter declined quickly and there was an ideal opportunity to determine whether there were differences in the rate of decay. Soil carbon was determined on a 0-23 cm depth sample, representative of the plough layer, in 1960, 1967, 1972 and 1980 and converted to kg C ha^{-1}.[12]

Soil Fertility and Soil Organic Matter 311

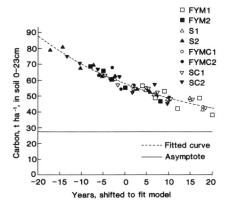

Figure 4 Fitted response curves to yields on Broadbalk, 1979-84, and the horizontal and vertical shifts required to bring the two curves into coincidence. Plots treated with fertilizer (Δ), FYM (O).

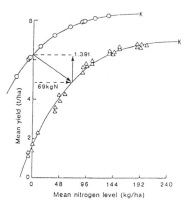

Figure 5 Decrease with time in soil organic carbon derived from different organic materials on a sandy loam, Woburn.

Initially % C was plotted against year (x axis) for each rate of the four manures and visual inspection suggested a family of similar decay curves. An exponential decay model was then fitted to all 16 sets of data, 4 manures x 2 rates x 2 blocks (Fig 5). Different exponential parameters for each treatment did not give statistically better fits. Horizontal shifts (in years) brought the curves into coincidence. These shifts were related to the different starting levels of organic matter in each soil. This apart, organic materials as diverse as FYM and sewage sludge when added to soil and allowed to equilibrate for some years produce organic matter which decays at the same rate, at least over a 20-year period.

There were separate decay curves for each block of the experiment and these, too, could be brought into coincidence but as yet we can offer no explanation for the difference. It was not related to clay content because these average % clay on blocks A and B was 9.4 and 9.4% respectively. From the fitted curve in Fig 5 and that for nitrogen (not shown) the half lives of C and N were calculated to be 20.1 and 12.4 years respectively. These estimates of half lives are strongly correlated with those of the asymptote and of the horizontal shifts and are subject to some error. However, for nitrogen it is possible to show just how large N losses can be. Starting with a total soil N content of 8000 kg ha^{-1} losses of nitrogen averaged 219 kg ha^{-1} each year during the first 5 years, and then

166, 125 and 95 kg N ha^{-1} each year during each succeeding 5-year period.

The successful outcome of the exercise described above prompted us to try the same equation for data from the silty clay loam at Rothamsted where old grassland was ploughed in 1959 and the site continuously fallowed since then. The same equation did not fit; that which fitted best gave a half-life for C and N of 10.1 and 6.9 years respectively. These values were significantly different to the half-lives obtained for the decay of organic matter on the sandy loam at Woburn. The difference is unlikely to be due to soil texture; decay should have been less rapid on the heavier textured soil. Probably cropping at Woburn but fallowing at Rothamsted was the principal reason. Clearly, however, the net rate of loss of organic matter depends on the type of farming system practised.

7 CONCLUSIONS

The results presented here have both an agronomic and an environmental significance. In recent years on both the silty clay loam at Rothamsted and the sandy loam at Woburn yields of arable crops, especially those which are spring sown, have often been larger on soils containing more humus. In part this is related to the introduction of cultivars with a high yield potential and the adoption of husbandry practices which protect that potential. In part it may also be related to the fact that such results tend to be on soils which have been in continuous arable cropping for more than 100 years. Thus the amount of humus may be lower than in soils on many farms. On the latter, soil humus levels probably increased when many fields were in semi-permanent grass during periods of agricultural depression as in the 1870s and 1930s. Agricultural prosperity since the 1950s and the polarization of farming enterprises to all animal husbandry or all-arable cropping means that on farms with a high proportion of arable crops soil organic matter levels are now probably declining, albeit slowly, to values similar to those on our soils for comparable soil types. Thus these results could be an "early warning" of possible future responses on other soils, especially on the lighter textured soils.

Equally important to the effects on yields is the observation that in a number of experiments fertilizer nitrogen was used inefficiently on soils low in humus. Clearly this is important whilst there is concern about losses of nitrate from arable land. The amounts of nitrogenous fertilizer applied should take much more cognizance of the yield potential of the site both to save unnecessary expense and help minimize the nitrate content of water draining from arable land.

However I cannot but be aware of other research at Rothamsted which has implicated the role of soil organic matter and its mineralization in producing large amounts of nitrate in soil in autumn. Subsequently this nitrate is at risk to loss by leaching during the winter period when soils are at field capacity and whenever through drainage occurs. Two examples of the size of these losses have been given here. Obviously the benefit arising from maintaining adequate levels of humus must not be overshadowed by the harm arising from the risk of nitrate leaching. Research must continue into ways of minimising nitrate leaching losses as another facet of research into the role of humus in soil fertility.

REFERENCES

1. MAFF, 'Modern Farming and the Soil'. Report of the Agricultural Advisory Council on Soil Structure and Soil Fertility (The Strutt Report) MAFF, HMSO, London, 1970.
2. E.W. Russell, *Phil. Trans. R. Soc. London. B*. 1977, **281**, 209.
3. A.E. Johnston, *Soil Use and Manag.*, 1986, **2**, 97.
4. P.J. Salter and J.B. Williams, *J. agric. Sci., Camb.*, 1969, **73**, 155.
5. R.J.B. Williams, 'Soil physical conditions and crop production', MAFF Tech. Bull. **29**, HMSO, London, 1975, p324.
6. W.B. Haines and B.A. Keen, *J. agric. Sci., Camb.*, 1925, **15**, 395.
7. A.E. Johnston, *Rothamsted exp. Stn., Rep. for 1972*, part 2, 1973, 131.
8. Russell, E.W. Russell's soil conditions and plant growth 11th ed. Longman Scientific & Technical, London, 1988, p599.
9. B.M. Church and R.J. Skinner, *J. agric. Sci., Camb.*, 1986, **107**, 21.
10. D.S. Jenkinson, *Rothamsted exp. Stn., Rep. for 1976* part 2, 1977, 103.
11. D.S. Jenkinson and A.E. Johnston, *Rothamsted exp. Stn., Rep. for 1976*, part 2, 1977, 87.
12. A.E. Johnston, S.P. McGrath, P.R. Poulton and P.W. Lane, 'Nitrogen in organic wastes applied to soils', Academic Press, London, 1989, p126.
13. A.E. Johnston, *Rothamsted exp. Stn., Rep. for 1974*, part 2, 1975, 102.
14. D.J. Eagle, 'Soil physical conditions and crop production', MAFF Tech. Bull. **29**, HMSO, London, 1975, p52.
15. G.E.G. Mattingly, M. Chater and P.R. Poulton, *Rothamsted exp. Stn., Rep. for 1973*, part 2, 1974, 134.

16. J.L. Short, *Exp. Husb.*, 1973, **28**, 103.
17. A.E. Johnston, *Proc. 4th ADAS Straw Util. Conf. Oxford 1978*, MAFF, HMSO, London, 1979, p42.
18. G.V. Dyke, B.J. George, A.E. Johnston, P.R. Poulton and A.D. Todd, *Rothamsted exp. Stn., Rep. for 1982* part 2, 1983, 5.
19. A.E. Johnston, *INTECOL Bull.*, 1987, **15**, 9.

Mineral Nitrogen Arising from Soil Organic Matter and Organic Manures Related to Winter Wheat Production

L.V. Vaidyanathan[1], M.A. Shepherd[2], and B.J. Chambers[1]

[1] MAFF/ADAS, SOIL SCIENCE DEPARTMENT, GOVERNMENT BUILDINGS, BROOKLANDS AVENUE, CAMBRIDGE CB2 2DR, UK
[2] GLEADTHORPE EHF, MEDEN VALE, MANSFIELD, NOTTS, NG20 9PF, UK

SUMMARY

Recent ADAS field experiments have evaluated the use of soil mineral nitrogen (SMN) measurements as an estimate of available soil nitrogen supply on organic/peaty soils, and following ploughed out grassland or organic manure applications.

The contribution of soil nitrogen supply to winter wheat and sugar beet crops growing on organic/peaty soils of the Black Fens in East Anglia was studied between March and September 1988. Nitrogen mineralisation during spring / summer meant that late autumn / early spring SMN measurements were not a good measure of soil nitrogen supply. Estimates of net mineralisation potential are required to predict soil nitrogen supply.

Field experiments in 1987 (23 sites) and 1988 (14 sites) studied the response of winter wheat to increasing additions of fertiliser nitrogen in potentially large nitrogen residue situations, following the ploughing out of grassland or organic manure applications. No response to fertiliser nitrogen was recorded where SMN levels were above 250-300 kg/ha N, measured in autumn. Below these amounts of SMN, responses to fertiliser nitrogen increased with decreasing SMN supply but were somewhat variable between sites with similar amounts.

Extension of the information from these response experiments has enabled ADAS to offer a SMN advisory service, which has provided farmers with the confidence to make full allowance for soil nitrogen supply in large and variable nitrogen residue situations.

INTRODUCTION

The top 30 cm of land growing winter wheat in the United Kingdom (2 million hectares) is estimated to contain about 12 million tonnes of organic nitrogen, compared with the 370,000 tonnes of fertiliser nitrogen (180 - 190 kg/ha) Survey of Fertiliser Practice 1988/89[1] applied to each season's crop by farmers. The annual wheat harvest of around 12 million tonnes removes up to 300,000 tonnes of nitrogen in grain and straw, about half of which is derived from soil organic matter.

Although only a small fraction (1 part in 80) of soil organic nitrogen reserves is accounted for in the crop, the rate and amount of nitrogen mineralised is important. In fact it remains the big, as yet unresolved, factor limiting our ability to determine the economic and environmentally appropriate fertiliser nitrogen requirement of crops.

Recent ADAS field experiments assessed the contribution of soil mineral nitrogen (SMN) supply to winter wheat crops growing on organic/peaty soils, and following ploughed out grassland or organic manure applications. Measurements of SMN (NH_4-N and NO_3-N) in the soil profile explored by roots when soils had reached field capacity in autumn and again in early spring (0-90 cm) were used to provide a surrogate estimate of available soil nitrogen, and to study nitrogen mineralisation from soil organic matter.

The ADAS Soil Nitrogen Index system[2], based on previous cropping, caters reasonably satisfactorily for crop response where soil nitrogen reserves are low or moderate. Index 0 fields are classed as having low soil nitrogen reserves (previous crops; cereals, sugar beet, maize) and Index 1 fields moderate nitrogen reserves (previous crops; peas/beans, potatoes, oilseed rape). Index 2 fields, having large and variable soil nitrogen reserves (following the ploughing out of intensively managed long leys/permanent pasture or fields receiving large/frequent organic manure dressings) are situations where SMN analysis can be useful.

NITROGEN MINERALISATION ON ORGANIC/PEATY SOILS

The organic/peaty soils of the Black Fens in East Anglia, Lancashire and elsewhere have large organic nitrogen reserves and contribute more nitrogen to crop requirement than mineral soils in long term arable cropping.

Nitrogen mineralised from soil organic reserves was estimated on Black Fen soils (organic matter 35%) between March and September 1988 under winter wheat and sugar beet crops. Under winter wheat given no fertiliser nitrogen, SMN status declined during spring from an initial value of 180 kg/ha N in March to 65 kg/ha N in June, due to crop uptake and increased again after anthesis to a value of 165 kg/ha N in August, similar to the initial spring value. The nitrogen released from soil organic reserves was utilised by the wheat crop between mid March and mid June (Figure 1).

In contrast under sugar beet, an initial SMN value of 180 kg/ha N in March increased to 280 kg/ha N by late May due to mineralisation of organic reserves. Uptake of nitrogen by sugar beet was later than for winter wheat, and led to a decline in SMN during July and August to a value of 70 kg/ha N in September, which was smaller than the initial spring value (Figure 2).

Estimates of net mineralisation were obtained by expressing changes in SMN status (0-90 cm) and crop nitrogen uptake as a function of time. The average net mineralisation rate under winter wheat over the period March to August was 1.3 kg/ha/day and under sugar beet over the period March to September 1.6 kg/ha/day. The apparently greater mineralisation rate under sugar beet is likely in part to be due to a greater rooting depth than for cereals.

Nitrogen mineralisation on these moisture retentive soils during the late spring/summer growing period means that late autumn or early spring SMN measurements are not a good measure of the amount of nitrogen that will be available to crops. Estimates of net mineralisation potential are required to predict nitrogen supply.

NITROGEN REQUIREMENTS IN LARGE RESIDUE SITUATIONS

Fields ploughed out of grassland or receiving organic manures represent two arable cropping situations that can benefit from field specific estimates of available soil nitrogen supply, which is not readily provided from the broad definition of ADAS N Index 2[3].

Substantial areas of land in permanent grassland were ploughed out during and after the 1939-45 war and more recently, following the introduction of dairy quotas. These areas now growing cereals and other arable crops have much larger organic nitrogen reserves than mineral soils in long term arable cropping. In addition, the Survey of Fertiliser Practice 1988/89[1] estimates that around 12% of the winter wheat area (240,000 ha) receives organic manures, which add to the organic nitrogen store. These large and variable organic nitrogen residue situations add further complexity to the uncertainty in determining fertiliser nitrogen requirement.

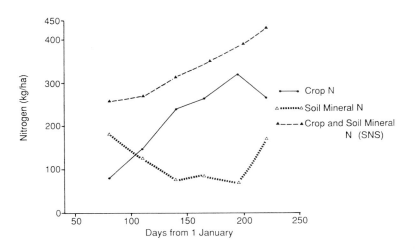

Figure 1 Nitrogen Mineralisation on Black Fen Soils in Relation to Winter Wheat Crop Demand

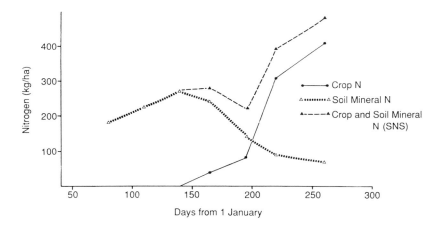

Figure 2 Nitrogen Mineralisation on Black Fen Soils in Relation to Sugar Beet Crop Demand

ADAS field experiments in 1987 (23 sites) and 1988 (14 sites) measured response to increasing additions of fertiliser nitrogen and crop nitrogen uptake, in these large and variable nitrogen residue situations on mineral soils. The results were used to evaluate the contribution of soil nitrogen supply and optimum fertiliser nitrogen requirement for winter wheat.

The sites had a wide range of SMN levels; 33 kg/ha N (a first wheat crop after sugar beet that had received 100 t/ha poultry manure) up to 470 kg/ha N (a field receiving frequent applications of pig slurry) measured in autumn (Table 1). This range of values for sites that would generally have been regarded as ADAS N Index 2 shows the inadequacy of the N Index system on an individual field basis in these situations.

Optimum fertiliser nitrogen requirement and autumn SMN supply were inversely related ($r = 0.571***$, Figure 3), as were yield response to fertiliser nitrogen addition and autumn SMN ($r = 0.511***$, Figure 4).

At SMN levels above 250-300 kg/ha N (measured in autumn) there was no response to fertiliser nitrogen (Figure 4). Below these SMN levels, responses increased with decreasing SMN supply but were somewhat variable between sites with similar levels. Sites where autumn SMN levels were low and optimum fertiliser requirement was also low were generally fields recently ploughed out of grassland (sites in SW England, 1987), or receiving large and frequent organic manure additions, the low fertiliser nitrogen optimum being a result of nitrogen mineralisation from soil organic matter during the season.

Optimum fertiliser nitrogen requirement and yield response to nitrogen addition showed a strong positive relationship ($r = 0.936***$, Figure 5), indicating that optimum fertiliser nitrogen rate is determined by yield response to nitrogen addition, which is dependent upon site and season.

SMN and crop nitrogen uptake were measured in spring (February) on all sites. Averaged over all sites, soil nitrogen supply (SNS, ie SMN and crop nitrogen) had decreased by 20% (177 to 142 kg/ha N) over winter 1986/7, and 21 % (113 to 89 kg/ha N) over winter 1987/8. The largest decreases were recorded for the 3 sites in Wales with excess winter rainfall over 1000 mm, on average 35% (198 to 128 kg/ha N) in 1986/7, and 47% (164 to 87 kg/ha N) in 1987/8.

Autumn SMN supply and spring SNS were strongly inversely related ($r = 0.863***$). Spring SNS also showed an inverse relationship with optimum fertiliser nitrogen requirement ($r = 0.544***$) which was not as strong as the relationship with autumn SMN supply.

Table 1. Winter Wheat Yields and Optimum Fertiliser Nitrogen Requirements in Relation to Measured Soil Mineral Nitrogen Supply on High Residue Sites. (37 trials, 1987 and 1988)

Site Number	Location*	Soil Supply (kg/ha, 0-90 cm)		Winter Wheat Yield (t/ha)		Fertiliser Nitrogen Optimum (kg/ha)	Recommended Fertiliser Nitrogen** (kg/ha)
		Autumn SMN	Spring SNS	Nil Fertiliser Nitrogen	Optimum Fertiliser Nitrogen		
1987:							
1	SW	126	73	6.55	7.59	70	119
2	SW	48	46	3.40	8.83	262	200
3	SW	75	65	3.97	7.05	224	170
4	SW	83	47	5.87	8.50	160	162
5	SW	101	116	8.33	8.63	50	144
6	W	220	137	6.92	7.82	110	40
7	W	177	114	7.29	8.98	134	60
8	W	198	133	7.47	8.44	98	47
9	E	258	262	6.24	6.78	25	40
10	E	33	49	5.22	7.66	160	200
11	E	414	268	7.16	7.16	0	0
12	E	470	327	7.06	7.06	0	0
13	E	82	74	6.05	7.38	138	163
14	E	365	331	6.86	6.86	0	0
15	E	54	67	5.97	7.74	90	191
16	E	136	105	6.03	7.21	75	109
17	E	124	99	6.65	8.91	133	121
18	E	95	107	6.34	7.34	120	150
19	E	196	247	5.47	5.60	25	40
20	E	245	237	5.81	6.13	25	40

Table 1. (continued)

Site		Soil Supply		Winter Wheat Yield (t/ha)		Fertiliser Nitrogen Optimum (kg/ha)	Recommended Fertiliser Nitrogen** (kg/ha)
Number	Location*	Autumn SMN	Spring SNS	Nil Fertiliser Nitrogen	Optimum Fertiliser Nitrogen		
		(kg/ha, 0-90 cm)					
21	MW	258	122	7.47	8.18	50	40
22	MW	179	142	8.33	8.77	25	66
23	MW	138	96	4.10	7.32	157	107
1988:							
24	SW	79	63	3.69	3.83	25	116
25	SW	29	76	5.95	6.25	30	150
26	SW	157	90	8.33	8.96	77	88
27	SW	62	90	6.76	8.25	93	183
28	SW	64	63	4.53	4.58	18	131
29	W	184	89	5.37	6.23	44	40
30	W	104	61	6.17	7.30	115	141
31	W	204	112	4.09	5.06	61	40
32	E	116	63	5.80	9.03	187	180
33	E	118	87	6.73	7.96	50	157
34	E	173	100	7.67	8.66	69	72
35	E	172	125	9.44	9.44	0	123
36	E	50	60	5.09	10.02	218	250
37	E	73	168	6.61	10.14	207	250

* ADAS Regions : SW, South West; W, Wales; E, Eastern; MW, Midlands and Western.
** Fertiliser nitrogen recommended from the ADAS SMN Service based on autumn SMN measurement.

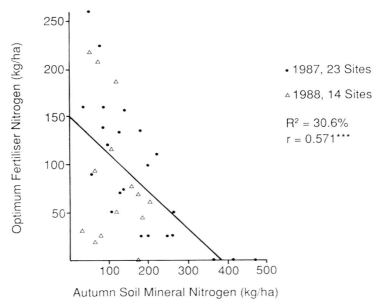

Figure 3　Optimum Fertiliser Nitrogen vs Autumn Soil Mineral Nitrogen (0-90 cm)

- 1987, 23 Sites
- 1988, 14 Sites

$R^2 = 30.6\%$
$r = 0.571^{***}$

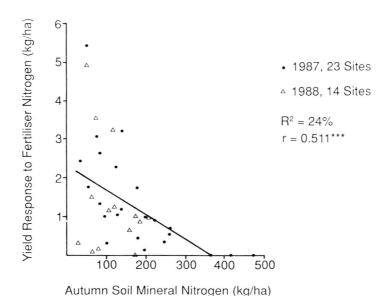

Figure 4　Yield Response to Fertiliser Nitrogen vs Autumn Soil Mineral Nitrogen (0-90 cm)

- 1987, 23 Sites
- 1988, 14 Sites

$R^2 = 24\%$
$r = 0.511^{***}$

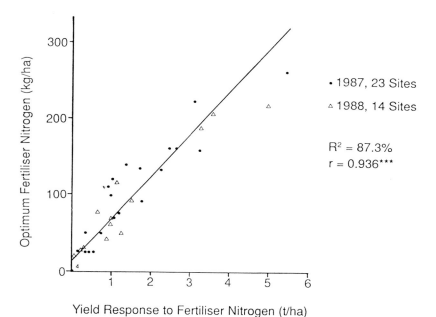

Figure 5 Optimum Fertiliser Nitrogen vs Yield Response to Fertiliser Nitrogen

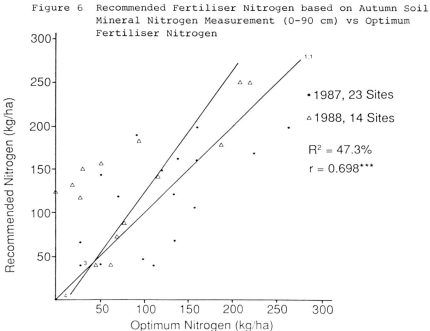

Figure 6 Recommended Fertiliser Nitrogen based on Autumn Soil Mineral Nitrogen Measurement (0-90 cm) vs Optimum Fertiliser Nitrogen

Using the response data from these experiments in relation to available soil nitrogen supply, ADAS has developed a field specific recommendation system based on autumn SMN measurement that is superior to the current ADAS N Index system in large and variable nitrogen residue situations for predicting optimum fertiliser requirement (r = 0.698***, Figure 6). Comparing the two fertiliser nitrogen recommendation systems using unrealised profit calculations as a measure of the system's success showed a lower unrealised profit and hence better recommendation based on autumn SMN measurement (£18.40/ha) than the N Index system (£27.30/ha) for years 1987/88.

Where the SMN system recommended less nitrogen than the experimentally determined optimum, this was on sites where over winter leaching losses were large (Wales and S.W. England). For this reason, spring sampling is advised in high rainfall areas. SMN system recommendations exceeded the experimentally determined optimum in some cases, probably reflecting the lack of allowance for organic nitrogen released in the later part of the growing season. This remains a source of inaccuracy that needs urgent improvement, despite the unpredictable effects of unforseeable weather factors on soil biological events.

ADAS SOIL MINERAL NITROGEN SERVICE

The ADAS SMN advisory service is targeted at fields with potentially large nitrogen residues, and aims to provide field specific nitrogen recommendations. The service has expanded greatly over the 4 years following its launch in 1986/87 (Table 2), and was used on over 700 fields in 1989/90 identifying SMN levels in the range 12-1750 kg/ha N.

Table 2. Summary of ADAS Soil Mineral Nitrogen Service Recommendations and Fertiliser Nitrogen Savings 1986/87 to 1989/90.

	1986/87	1987/88	1988/89	1989/90
Samples taken	103	119	250	720
Average SMN supply (kg/ha)	146	118	183	217
Average N recommended (kg/ha)	–	125	94	89
Net saving per field (£)	–	150	160	170
Average N saved (kg/ha)	40	38	44	46

The larger SMN supplies measured during the last two years are undoubtedly a reflection of better identification of high residue fields by farmers and ADAS advisers.

Figure 7 Distribution of Soil Mineral Nitrogen Reserves from 286 fields receiving Organic Manures (ADAS SMN Service 1989/90)

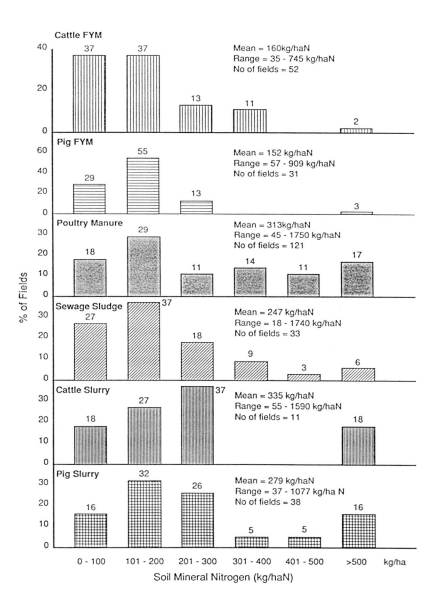

Figure 8 Distribution of Soil Mineral Nitrogen Reserves from 92 fields Ploughed Out of Grassland (ADAS SMN Service 1989/90)

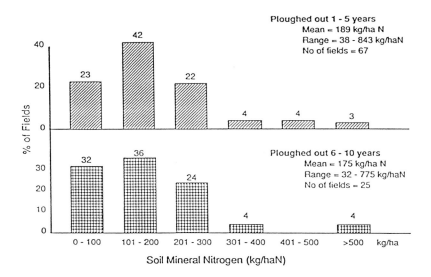

The distribution of results from the 286 fields (1989/90) receiving different types of organic manures is shown in Figure 7. Large SMN reserves (>300 kg/ha N) were identified in 75 fields (26% of total), of which 50 fields had received poultry manure and 9 fields pig slurry applications. No nitrogen fertiliser was recommended for 83 fields (29% of total) receiving organic manures, and only 40 kg/ha N for 40 fields (14% of total).

The distribution of results from 92 fields (1989/90) ploughed out of grassland 1 to 5, and 6 to 10 years ago is shown in Figure 8. Large SMN reserves (>300 kg/ha N) were identified in 10 fields (11% of total), of which 8 were ploughed out between 1 and 5 years ago. No nitrogen fertiliser was recommended for 10 fields (11% of total) and only 40 kg/ha N for 25 fields (27% of total). Stopes 1989[4] identified SMN reserves up to 200kg/ha N following the ploughing out of a grass ley.

The ADAS Soil Mineral Nitrogen service has provided farmers with the confidence to make full allowance for soil nitrogen supply in large and variable residue situations.

ACKNOWLEDGEMENTS: We are grateful for support from colleagues and cooperation by farmers, which was an essential part of the experimental programme. Thanks are also due to A. E. Johnston, Rothamsted Experimental Station for helpful comments on the text.

REFERENCES :

1. Survey of Fertiliser Practice. (1988/89) Fertiliser use on farm crops in England and Wales. Agricultural Development and Advisory Service; ICI fertilisers; Hydro fertilisers; Kemira Ltd;

2. MAFF (1988). Fertiliser recommendations for agricultural and horticultural crops. MAFF/ADAS Reference Book 209. HMSO London.

3. Sylvester-Bradley, R., Addiscott, T.M. Vaidyanathan, L.V. Murray, A.W.A. and Whitmore, A.P. (1987). Nitrogen advice for cereals; present realities and future possibilities. Proceedings of the Fertiliser Society, 263, 1-36.

4. Stopes (1989). The supply and utilisation of nitrogen in organic farming systems. J. Sci. Food Agric, 48, 120-121.

The Effect of Long-term Applications of Inorganic Nitrogen Fertilizer on Soil Organic Nitrogen

M.J. Glendining and D.S. Powlson

SOIL SCIENCE DEPARTMENT, AFRC INSTITUTE OF ARABLE CROPS RESEARCH, ROTHAMSTED EXPERIMENTAL STATION, HARPENDEN, HERTS. AL5 2JQ, UK

1 INTRODUCTION

Many arable soils have now received applications of inorganic nitrogen (N) fertilizers for several decades. It is well known that long-term applications of organic N fertilizers can lead to large increases in soil organic N content, and N mineralization[1,2]. However, the effects of long-term applications of inorganic N fertilizers on soil organic N are generally not considered, although they may also be important.

The amount of organic N in a soil is determined by inputs of organic and inorganic N and outputs of mineralized N, removed from the soil by plant uptake, leaching and gaseous losses. Applications of inorganic N may increase soil organic N content in two ways: directly by the immobilization of fertilizer N, and indirectly by increasing inputs of organic N in the form of crop residues (roots, root exudates, stubble, straw, etc.). Nitrogen fertilizer may increase both the amount of crop residues and their N concentration. The N content of the residue will also influence its rate of mineralization, and thus the output of mineralized N[3].

These additional inputs are small but cumulative and may take many years to show up against the huge background of organic N already in the soil. However, in the long term the continuous use of inorganic fertilizer N may increase both the total soil N content, and the supply of mineralized N. The latter may benefit subsequent crops but may also increase the risk of nitrate leaching and other losses of N.

Results are presented from a number of experiments in Europe and North America given inorganic N for many years, which provide an opportunity to investigate these effects.

2 THE EXPERIMENTS

Broadbalk Continuous Wheat Experiment

This is located at Rothamsted Experimental Station in South East England, and was started in 1843 by Lawes and Gilbert. Winter wheat is grown every year, except for occasional fallow years to control weeds. The straw is removed, so the only crop residues are roots, root exudates, stubble and part of the chaff. The experiment[5,6] includes plots given the same rate of inorganic fertilizer N each year since 1852 - 0, 48, 96 and 144 kg N ha^{-1} yr^{-1}, referred to as the N_0, N_1, N_2 and N_3 treatments respectively. Another plot has received a larger rate of N (192 kg N ha^{-1} yr^{-1}, the N_4 treatment) since 1968; previously it received the N_1 treatment. All plots receive P, K, Mg and lime as required. The soil is a flinty silty-clay loam overlying clay-with-flints (Batcombe series), with 28% clay in the topsoil (0-23 cm). The plot size is 28 x 5.3 m, and there are no replicate plots.

Hoosfield Continuous Barley Experiment

This is also at Rothamsted, on the same soil series as Broadbalk. Spring barley has been grown since 1852, except for occasional fallow years, and the straw is removed. The experiment includes[6,7] two long-term inorganic fertilizer treatments - the PK treatment (no inorganic N, 1852-1968) and the NPK treatment (48 kg N ha^{-1} yr^{-1}, 1852-1968), both receiving adequate P, K, Na, Mg and lime. Since 1968 both treatments have been divided into four sub-plots receiving 0, 48, 96 or 144 kg N ha^{-1} yr^{-1} as inorganic N. The main treatment plot size is 42 x 17.6 m and like Broadbalk there are no replicate plots.

Dehérain Experiment

This experiment began in 1875, at Grignon, Northern France[8]. It includes a range of inorganic fertilizer treatments repeated each year, but started at different times. The two treatments considered here are the NPK (87 kg N ha^{-1} yr^{-1}) treatment, started in 1902, and the PK treatment, which began in 1929. The cropping was a winter wheat/sugar beet rotation between 1928-1987; the straw and tops were removed. Previously a range of crops were grown. The soil is a loam with 22-30% clay. There are eight replicate plots, each 10 x 40 m.

Swedish 'M' Series of Soil Fertility Experiments

This consists of six different experimental sites in Skåne, Southern Sweden[9], representing different climatic and soil conditions (mainly till soils, with 8-30% clay in the topsoil). There is a four year crop rotation of barley/oilseed/winter wheat/sugar beet; the straw and tops are incorporated. The present inorganic fertilizer treatments started in 1962. They include PK and NPK (150 kg N ha^{-1} yr^{-1}) treatments, both receiving all other nutrients and lime as required. There are two replicate plots, 20 x 6.3 m at each site.

Lethbridge Rotation Experiment

This is at the Agriculture Canada Research Station, Lethbridge, Alberta, Canada[10,11]. The experiment began in 1912 and includes two cropping treatments - continuous spring wheat, and a three year wheat/wheat/fallow rotation. Straw has been incorporated since 1943. Two inorganic N fertilizer treatments were started in 1967, 0 and 45 kg N ha^{-1} yr^{-1}, with and without 20 kg P ha^{-1} yr^{-1} (since 1972). The results given are the mean of the two P treatments. Nitrogen is applied every year, including fallow years. Each plot is 0.1 ha, with no replicates. The soil is a calcareous sandy loam (pH = 7.7).

3 RESULTS

Broadbalk Continuous Wheat Experiment

Table 1. Changes in total N content (%N) of topsoil (0-23 cm) in the Broadbalk Continuous Wheat Experiment between 1865 and 1987

Year sampled	Plot number, fertilizer treatment and date started				
	5 N$_0$PK 1843	6 N$_1$PK 1852	7 N$_2$PK 1852	8 N$_3$PK 1852	9 N$_4$PK 1968
1865	0.111	NA	0.122	NA	NA
1881	0.102	0.115	0.126	0.131	NA
1893	0.101	0.111	0.122	0.119	NA
1914	0.103	0.112	0.120	0.129	NA
1936[a]	0.105	NA	0.120	NA	NA
1944[a]	0.105	0.112	0.121	0.123	0.114
1966[a]	0.107	0.113	0.115	0.118	0.111
1966[b]	0.111	0.121	0.124	0.123	0.116
1980/81[c]	0.109	0.117	0.127	0.131	0.124
1987[b]	0.104	0.113	0.124	0.126	0.123

NA = not available. %N in air-dry soil. Fertilizer treatments (ha^{-1} yr^{-1}): 'PK' = 35 kg P, 90 kg K, 15 kg Na, 10 kg Mg; N$_0$ = 0, N$_1$ = 48, N$_2$ = 96, N$_3$ = 144, N$_4$ = 192 kg N respectively. Before 1968 plot 9 received N$_1$PK. [a] = Mean of all 5 sections; [b] = Mean of continuous wheat sections (sections 1 and 9) only; [c] = section 1 only, mean of 1980 and 1981. Sources: [4,12,13] and Poulton (unpublished).

Total soil N content. All plots receiving inorganic N now contain more total N in the soil than the treatment never given fertilizer N (Table 1). However, the maximum increase is only 20%, after applying 144 kg N ha^{-1} yr^{-1} for 135 years. This is equivalent to an additional 0.5 t N ha^{-1} in the plough layer (0-23 cm), less than 3% of the total amount of fertilizer N applied. This difference appeared to be established by 1881, when the N_3 treatment contained an extra 0.029% N. It is not possible to calculate standard errors, as there are no replicate plots in the experiment. However, one of the major features of Table 1 is the consistency of the differences between the treatments over the last 100 years. For example, with the exception of 1966, the difference between the N_0 and the N_3 treatment ranged from 0.018 to 0.029% N, with no consistent change over time. The 1980/81 measurements are from microplots given ^{15}N-labelled fertilizer that were established within Section 1 (continuous wheat). The values given in Table 1 are the mean of 3 replicate microplots, for which the standard error of the difference, for all rates of N, was 0.0040% N. The plot now receiving the N_4 treatment (plot 9) contains less total N than the N_3 treatment; however it has only received this high rate since 1968: previously it received the N_1 treatment. Plot 9 now contains 9% more total N than plot 6, which has continued to receive the N_1 treatment. Both plots had similar total N contents when measured in 1944 and 1966. These results suggest that measureable changes in soil total N content can occur within 15-20 years, at least when large amounts of inorganic N are applied.

N mineralization. Changes in the amount of readily mineralizable N have been measured directly, as the amount of inorganic N released during aerobic incubation at 25°C for 20 days[14]. More N was mineralized from treatments with a history of fertilizer N additions than from the N_0 treatment (Table 2(a)). Soil given the N_3 treatment since 1852 mineralized almost 60% more N during incubation than the unfertilized plot. However, in an earlier incubation experiment[15] the increase in mineralization was much less. In both experiments the soil samples had been stored frozen before incubation and it is not known whether the flush of mineralization resulting from freezing and thawing would have affected all soils equally, or amplified any differences.

Nitrogen mineralization has also been measured indirectly, as crop uptake of N from the soil. This N will be from (a) non-fertilizer inputs (rain, seed, dry deposition, etc.), assumed to be the same for all treatments; (b) inorganic residues of fertilizer N, thought to be negligible at this site and these application rates[16], and (c) mineralization of organic N. Differences in the uptake of soil N are assumed to reflect differences in N mineralization. ^{15}N-labelled fertilizer was applied at the usual rates to distinguish

Table 2. The effects of long-term applications of inorganic N fertilizer on mineralizable N in soil, the Broadbalk Continuous Wheat Experiment, Section 1.

N fertilizer kg N ha^{-1} yr^{-1}	Applied since	a) N Mineralized during incubation kg ha^{-1} d^{-1}	b) Crop uptake of unlabelled N kg ha^{-1}
0	1843	0.69	30.5
48	1852	0.74	45.0
96	1852	0.94	66.8
144	1852	1.10	73.8
192	1968	1.20	73.7
SE		0.141	3.76

SE = standard error of difference. See text for details.

between fertilizer and soil N[4]. Crop uptake of unlabelled N increased from about 30 kg N ha^{-1} from soil never given inorganic N, to more than 70 kg N ha^{-1} from the N_3 treatment (Table 2b). Even if the N in the weeds (present only in the N_0 and N_1 treatments) is included, these results suggest that a minimum of 30 kg ha^{-1} of additional N was mineralized and available to the crop in the N_3 treatment. This difference may have been even greater, as some of the mineralized N may have been lost. Although there can be artefacts due to pool substitution effects when using ^{15}N-labelled fertilizer[17], there was no evidence of this in this particular experiment[4].

After only 15 years at the higher N rate, soil receiving the N_4 treatment mineralized at least as much N as that under the N_3 treatment, although it contained less total N.

Hoosfield Continuous Barley Experiment

Total soil N. In contrast to the Broadbalk experiment, there has been no increase in total soil N content on the Hoosfield experiment as a result of applying 48 kg N ha^{-1} yr^{-1} since 1852 (Figure 1a). The reasons for this difference between the experiments are not clear, as both are on the same soil series, and crop yields (and presumably inputs of organic N) were much greater from the NPK treatments than from the PK treatments[7,18]. A major limitation to both the Broadbalk and Hoosfield experiments is the lack of replication - there may be long-lasting soil differences between the plots which existed before the fertilizer treatments began.

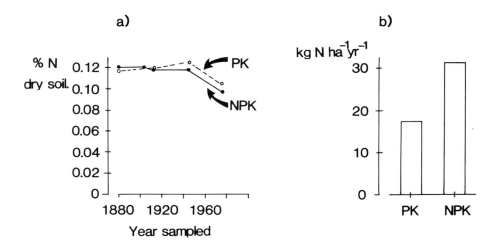

Figure 1. The effects of very long-term applications of inorganic N fertilizer to the Hoosfield Continuous Spring Barley Experiment, Rothamsted. a) Total soil N content, 1852-1975; b) Mean crop uptake of N, 1970-72. No N applied 1967-73. From [19]. N = 48 kg N ha^{-1} yr^{-1}.

N mineralization. Despite the absence of a difference in total soil N content, there appears to have been a large increase in the supply of mineralized N (Figure 1b). Part of the NPK treatment received no inorganic N between 1967 and 1973. During this period the mean yield was 2.83 t ha^{-1} (grain plus straw), 60% more than from the PK plot, which had never received inorganic N. There was no obvious decline in the difference with time. The amount of N taken up by the crop was also much greater: 31.3 and 17.7 kg N ha^{-1} yr^{-1} respectively, mean of 1970-72[19]. Presumably this additional N came from greater mineralization, as there would be no inorganic residues of fertilizer N remaining four years after the last application of inorganic N in 1966.

Dehérain Experiment

Plots receiving the NPK treatment for 30 to 60 years contained 10-15% more total soil N than those given only PK fertilizer for the same number of years (but sampled at different dates, Figure 2). All treatments are still showing a decline in total soil N content, unlike the two long-term experiments at Rothamsted. This is possibly because the initial soil N content at the start of the experiment was much higher - 0.204% N in 1875 in the Dehérain Experiment[8], compared to an estimated 0.115% at the start of the two Rothamsted experiments, around 1850[12,19].

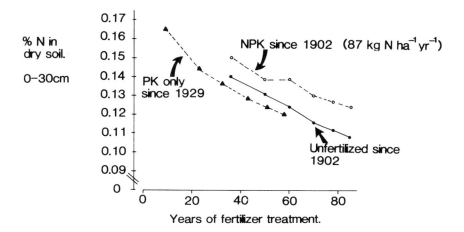

Figure 2. Changes in total soil N content, the Dehérain Experiment, Grignon, France. From [8] and R. Chaussod, pers. comm.

Nitrogen mineralization has also been measured[20], however, it is not possible to draw any conclusions as the PK fertilizer treatment was started 27 years after the NPK treatment.

Swedish 'M' Series of Soil Fertility Experiments

The application of 150 kg ha^{-1} inorganic N significantly increased total soil N content (P = 0.05) after only 22 years[21], with a mean increase of 12% or 0.4 t N ha^{-1} in the topsoil (0-20 cm). Changes in N mineralization were even greater: 21% more in the N fertilized soil than in those that had not received fertilizer N (P = 0.05)[21], a mean of 68.4 and 56.6 µg N g^{-1} soil, respectively. Soils from all the sites were incubated under aerobic conditions at 25°C for 17 weeks. Mineralization in the first two weeks was discounted to exclude the effects of drying and rewetting the soil.

Annual additions of organic N are greater than in the three previous experiments, as the straw and sugar beet tops are incorporated into the soil each year. This may well account for the more rapid establishment of measureable differences in soil organic N content.

Lethbridge Rotation Experiment

The soil was sampled after 18 years of the N fertilizer treatments[11]. N fertilizer significantly (P = 0.001) increased soil organic N content in both rotations, by 15 and 11% in the continuous wheat and wheat/wheat/fallow rotations respectively (Figure 3). N mineralization showed an even greater response, with increases of 32% in the continuous wheat rotation and 23% in the wheat/wheat/fallow rotation. The soil was aerobically incubated for 18 weeks at 35°C, with periodic leaching with 0.001 M CaCl$_2$ under suction[11]. Carbon mineralization followed a similar trend.

Organic N contents and levels of mineralizable N were significantly higher under continuous cropping than under the wheat/wheat/fallow rotation, presumably because of greater inputs of organic N. However, there was no interaction between N treatment and cropping system.

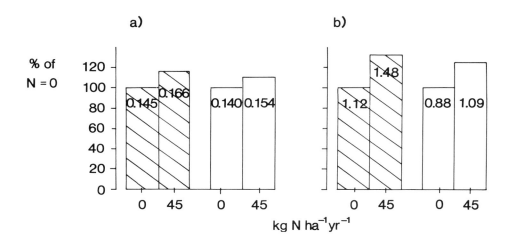

Figure 3. 18 years of N fertilizer applications to the Lethbridge Rotation Experiment, Alberta, Canada. a) Total soil N content, %N, 0-10 cm; b) N mineralization rate, mg N kg^{-1} soil day^{-1}. From [11]. Hatched columns = continuous spring wheat, open columns = wheat/wheat/fallow rotation. Expressed as a percentage of the treatment receiving no inorganic N. Figures in columns are actual values.

4 DISCUSSION

All experiments, except Hoosfield, showed a small increase in <u>total</u> soil N content, as a result of long-term (18-135 years) applications of inorganic nitrogen fertilizer. The maximum increase was shown by the treatment given the greatest amount of N - the N$_3$ plot of the Broadbalk experiment, given 144 kg N ha^{-1} yr^{-1} for 135 years. This now contains 20% more total organic N than the plot never given fertilizer N, a difference of 0.5 t ha^{-1} (0-23 cm), though less than 3% of the fertilizer N applied. Changes were apparent after only 15-20 years when straw and sugar beet tops were incorporated into the soil, as in the experiments at Lethbridge and in Sweden, or when large amounts of fertilizer N were applied as in the N_4 treatment on Broadbalk.

All experiments showed increases in the amount of readily mineralizable N, after receiving inorganic N for many years. These increases were always greater than the changes in total N content, and were often considerable. This suggests that the small amount of additional organic N returned to the soil as a result of using inorganic N fertilizer is in a fraction that turns over more rapidly than the N in older organic matter. Shen et al. (1989)[14] show that recently added residues from wheat in the Broadbalk experiment are about seven times more mineralizable than the native (old) organic matter.

However, the results from Broadbalk also illustrate two problems with this type of study. First, the lack of replicate plots in many of the older long-term experiments. Second, the difficulty of obtaining absolute values for differences in mineralization: a difference obtained in a laboratory incubation on a soil sample taken at a particular time may not reflect differences occurring over the whole year. In situ field incubation of soil cores may be useful in this respect[22].

These results have important agricultural and environmental implications. Many agricultural soils have now received inorganic N fertilizer for several decades. This is likely to have increased the amounts of total and mineralizable N in the soil, particularly when large amounts of crop residue are returned to the soil. This will obviously benefit subsequent crops, and should be considered when predicting the fertilizer N that crops require. However, a greater supply of mineralizable N will also have the disadvantage of an increased risk of nitrate leaching. Much of the nitrate leached out of arable soils in the autumn is from the mineralization of soil organic matter. Any process which leads to an increase in the supply of mineralizable N will inevitably increase the risk of nitrate leaching. Although in the short term accurate and well timed applications of fertilizer N may make little direct contribution to nitrate leaching[16], in the long term the continuous use of inorganic N may indirectly increase the risk.

Acknowledgements
This work was partially funded by a grant from the Commission of the European Communities. The authors thank Mr P.R. Poulton and Dr D.S. Jenkinson for helpful discussion.

REFERENCES
1. C.M. Woodruff, Soil Sci. Soc. America Proc., 1949 14, 208.
2. F.J. Stevenson, 'Soil Nitrogen' eds. W.V. Bartholomew and F.E. Clarke, American Society of Agronomy, Madison, USA, 1965, Monograph 10, Chapter 1, p1.
3. D.S. Jenkinson, 'The Nitrogen Requirement of Cereals', MAFF/ADAS ref. book 385, HMSO, London, 1984, p79.

4. D.S. Powlson; G. Pruden; A.E. Johnston and D.S. Jenkinson, J. Agric. Sci., Camb., 1986, 107, 591.
5. G.V. Dyke; B.J. George; A.E. Johnston; P.R. Poulton and A.D. Todd, Rothamsted Report for 1982 Part 2, 1983, p5.
6. Anon, 'Details of the Classical and Long-Term Experiments 1968-73', Rothamsted Experimental Station, Harpenden, 1978.
7. R.G. Warren and A.E. Johnston, Rothamsted Report for 1966, 1967, p320.
8. R. Morel; T. Lasnier and S. Bourgeois, 'Les essais de fertilisation de longue durée de la station agronomique de Grignon', INRA, Paris, 1984.
9. K. Ivarsson and S. Bjarnason, Acta. Agric. Scand., 1988, 38, 137.
10. S. Freyman; C.J. Palmer; E.H. Hobbs; J.F. Doormaar; G.B. Schaalje and J.R. Moyer, Can. J. Plant Sci., 1982, 62, 609.
11. H.H. Janzen, Can. J. Soil Sci., 1987, 67, 165.
12. A.E. Johnston, Rothamsted Report for 1968 Part 2, 1969, 93.
13. D.S. Jenkinson, Rothamsted Report for 1976 Part 2, 1977, 103.
14. S.M. Shen; P.B.S. Hart; D.S. Powlson and D.S. Jenkinson, Soil Biol. Biochem., 1989, 21, 529.
15. J.K.R. Gasser, Plant and Soil, 1962, 17, 209.
16. A.J. Macdonald; D.S. Powlson; P.R. Poulton and D.S. Jenkinson, J. Sci. Food Agric., 1989, 46, 407.
17. D.S. Jenkinson; R.H. Fox and J.H. Rayner, J. Soil Sci., 1985, 36, 425.
18. H.V. Garner and G.V. Dyke, Rothamsted Report for 1968 Part 2, 1969, p26.
19. D.S. Jenkinson and A.E. Johnston, Rothamsted Report for 1976 Part 2, 1977, p87.
20. S. Houot; J.A.E. Molina; R. Chaussod and C.E. Clapp, Soil Sci. Soc. Am. J., 1989, 53, 451.
21. S. Bjarnason, Acta Agric. Scand., 1989, 39, 361.
22. D.J. Hatch, S.C. Jarvis and L. Philipps, Plant and Soil, 1990, 124, 97.

Straw Incorporation into Soils Compared with Burning during Successive Seasons - Impact on Crop Husbandry and Soil Nitrogen Supply

J.S. Rule[1], D.B. Turley[2], and L.V. Vaidyanathan[2]

[1] BOXWORTH EHF, MAFF/ADAS, BOXWORTH, CAMBS., UK
[2] MAFF/ADAS, SOIL SCIENCE DEPARTMENT, GOVERNMENT BUILDINGS, BROOKLANDS AVENUE, CAMBRIDGE CB2 2DR, UK

ABSTRACT

Concern about decline in soil fertility during the early fifties due to loss of organic matter was attributable to ploughing up grassland for arable cropping, but the extensive burning of surplus straw was also implicated. Experiments at MAFF EHFs on the effects of straw on soil fertility showed, after 18 successive years, that soil organic matter levels were hardly different irrespective of regular ploughing in, baling and removing or straw burning. Yields were also unaffected when adequate spring fertiliser nitrogen was added.

A new series of straw disposal experiments was started by ADAS and AFRC in 1982-83 because of the atmospheric pollution from smoke and smuts and the impending ban on straw burning. These evaluate the economics and practicability of incorporating straw using the conventional mouldboard plough, discs and tines, and a range of custom designed implements. This paper presents results from these recent studies describing effects on crop establishment and yield and influence of retained straw on soil organic matter and nitrogen supply.

Effects on yield were usually small (1-3%), except where inappropriate cultivations were used or where grass weed infestation was serious with reduced cultivation and continuous winter cereal cropping.

Straw incorporation did not exacerbate pests or diseases. The major problem identified is the grass weed risk, especially on heavy clay soils when shallow, non-plough cultivation systems are practised. Ploughing when soil is dry causes serious difficulties in achieving a seedbed for autumn sowing and may lead to a substantial yield penalty.

Non-cereal break crops established in the presence of shallowly incorporated straw residues help reduce the need for chemical weed control. Moisture conservation and reduced tillage techniques are successful for the early sowing required for winter oilseed rape.

Soil organic matter and mineral nitrogen measurements show that incorporated straw immobilises soluble nitrogen only occasionally in amounts comparable to those expected from C:N ratios; nor has retained straw systematically influenced mineral nitrogen levels per se. No fertiliser nitrogen is required in autumn seedbed to counteract the alleged immobilisation of nitrogen during the early stages of straw decomposition.

At Boxworth EHF, following three successive years, optimum yields and fertiliser nitrogen requirements were similar when straw was burnt or shallowly incorporated with tines, provided chemical weed control was successful. In contrast, ploughing in straw resulted in larger optimum yields with notable reduction in fertiliser nitrogen requirement. However, where uncontrolled grass weeds were competing, especially after the first year of non-plough incorporation, optimum yield was greatly reduced and fertiliser nitrogen requirement increased. These yield and fertiliser nitrogen effects were not related to soil mineral nitrogen supply.

INTRODUCTION

The improvement or maintenance of soil fertility by the return of cereal residues to the soil was the main objective of an experiment at Rothamsted in 1934, and another experiment at MAFF Experimental Farms that started in the early 1950s. The latter experiment was also started in response to concern at falling levels of soil organic matter particularly after ploughing up grassland for intensive production of cereals. Straw burning for disposal of residues was also implicated as a causal factor in this organic matter decline. However, after eighteen years of ploughing in residues from a six course rotation, four of which were white straw crops, there were no significant changes in levels of organic matter attributable to straw incorporation or burning[1]. With only four cereal crops with relatively low yields in the rotation, and cultivation by ploughing, perhaps this result was not surprising. Ploughing can cause organic matter to decompose and nitrogen to mineralise faster than reduced cultivation methods[2]. Microbial activity is also promoted by ploughing due to the admixture of soil and organic material, and the exposure of fresh organic material to aerated microbially active soil[3].

Yield effects in the Rothamsted experiment were attributed to the return of potash, and nitrogen immobilisation during the straw rotting process rather than to any effect on soil structure. Similar conclusions were drawn from an experiment at the Norfolk Agricultural Station[4].

Soil Fertility Effects

In the MAFF experiments, yields were unaffected by eighteen years of straw incorporation provided that adequate nitrogen was applied in spring. Soil structure differences as determined by draught measurements with a dynamometer showed no differences between burnt or incorporated treatments. Organic matter declined as cropping practice changed at the start of the trial except at

Gleadthorpe where no such change occurred. There was relatively little difference in organic matter between disposal methods over the period of the trial. The use of farmyard manure for potatoes in the rotation provided additional organic matter, which tended to reduce the overall decline that occurred[1].

Non Biological Effects

The effects of straw incorporation may be immediate or long term. Immediate effects include the physical blockage of implements or drills if the straw is inadequately chopped and spread; also, moisture loss during cultivation and subsequent looseness of the seedbed may be exacerbated by the pressure of straw residues and may adversely affect establishment. However, if additional or inappropriate cultivations are used, then any consequent yield loss must be attributed to this cause rather than the presence of straw residues.

Biological effects

Scepticism about successful straw incorporation was encouraged in the mid 1970's when the Letcombe Laboratory[5] reported adverse effects on seed germination and seedling growth. This was attributed to the production of aliphatic acids during straw incorporation under anaerobic conditions. However, the drilling of seeds in anaerobic conditions is relatively rare, and the development of toxicity positively attributed to straw residues in field scale trials and commercial practice has not been observed to be a problem[6].

To evaluate this risk during the early years of the current ADAS series of experiments, incorporated straw was dug up at intervals for anaerobic incubation in the laboratory to determine the potential for producing aliphatic acids. This was done at the Wheat '83 site at Trumpington, and at Boxworth, Terrington and Drayton EHFs in 1984/85. There was a rapid decline in the production of acetic acid from about 18 mg/g straw in September (fresh straw) to 4-5 mg/g (incorporated straw) by mid November.

Nitrogen immobilisation may occasionally cause temporary crop yellowing which may tempt remedial action; but autumn nitrogen for this or other reasons has not been effective or economically justified.

Other factors adversely affecting germination and establishment are often attributed to the presence of straw, but are more likely to be due to cultivation effects and moisture loss associated with seasonal differences, drill coulter type, nitrogen availability or disease incidence. Weed effects are because of the survival of weed seeds and seedlings in the absence of burning rather than to the effect that straw residues may have on herbicide efficacy.

Current Experiments

In 1983, responding to public concern at atmospheric pollution by smoke and smuts from burning straw, the current series of ADAS straw disposal experiments started to investigate methods of incorporating straw. Existing equipment and new and novel implements were used, and results from all incorporation methods were compared with those after burning straw. Most of the sites were on MAFF Experimental Husbandry Farms covering a range of soil types. Non-cereal, combinable break crops are included in the rotation to provide information on their husbandry requirements. This paper presents results from conventional methods only. There were almost one hundred experiments between 1983 and 1989, with just over one half on soils classified as difficult by the Soil Survey of England and Wales[7].

RESULTS

Organic Matter Changes

Small but fairly consistent increases in level of soil organic matter occurred except at Drayton EHF (Table II). These on average approached the Rothamsted model[8], an increase of 0.2% in organic matter after five years when straw was returned to the soil annually. However, changes were not significantly different except at Terrington ($P<.001$). The reason for absence of any change in organic matter at Drayton remains obscure. There were no interactions with method of incorporation. Small further increases could occur slowly during subsequent decades. Soil mineral nitrogen levels have shown little increase so far.

Seedbed Conditions

Good crop establishment for autumn sown crops is vital. The requirement to incorporate straw may result in cloddy conditions in dry seasons, increased soil moisture loss, and puffiness or looseness of the seedbed. These factors will be detrimental to drilling and good establishment, and may exacerbate root diseases such as takeall, or increase slug damage if not well rolled to consolidate.

On light and medium soils ploughing has been an effective method for straw incorporation, with no appreciable increases in costs other than the cost of chopping the straw, and with least risk of grass weeds developing. In contrast, shallow cultivations with tines or discs have occasionally led to severe brome grass infestation.

On heavy soils ploughing is readily feasible but at greater cost due to the secondary tillage that is required to obtain a drillable seedbed particularly in dry or wet autumns. On the heaviest soils tined cultivation has proved effective and least expensive, but also at risk from grass weed infestation.

Table I Effects of Straw Disposal Method on Organic Matter concentration in Cultivated Layer (% air dry soil), (average of 3-4 replicates)

Site		Tined cultivation		Plough	
		burnt	incorporated	burnt	incorporated
Boxworth clay loam	1983-90	3.87	4.20	3.40	3.40
Bridgets chalky silty loam	1982-90	4.52	4.57	4.17	4.55
Drayton clay	1983-90	4.93	4.83	NT	4.90*
High Mowthorpe silty clay loam	1982-90	NT	4.30*	3.75	4.05
Terrington silt loam	1982-90 (P<.001) SED ± 0.026	2.85	2.97	NT	2.72*
7-8 year mean*		4.04	4.14	3.77	4.00
			+ 0.10		+ 0.23
Sproatley clay loam	1984-90	4.50	4.63	4.57	4.60
Rochford clay	1984-90	4.17	4.43	3.90	4.07
5 year mean		4.33	4.53	4.23	4.33
Change			+ 0.20		+ 0.10
Overall mean		4.14	4.27	3.96	4.13
Change			+ 0.13		+ 0.17

NB. All the sites had a ripening crop of winter wheat when sampled at end of July 1990.

NT: these treatments were not tested at these sites.

*   not included in 7-8 year mean.

Yield Effects

Yield losses attributable to various straw incorporation methods have usually been quite small (table II), except where grass weed problems developed. (table III).

Incorporation by tines or discs led to yield penalties on light soils, where grass weeds frequently added to the problem. In contrast ploughing in straw gave similar yields to where straw was burnt on all soil types, but at a higher cost on the heavier soils.

Results averaged for the range of soil types are in Table II.

Table II Effect of Straw Incorporation on Winter Wheat Yields: ADAS Sites 1983-1989. Values Expressed as a Percentage of Yields from Straw Burnt Plots.

	No. of sites	Straw: burnt tines yield t/ha	incorporated tines/discs	plough
Light soils	26	6.91	89	99
Medium soils	15	8.45	97	98
Heavy	18	7.71	97	100
V Heavy	25	6.99	100	99

Data in Table III are from one of the sites fairly typical of experience with straw incorporation on heavier soils.

Table III Effect of Straw incorporation on Winter Wheat Yields. Bescaby (Leics) 1985-88. Soil: decalcified clay loam. Values Expressed as a Percentage of Yields from Straw Burnt Plots.

	Straw: burnt tines Yield t/ha	incorporated tines 10 cm	incorporated tines 15 cm	plough
1985	7.51	104	101	99
1986	6.77	98	92	106
1987	5.47	99*	94*	85
Mean 3 yrs		100	96	97
1988	7.00	87**	81**	103
Mean 4 yrs	6.69	97	92	98

* Barren brome infestation.   ** Very severe Barren brome infestation.

The results show:
1. The variation between years in yield is associated with difficulties with ploughing on heavy soils.

2. Grass weed competition increases with runs of continuous cereals and continuous shallow cultivations for incorporation.

3. The better dispersal of straw residues expected from deeper cultivations did not improve yield in any year.

With the benefit of hindsight a change of cultivation or rotation following the 1987 harvest would have been the correct decision. Thus, yields of first wheats after breakcrops have been maintained where straw has been incorporated continuously (Table IV).

Table IV Effect of Straw Incorporation on Winter Wheat Yield: 1st Wheats after Breakcrops, 1989 and 1990. Expressed as a Percentage of Yield from Straw Burnt Plots.

Site	Year	burnt tines Yield t/ha	incorporated tines	plough
Rochford	1990	9.00	102	101
Boxworth	1989	7.98	105-108	105
Sproatley	1989	11.31	104	102
Terrington	1990	10.42	98	97

Effects of Straw Incorporation on Break Crops

Results with breakcrops, such as peas and winter oilseed rape, are shown in Table V.

Table V Yields of Winter Oilseed Rape (OSR) and Peas Following Successive Years of Cereal Straw Incorporation. Expressed as a Percentage of Yields from Straw Burnt Plots.

Site	Crop	Year	Straw: burnt tines Yield t/ha	incorporated tines	plough	other implements
Rochford clay	Peas	1988	4.64	104	104	96-103
Boxworth clay loam	OSR	1988	3.42	102	102	108-116
Sproatley clay loam	OSR	1988	3.22	53*	114	
Terrington silt loam	OSR	1989	3.20	105	95	

* poor establishment and weed control.

Breakcrops were successfully grown except after tined tillage at Sproatley where poor establishment and weed growth seriously affected yield. The break from cereals normally allows the use of herbicides with a different mode of action, enabling good control of some difficult grass weeds. Yields obtained were quite encouraging with no adverse effects from straw residues.

Effect of Straw Incorporation on Diseases, Pests and Weeds

Diseases

Root diseases such as Gaeumannomyces graminis syn Ophiobolus (takeall), and Pythium spp thought likely to increase with non burning cultivation regimes[9] have not increased to serious levels. In some instances higher levels of takeall have occurred with straw incorporation by shallow cultivation methods, and under-consolidated seedbeds. Stem base and leaf diseases such as Pseudocercosporella herpotrichoides (eyespot), Fusarium + Septoria spp, Rhynchosporium secalis (leaf blotch), or Pyrenophora teres (net blotch) are most likely to occur in the presence of trash but levels have not increased markedly with the sensible use of appropriate fungicides.

Pests

Slug numbers are thought to increase inevitably with straw incorporation. Where slug numbers have increased frequently, crop damage has not increased proportionally. Adequate depth of drilling and seedbed consolidation are important factors contributing to this result[10]. Other insect pests were not differentially affected by the presence of straw.

Weeds

Grass weeds have been the main problem with the cessation of straw burning, particularly with reduced cultivation techniques. The extent to which grass weeds could be reduced by straw burning was shown by the AFRC Weed Research Organisation[11].

	Control by burning
Alopecurus myosuroides (black grass)	40-80%
Avena fatua, A ludoviciana (wild oats)	32%
Bromus sterilis (barren brome)	40-97%

Boxworth data 1983-1989 shows that the control of Bromus sterilis, and Bromus commutatus by regular straw burning has consistently decreased these to between 78-99% lower than where straw has not been burnt. These grass weeds are proving difficult to control with existing herbicides. Furthermore the spread of herbicide tolerant strains of blackgrass is adding to the difficulties which will be even more serious when straw burning becomes proscribed after the 1992 harvest.

Yields from all methods of straw disposal have been similar where grass weed control was good (Table VI).

Grass weed control may be affected by large levels of soil organic matter, which reduce the efficacy of soil acting herbicides[12]. The effect of straw residues (Table VI) or ash after burning on a selection of grass weed herbicides has been studied during 1983-1989 at Boxworth.

Table VI Effect on Winter Wheat Yield of Grass Weed Control by Herbicides where Straw was Burnt or Incorporated Yield t/ha @ 85% DM.

Straw:	1986 (3rd year) burnt	incorporated	1989 (6th year) burnt	incorporated
Untreated	6.74	4.58	3.17	2.55
Avadex + Dosaflo	7.13	6.66	6.02	6.62
SMY 1500 in autumn	7.03	6.97	6.35	6.80
Soil Kd values*			(3.9)	(3.2)

Effect of Cultivation Methods on Percentage Organic Matter, pH and Soil Kd value* after Straw Burning

Contrasting cultivations during the period 1974-1981 where straw was burnt, show the cumulative difference in gradients with depth of soil organic matter and pH, and soil adsorption factor - Kd value*[12]. Accumulation of ash in the shallow surface layer where soil acting herbicides are placed results in enhanced adsorption of the chemical and loss of its effectiveness. Unburnt straw residue is not as adsorptive as ash. These data emphasise the reasoning behind the concept of rotational ploughing to avoid large concentrations of organic matter building up at the surface (Table VII).

Table VII Effect of Cultivation Methods at Boxworth EHF 1974-1981, on Soil OM%, pH and Kd in the surface layer.

Cultivation:	direct drill			tine			plough		
Sampling depth cm	OM%	pH	Kd	OM%	pH	Kd	OM%	pH	Kd
0-2.5	5.4	6.5	10.8	4.3	7.3	-	3.6	7.7	2.5
2.5-5.0	4.0	7.2	-	4.3	7.4	-	3.5	7.7	-
5.0-7.5	3.6	7.5	-	4.2	7.5	-	3.5	7.7	-
7.5-10.0	3.6	7.5	-	3.8	7.6	-	3.5	7.7	-
10.0-20.0	3.6	7.6	-	3.7	7.6	-	3.6	7.7	-

- not determined

Effect of Straw Incorporation on Soil Nitrogen

Mineral Nitrogen

Jenkinson[8] using C:N ratios predicted, assuming one third of straw carbon remained after one year's decomposition, that 1 tonne of straw containing 4-8kg nitrogen (N), would immobilise 9-5 kg/ha N. However, such levels of N immobilisation have rarely been encountered in recent ADAS field experiments where effects of straw on soil mineral N have been studied. Shallow disc/tine incorporation of straw, where greater N immobilisation is expected[13] has had variable and inconsistent effects on spring soil mineral N level; similar results are obtained when ploughing in straw (Table VIII).

Table VIII Effect of Straw Disposal Methods on Mineral Nitrogen in the Cultivated Layer of Soils. (average of 3-4 replicate measurements by 2M KCl extraction of fresh soil)

Mineral nitrogen kg/ha

Site, duration	Cultivation method	Straw Disposal Method burnt	incorp.	Mean	Analysis of Variance Factor	SED	P=	CV%
Bridgets 1982-90	Tine	54.4	79.4	66.9	Cultivation	10.67	0.185	36.0
	Plough	37.8	65.4	51.6	Straw disposal		0.036	
	Mean	46.6	68.4	59.3	Interaction	15.09	0.092	
High Mowthorpe 1982-90	Tine	29.1	24.8	27.0	Cultivation	1.59	0.152	13.1
	Plough	22.0	24.3	23.2	Straw disposal		0.879	
	Mean	25.6	24.6	25.1	Interaction	2.26	0.231	
Boxworth 1983-90	Tine	102.0	122.9	112.5	Cultivation	7.49	0.155	12.2
	Plough	104.7	95.8	100.3	Straw disposal		0.456	
	Mean	103.4	109.4	106.4	Interaction	10.59	0.094	
Sproatley 1984-90	Tine	24.5	26.9	25.7	Cultivation	7.65	0.063	38.5
	Plough	46.7	39.6	43.2	Straw disposal		0.768	
	Mean	35.6	33.3	34.4	Interaction	10.82	0.562	
Terrington 1982-90	Tine	12.3	18.0	15.2	Cultivation	4.70	0.179	37.8
	Plough	NT	22.4	NT				
	Mean		20.2					
Drayton 1983-90	Tine	34.3	33.8	34.1	Cultivation	6.19	0.365	24.3
	Plough	NT	25.5	NT				
	Mean		29.7					

NB. All the sites had a ripening crop of winter wheat when sampled at the end of July 1990.

NT: these treatments were not tested at these sites.

Total Organic Nitrogen

Small, but significant, increases in total organic N (%N) occurred in 3 sites - Bridgets, High Mowthorpe and Boxworth EHF's - where straw had been incorporated by shallow (10-15 cm deep) disc/tine cultivation, compared with burning (Table IX).

Ploughing the straw in to a greater depth of about 20 cm had negligible effect on %N accumulation.

Table IX Effect of Cultivation and Straw Disposal Methods on Total Organic Nitrogen % N in Cultivated Layer of Soils.

Site	Cultivation Method	Straw Disposal Method burnt	incorp.	Mean	Analysis of Variance Factor	SED	P=	CV%
Bridgets	Tine	0.32	0.33	0.33	Cultivation	0.0098	0.046	6.4
	Plough	0.29	0.28	0.29	Straw disposal		0.621	
	Mean	0.31	0.31	0.31	Interaction	0.0138	0.332	
High	Tine	0.29	0.27	0.28	Cultivation	0.0058	<0.001	4.5
Mowthorpe	Plough	0.23	0.25	0.24	Straw disposal		0.675	
	Mean	0.26	0.26	0.26	Interaction	0.0082	0.007	
Boxworth	Tine	0.22	0.24	0.23	Cultivation	0.0052	<0.001	4.3
	Plough	0.19	0.19	0.19	Straw disposal		0.028	
	Mean	0.21	0.22	0.21	Interaction	0.0073	0.065	
Sproatley	Tine	0.27	0.28	0.28	Cultivation	0.0103	0.301	6.7
	Plough	0.26	0.26	0.26	Straw disposal		0.645	
	Mean	0.27	0.27	0.27	Interaction	0.0146	0.645	
Terrington	Tine	0.16	0.16	0.16	Cultivation	0.0039	0.096	3.5
	Plough	NT	0.15	NT				
	Mean		0.16					
Drayton	Tine	0.28	0.28	0.28	Cultivation	0.0253	0.095	11.0
	Plough	NT	0.27	NT				
	Mean		0.28					

NB. All the sites had a ripening crop of winter wheat when sampled at end of July 1990.

NT: these treatments were not tested at these sites.

Terrington showed no change in %N, despite the highly significant increase of organic matter from 2.85 to 2.97% (Table I) after 8 years of tine incorporation; this site had the smallest level of organic matter. Soil mineral N levels in this site are usually much smaller than elsewhere; it is likely that much loss of nitrogen may be lost by denitrification in this marine alluvial silt loam.

Effect of Straw Incorporation on Fertiliser Nitrogen for Optimum Yield

Fertiliser N requirement for economic optimum yield, derived from a linear plus exponential function[14] fitted to responses measured in experiments testing increasing amounts of applied N, were similar irrespective of straw disposal method (Table X). The site was on a chalky boulder clay field with clay loam topsoil at Boxworth EHF growing winter wheat continuously since 1970.

Data are averages of 6 sets of responses in experiments during 1985-86 to 1987-88 and represent effects of 1, 2 and 3 successive seasons of incorporation. Analysis of variance on the averages of optimum yield and corresponding amount of fertiliser N was made using the values derived for the successive seasons as replicates based on the absence of any significant interactions.

Table X   Effect of Straw Incorporation Method on Winter Wheat Yield - Optimum* Yield and Fertiliser N.

Straw disposal method	Nil N	Yield @ 85% DM t/ha With optimum* fertiliser N			Optimum* fertiliser N kg/ha^{-1}		
		None in autumn	40 kg/ha N in autumn	Mean	None in autumn	40 kg/ha N in autumn	Mean
Burnt	3.09	5.51	5.51	5.51	196	181	189
Tine+	2.91	5.56	5.65	5.61	207	220	213
Plough	3.42	6.07	6.21	6.14	189	203	196
SED	0.627	0.408	0.423	0.412	14.87	14.85	11.16
P =	0.516	0.688	0.607	0.642	0.791	0.344	0.442

* assuming grain to N price ratio of 3:1
\+ severe brome grass infestation reduced yields in the second and third seasons - see section on weeds.

Seasonal effects of drought and increasing incidence of grass weed patches influenced crop performance far more than the effect of straw incorporation methods; these adverse factors resulted in large variability and contributed to the large errors of measurement. 1985 autumn was very hot and dry when primary cultivations were done. Ploughing resulted in harsh, cloddy seedbed and crop emergence was protracted and early growth seriously penalised.

Responses to autumn applied N have not been consistently larger where straw is incorporated; similar yield responses could have occurred if this N was applied in Spring.

Similar to the above effects, additional autumn fertiliser N dressings (applied to counteract any possible effects of N immobilisation by straw) have only provided an economic yield return in half of the cases studied (Table XI); significant financial losses occurred in the remaining cases.

Table XI Responses to Additional Autumn N Dressings in 18 Recent*
ADAS Studies where About 200 kg/ha N was Applied in Spring.

Straw Disposal Method	% Cases Giving an Economic Yield Response+ to Autumn N	% Cases Where Yield Responses were Greater Where Straw was Incorporated
Straw burnt Shallow tine	44%	-
Straw incorporation Shallow tine	50%	50%

*. 6 sites in 1985, 6 in 1986, 2 in 1987, and 4 in 1988.
+. A yield return in excess of 3 kg of grain/kg autumn N.

The above effects are also reflected in the inability of incorporated straw to consistently retain autumn applied N-fertiliser in soil in excess of that found where straw is burnt. 40 kg/ha N as $^{15}NH_4$ $^{15}NO_3$ (5 atom % excess ^{15}N) was applied (22.11.85 and 27.11.86 to crops drilled 5-6 weeks earlier) to duplicate 1m x 1m areas located in main plots of a nitrogen response experiment at Boxworth EHF. Recovery of ^{15}N in above ground dry matter of crops and that left in 0-30cm soil layer was measured in samples collected from 0.4m x 0.4m in the middle of ^{15}N labelled areas. Sampling was at GS22 & 23[20] on 5.3.86 and 13.2.87 respectively. $^{15}N:^{14}N$ ratios were determined after Kjeldahl digestion for % N measurements.

Although recovery of the autumn applied fertiliser N was better with straw incorporation in the 1986-87 season, this benefit was not significant at the 95% probability. Some retention of the added fertiliser N during decomposition of straw is implicated. This is in accordance with other observations in ADAS studies[15] that immobilisation of soil mineral N by incorporated straw, while not an invariable or consistent occurrence, is most frequent in soils with inherently large organic matter levels and correspondingly large mineral N in the cultivated layer. Luxury uptake of soluble N by decomposer organisms has been suggested as a possibility in these circumstances[16] - a response to variation in N availability changing the N concentrations of newly synthesised microbial cells, as proposed by Allison[17].

There is an obvious environmental benefit from net N immobilisation by straw incorporated in the autumn and thereby a reduction in leachable nitrogen. Such benefit will continue only if immobilised N and N accumulating from incorporated straw is mineralised only after cessation of excess rainfall.

The N requirements of decomposing straw will be met not only by soil and fertiliser N sources, but also through mineralisation and solubilisation of N from straw itself and through recycling of N in the decomposer biomass. In studies following changes in straw N content during decomposition, in most cases net

Table XII Spring Recovery from Crop and Soil (0-30 cm) of 40 kg/ha 15-N Labelled Fertiliser Applied in Autumn. Boxworth EHF, Hanslope Soil Series.

Straw Disposal Method	Number of Successive Years of Disposal	% Recovery Spring 1986		% Recovery Spring 1987	
		Soil	Crop + Soil	Soil	Crop + Soil
Straw burnt Shallow tine	1	30.5	34.0	40.1	43.9
	2			43.8	48.4
Straw Incorporated Shallow tine	1	29.7	31.5	65.2	68.5
	2			57.0	61.9
Plough	1	28.4	29.3	41.2	49.9
	2			56.3	65.5
SED (6 df for 1986, 11 for 1987)					
Straw disposal method		9.69	9.16	8.29	8.43
P =		0.98	0.87	0.11	0.12
Years of incorporation				6.77	6.88
P =				0.61	0.53
Disposal method x year				11.72	11.92
P =				0.40	0.45

mineralisation of N from straw occurred within a single season (16, 18, 19) counteracting any earlier immobilisation.

The findings of this and other studies suggest that, with little net N immobilisation occurring in a season during straw decomposition, there is little to release during future seasons in the short to medium term. Thus, a number of successive years of incorporation maybe required before any effect of straw increasing soil N supply and reducing fertiliser N requirement becomes noticeable.

ACKNOWLEDGEMENTS

Many colleagues in ADAS helped in making the experiments and gather data. We are grateful to them and to the farmers who provided facilities. Special thanks are due to Mr. J.T. Clapp, P.J. Skeels and Sylvia Smith of Boxworth EHF, and to the Analytical Chemists of ADAS, Cambridge Regional Office for their support. Thanks are also due to Ransomes of Ipswich for their financial support to David B. Turley during 1985-9, and to MAFF for his post-graduate grant. Support from the H-GCA for isotope ratio analysis is gratefully acknowledged.

REFERENCES

1. J L Short. <u>Experimental Husbandry, 1973</u>, Vol 25, p 103-106.
2. K Baeumer and W A P Bakermans. `Zero Tillage`, <u>Advances in Agronomy</u>, 1973, Vol 25, p 77-122.
3. P L Brown and D D Dickey. `Losses of Wheat Straw Residue Under Simulated Field Conditions`, <u>Proceedings of the Soil Science Society of America</u>', 1970, Vol 34, p 118-124.
4. F Rayns and S Culpin. `Rotational Experiments and Straw Disposal`, <u>Royal Agricultural Society</u>, 1948, Vol 109, p 128-137.
5. S Harper. `Straw Breakdown and Toxin Production in the Field`, <u>AFRC Publication - Straw, Soils and Science</u>, 1985, p 10.
6. `Changing Straw Disposal Practices`, <u>HGCA Research Review</u>', 1988, Vol 11, Editors R D Drew and B D Smith.
7. D B Davies and A J Thommasson. 'Soil Survey of England and Wales`, 1984.
8. D S Jenkinson. `How Straw Incorporation Affects the Nitrogen Cycle`, <u>Straw Soils and Science-current Research into Viable Alternatives to Disposing of Straw by Burning</u>, Sponsored by the AFRC', 1985, AFRC, London, p 14-15.
9. R J Cook, J W Sitton and J T Waldher. `Evidence for Pythium as a Pathogan of Direct Drilled Wheat in the Pacific Northwest`, <u>Plant Disease, 1980</u>, Vol 64, p 102-103, 'Farmers Weekly', 1986
10. D M Glen, N F Milsom and C W Wiltshire 1989. `Effect of Seedbed Conditions on Slug Numbers and Damage to Winter Wheat on a Clay Soil`, <u>Annals of Applied Biology 115</u>, p 177-190.
11. S Moss. `Straw Burning and Effects on Weeds`, <u>AFRC Publication - Straw, Soils and Science</u>' 1985, p 18.
12. G W Cussons, S R Moss, R J Harce, S J Embling, D Caverley, T G Marks and J J Palmer `The Effect of Tillage Method and Soil Factors on the Performance of Chlortoluron and Isoproturon', <u>Proceedings of the 1982 British Crop Protection Council - Weeds. 1982</u>, Vol 1, p 153-160.
13. E Lord, B Davies and J S Rule. `Guidelines for Straw Incorporation in England`, <u>Agricultural Progress, 1986</u>, Vol 61, p14-25.
14. B J George `Design and Interpretation of Nitrogen Response Experiments', In: <u>The Nitrogen Requirement of Cereals</u>, MAFF Reference Book 385. Pub. HMSO, London, p 133-150. 1984.
15. D B Turley. `Effect of Disc/Tine and Plough Based Straw Disposal Systems on Wheat Production in Clay Soils, 1990`, <u>Ph D Thesis, Cranfield Institute of Technology, Silsoe College</u>.
16. B T Christensen. `Barley Straw Decomposition Under Field Conditions: Effect of Placement and Initial Nitrogen Content on Weight Loss and Nitrogen Dynamics`, <u>Soil Biology Biochemistry, 1986</u>, Vol 18, p 523-529.
17. Allison. 1973. `Soil Organic Matter and its role in Crop Production` <u>Developments in Soil Science III</u>, Elsevier Scientific Publishing, Amsterdam, London, New York.

18. J H Smith and C L Douglas. `Wheat Straw decomposition in the Field`, <u>Soil Science Society America Proceedings</u>. 1971, <u>Vol 35</u>, p 269-272.
19. C L Douglas, R R Allmaras, P E Rasmussen, R E Ramig and N C Roager. `Wheat Straw Composition and Placement Effects on Decomposition in Dryland Agriculture of the Pacific Northwest`, <u>Soil Society America Journal</u>, 1980, <u>Vol 44</u>, p 833-837.
20. D R Totman and H Broad. `Decimal Code for the Growth Stages of Cereals`, <u>Annals of Applied Biology</u>, 1987, <u>Vol 110</u>, p 683-687.

Soil Organic Matter: Its Central Position in Organic Farming

R.D. Hodges

DEPARTMENT OF BIOCHEMISTRY AND BIOLOGICAL SCIENCES, WYE COLLEGE (UNIVERSITY OF LONDON), WYE, ASHFORD, KENT, TN25 5AH, UK

1 INTRODUCTION

Organic farming systems developed in parallel with intensive, conventional farming systems from the 1940's onwards. They developed as a response to what was seen by the pioneers of organic systems as a retrograde development in agriculture; the increasingly widespread use of synthetic chemicals as the essential basis of chemical fertility renewal and pest and disease control.

The term organic, as applied to farming systems, refers to the development of a farm as an organic whole, a carefully integrated and balanced system, rather than to a unit which specifically utilises recycled organic matter; although the latter interpretation also applies. During the 1950's and 60's organic farming was generally rejected as being an eccentric system with no place in the mainstream of agriculture. More recently, particularly as the inherent weaknesses of conventional, chemically-based systems have begun to appear,[1-3] organic farming is increasingly being seen as an alternative to conventional farming. Studies are showing that organic farming, or similar variations on a basic theme, is economically viable, productive, environmentally-benign and scientifically sound.[4-10]

If organic farming can be as productive and profitable as its equivalent conventional counterpart, how can it manage this without making use of the essential underpinnings of conventional farming, moderate to large inputs of fertilizers and pesticides? This review concentrates upon the central role that organic matter plays in organic farming, in particular in soil structure and fertility maintenance and pest and disease control.

2 SOIL ORGANIC MATTER IN AN ORGANIC SYSTEM

In conventional agriculture the ability to modify soil, in particular by the addition of chemically-formulated plant nutrients, has meant that biological or inherent soil fertility,[11] generated by complex interactions between the soil minerals, microflora and fauna, organic matter and crop roots, has become of only secondary importance. The final stage in this process was reached with the development of soil-less culture, as in hydroponics or nutrient film technique, where even the mechanical support that soil provides for plant roots is no longer needed, and the crop grows in a nutrient solution.

On the other hand the development and maintenance of a soil of high biological activity, which then develops improved fertility and a whole range of other important positive characteristics, is the primary concern of the organic farmer. The concept of the soil as a biologically-active medium, the "living soil", was first put forward by Balfour[12] and developed further by Matile[13], who likened a fertile topsoil containing about 25 tonnes per hectare of organic material to a gigantic, complex organism. The U.S. Department of Agriculture report on organic farming in 1980[4] considered that the concept of the soil as a living system was central to their definition of organic farming. In seeking to enhance the level of biological activity in a soil the organic farmer relies largely upon the maintenance of an optimum level of organic matter and the activity of microflora and fauna. An "organic" soil can only be maintained by regular inputs of fresh or partly-decomposed organic matter. In organic farming organic matter is the major source of soil fertility, so its inputs and optimum levels tend to be higher than in conventional systems reliant on fertilisers for nutrient inputs.

Fresh organic matter can be added to the soil directly, via the incorporation of straw and other residues or by the deposition of animal excreta on grassland. This approach may, however, be inefficient and unsatisfactory; consequently organic farmers, where possible, prefer to compost manures and plant residues, a process which has a number of advantages.[14] Among these are: a considerable reduction in both weight and volume without loss in quality; the stabilisation of plant nutrients via incorporation into biomass or humus complexes; the optimisation of the carbon:nitrogen ratio; the destruction of pathogens and weed seeds; and the production of a material easily assimilated into the biological activity and physico-chemical structure of the soil following incorporation.

The heterogeneous nature of soil organic matter, or soil humus, makes its definition and classification complex. A number of simple classifications have been attempted and, more specifically in relation to organic farming systems, organic matter has been divided by Koepf et al.[15] into three components:

a. "Effective humus". This consists of a range of plant residues and wastes, including undecomposed plant materials and manures, together with added composts, green manures and dead soil organisms. These materials interact within the soil and a large proportion is rapidly decomposed to carbon dioxide, water and plant nutrients. "Effective" humus has a half-life measured in months to years and only a small proportion is transformed into a stable form.

b. "Stable" humus results from the interaction between "effective" humus and soil microorganisms. It is dark brown to black in colour and its presence in soils in relatively small amounts greatly improves their physico-chemical properties. As its name implies it has a long half-life measured in decades or centuries.

c. The soil biomass or edaphon. This is composed of a wide range of bacteria, actinomycetes, fungi and algae, plus the soil fauna - protozoa, nematodes, annelids, arthropods, molluscs, etc.

Organic farming methods aim to raise the soil organic matter above a threshold level and thereafter to maintain this level together with the right balance of the three components. Koepf <u>et al</u>.[15] suggest that sandy soils should contain more than 2%, loamy soils 2.2-3.5% and clayey soils 3-4.5% "stable" humus and this requires fairly regular inputs of "effective" humus to soils. It is the continuous turnover of "effective" humus by the soil biomass which maintains soil fertility in organic systems.

The Maintenance of Organic Matter Levels

Soil organic matter levels are maintained on organic farms at or close to the above-mentioned levels through the application of the following techniques.

1. <u>Mixed Farming</u>. Organic farming has traditionally been based on mixed farming systems; a combination of livestock and cropping being seen as the most efficient way of utilising and recycling nutrients and organic matter within the system. Successful all-arable organic systems have, however, been developed.

2. <u>Rotations</u>. Complex rotations are an essential part of organic farming, in order to balance crops which use up fertility, such as grains, with crops which help to build up fertility, such as legumes and leys. Organic rotations may extend over six or eight years, as distinct from the often truncated rotations found in conventional farming and this approach also helps to

control pests and diseases. Two important components of organic rotations in a mixed farming system are the mixed ley and one or more grain cash-crops. A ley builds up organic matter and fertility as a result of the interactions of the grass roots with soil microbes and earthworms, together with the nitrogen fixation ability of the legumes plus the excreta of stock grazing on the sward.

3. Recycling. Organic matter and plant nutrients are recycled by the efficient re-utilisation of all plant residues, manures and other on-farm waste materials. These wastes are carefully conserved by composting and other processes to reduce losses of mobile nutrients such as nitrogen and potassium.

4. Green Manures. Green manure crops may be grown, either undersown or used to fill spaces between main crops in a rotation. The crop is turned in whilst still relatively young to add green organic matter to the soil which breaks down quickly releasing nutrients for a following crop. Green manures have a range of valuable functions,[16] particularly providing moderate to large amounts of "effective" humus.

5. Bought-in Fertility. The off-take of produce causes a constant drain on the fertility of farm soils. With conventional farms a high proportion of the chemical fertility required to continue production is bought-in in the form of fertilisers. Organic farms also import fertility, largely in the form of organic materials acceptable to the organic standards.[17] Among such materials are seaweed meals and extracts, manures and composts and sewage sludges, provided the latter are relatively uncontaminated with heavy metals.[18]

Functions of Organic Matter

The functions of soil organic matter are complex. They are important in all agricultural systems but particularly so in organic farming where inputs of chemicals are replaced by techniques aimed at regulating and enhancing biological processes in the soil. Soil organic matter is essential for many of these processes and the focal point from which they largely originate. The following general summary of functions is based upon a number of recent major reviews;[15, 16, 19-22] a comparison of the impact of conventional and organic agriculture on soil organic matter and its functions has been made by Arden-Clarke and Hodges.[23]

1. Energy for Soil Organisms. This function is given priority because many soil properties related to a sound structure and high inherent fertility are closely associated with the biological activity of the soil. Regular inputs of "effective" humus provide a source of

energy and nutrients for soil organisms, helping to build up populations and to improve their positive activities. A fresh addition of organic matter usually results in an outburst of microbial activity leading to improved structure and fertility and to increased numbers of soil animals.

2. Soil Structure and Stability. A well-developed structure, or tilth, is essential for optimal crop growth and production and soil organic matter significantly influences the development and maintenance of structure and stability in soils. A well aggregated and structured soil shows a positive correlation with a range of important properties.

Organic substances of plant and microbial origin cement soil particles together to form aggregates.[24] The addition of "effective" humus - largely producing relatively transient microbial polysaccharides - generates less stable aggregates than the interactions occurring between "stable" humus and the soil particles. Thus with regular inputs and turnover of organic matter soil structure will be improved, stabilised and maintained with light soils becoming more aggregated and heavy soils becoming more porous.[25] With many stable aggregates present, a soil becomes more easily tillable and less easily damaged in wet conditions; and the presence of numerous, well-developed pores facilitates good gaseous exchange, reduces the bulk density, and roots penetrate more deeply.[25-27]

3. Water Relations. Soil water characteristics are also improved as a result of increasing organic matter levels. Improved soil structure results in better permeability and drainage and less likelihood of slaking and capping. Soil organic matter can absorb and hold substantial amounts of water, much of which is available to crop roots.

4. Soil Erosion Control. Where inappropriate or intensive farming practices have been continued for any length of time there is a tendency for soil organic matter levels to fall and soil structure may deteriorate. The risk of soil erosion may then increase. In organic farming, the essential maintenance of organic matter levels together with rotations and other techniques result in a continuing protection of the integrity of the soil.[28]

5. Soil Warming. Organic matter and particularly the stable, humified fraction imparts a dark brown colour to the soil which speeds up soil warming. Earlier crop growth may be promoted as a result.

6. Availability of Plant Nutrients. Depending on soil type, microbial mineralisation of organic matter will release variable amounts of a whole range of plant

nutrients, in particular N, P and S but also minor elements. Microbial activity also helps to mobilise nutrients from the soil minerals.

The supply of plant nutrients must depend heavily on the relatively rapid breakdown of "effective" humus, since "stable" humus is mineralised slowly and only gradually releases its nutrient content. This is a further reason for organic farmers to maintain inputs of "effective" humus.

7. Ion Exchange, Chelation and Buffering. In addition to supplying plant nutrients soil organic matter, because of its electro-chemical properties, can retain nutrients which are thus protected from loss by leaching though still available for plant use. This property results in part from the considerable cation exchange capacity (C.E.C.) of humus. Both clay minerals and organic matter contribute to the soil C.E.C. but humus has a C.E.C. considerably in excess of most clays such that between 20 and 70% of the C.E.C. of soils can be directly attributable to their content of organic matter.

Another property of organic matter is the ability of certain components to chelate with metal ions and, of particular importance, with trace elements. In the absence of chelation trace elements can become insoluble at the usual pH values found in agricultural soils; when chelated they remain soluble and available to plant roots.[16,29].

Organic matter also has a substantial buffering capacity in the soil. In combination with clay minerals, humus forms negatively-charged colloidal complexes which buffer changes in pH.

8. Effects on Crop Growth. Soil organic matter through its interactions with microorganisms can also influence plant growth in a number of ways. These influences may be positive or negative.

Vaughan and Malcolm[21] have reviewed an extensive literature which shows that, under laboratory conditions, many humic substances can have a positive effect on the growth of plants as measured by a wide range of parameters. There is also considerable evidence that humic materials can beneficially affect the plant content and thus presumably the uptake of many nutrients. Similarly these substances may also positively affect the growth of microorganisms. In general, however, the influence of humic substances on plants in the field is much less certain.

Humic substances influence plant roots indirectly or directly.[16,30] Indirect effects, at the root surface or actually on the root cell membranes, optimise the availability or uptake of nutrients. Direct effects involve the actual uptake of the humic substance by the plant, which may then influence metabolic processes within the cells. Many experimental studies have shown that soluble humic acids, or humates, have a positive effect on biochemical mechanisms and processes within plant cells such as membrane permeability and transport, respiration activation and ATP production, chlorophyll content and photosynthesis, and nucleic acid synthesis.[31]

Direct effects of humic substances occur through uptake by the plant. For many years theories of plant nutrition have concentrated on the supply and uptake of nutrient ions, with the involvement of more complex molecules such as humus materials being very largely ignored. It is clear, however, that plant roots are capable of taking up humic substances of low molecular weight via membrane transport mechanisms and that both low and high molecular weight substances can be taken up by pinocytosis.[16,30]

Other products of the interactions between soil organic matter and microorganisms influencing crop growth and composition are plant growth hormones, enzymes and antibiotics,[20,32-34] which can be taken up and utilised by higher plants.[35,36]

Negative influences on plant growth and development have also been reported, due to the production of phytotoxic substances by microbes breaking down organic residues.[16,20,21,33]

9. Incidence of Plant Pathogens and Diseases. It is now well understood that additions of organic matter to soils can directly influence and often control soil-borne plant diseases and this is supported by an extensive literature.[16,20,37] Most recently Schueler et al. have demonstrated the antiphytopathogenic properties of composted organic household wastes when used against Pythium and Rhizoctonia infections of beans and peas.[38] When inputs of effective humus are increased, all aspects of biological activity in the soil increase and this enables a whole range of biological "checks and balances" to develop. Among these are the production of antibiotics which may help to control pathogenic bacteria associated with plants; the increase in naturally-occurring populations of antagonistic organisms to control pathogens; and a general increase in saprophytic organisms over parasitic ones, thus reducing populations of the latter.

3. DISCUSSION AND CONCLUSIONS

Very little of what has been described in the preceding sections is new; the techniques for maintaining organic matter levels used by organic farmers are essentially traditional husbandry methods, whilst the summary of the functions of soil organic matter provides knowledge of value applicable to any agricultural system. Both the practice of fertility renewal and the knowledge of organic matter functions are, for the organic farmer, part of what can be called the biological approach to agriculture. These techniques appear to work and both production in the field and research results lend support to the value of the biological/organic approach.

The properties of organic matter are utilised by organic farmers largely on a practical basis and the "fine tuning" of the system can often produce results which are as profitable and productive as the conventional counterpart.[5,10] The produce may be of higher quality,[39,40] and in terms of soil conservation it would seem to be a more sustainable system.[28,41]

The biological approach centres round the optimisation of factors developing from the interaction between the soil, organic matter, plant roots and microorganisms; a combination of complex processes which is supported by some experimental evidence. A comparison by Bolton et al.[42] suggests that organic farms have a larger and more active soil microflora than their conventional counterparts, and Elliott and Papendick[43] state that organic methods do appear to stimulate the soil microflora.

Together with improvements in crop varieties and mechanisation, conventional farming has developed over the past 40 years largely on the basis of increasing inputs of agricultural chemicals, fertilisers and pesticides, in some cases augmenting but frequently replacing the biologically-based processes of fertility maintenance and pest control found in its traditional predecessor. These developments have been based upon modern theories of plant nutrition linked to the use of a range of simple ionic nutrients alone. Over 150 years ago the Humus Theory suggested that plants were nourished by direct uptake of humic materials from the soil. This was apparently disproved by the work of Liebig and others and was replaced by the Mineral Theory. The modern theories leave little place for organic matter, apart from its improvement of soil structure plus its nutrient supply, and have led to modern textbooks saying little or nothing about the place of humic substances in plant nutrition.[44] Thus there seems to be a dichotomy in modern, conventional agriculture. On the one hand there is the theory and practice of crop production based upon knowledgable soil management plus the use of soluble nutrient ions; on the other hand there are extensive data which suggest that soil organic matter is more directly involved in plant nutrition and crop protection than just as a source of nutrient ions. The successful organic farmers are demonstrating the latter to be so on a practical basis.

REFERENCES

1. R. D. Hodges, 'Biological Husbandry', B. Stonehouse, ed., Butterworths, London, 1981, p. 1.
2. R.D. Hodges and A.M. Scofield, Biol. Agric. Hortic., 1983, 1, 269.
3. C. Arden-Clarke, 'The Environmental Effects of Conventional and Organic/Biological Farming Systems', Research Reports RR-16 and RR-17, Political Ecology Research Group, Oxford, 1988.
4. United States Department of Agriculture, 'Report and Recommendations on Organic Farming', U.S. Government Printing Office, Washington, D.C., 1980.
5. W. Lockeretz, G. Shearer and D.H. Kohl, Science, 1981, 211, 540.
6. D.F. Bezdicek and J.F. Power, 'Organic Farming: Current Technology and its Role in a Sustainable Agriculture', American Society of Agronomy, Madison, Wisconsin, 1984.
7. T.C. Edens, C. Fridgen and S.L. Battenfield, 'Sustainable Agriculture and Integrated Farming Systems', Michigan State University Press, East Lansing, MI, 1985.
8. National Research Council, 'Alternative Agriculture', National Academy Press, Washington, D.C., 1989.
9. C.A. Edwards, R. Lal, P. Madden, R.H. Miller and G. House, 'Sustainable Agricultural Systems', Soil and Water Conservation Society, Ankeny, Iowa, 1990.
10. N. Lampkin, 'Collected Papers on Organic Farming', 2nd. edn., Centre for Organic Husbandry and Agroecology, Aberystwyth, 1990.
11. R.D. Hodges, Biol. Agric. Hortic., 1982, 1, 1.
12. E.B. Balfour, 'The Living Soil', Faber and Faber, London, 1943.
13. P. Matile, Wasser und Abwasser, July 1973 Supplement. Reported in Soil Association Journal, 1973, 1, No. 8, 5.
14. H.W. Dalzell, A.J. Biddlestone, K.R. Gray and K. Thurairjan, 'Soil Management: Compost Production and Use in Tropical and Subtropical Environments', Soils Bulletin No. 56, U.N. Food and Agriculture Organisation, Rome, 1987.
15. H.H. Koepf, B.D. Petterson and W. Schaumann, 'Bio-Dynamic Agriculture. An Introduction', The Anthroposophic Press, Spring Valley, New York, 1976.
16. F.E. Allison, 'Soil Organic Matter and its Role in Crop Production', Elsevier, New York, 1973.
17. The Soil Association, 'Standards for Organic Agriculture', The Soil Association, Bristol, 1989.
18. K.E. Giller and S.P. McGrath, New Scientist, 1989, 124, No. 1689, 31.
19. M. Schnitzer and S.U. Kahn, 'Soil Organic Matter', Elsevier, Amsterdam, 1978.
20. F.J. Stevenson, 'Humus Chemistry. Genesis, Composition, Reactions', John Wiley, New York, 1982.
21. D. Vaughan and R.E. Malcolm, 'Soil Organic Matter and Biological Activity', Martinus Nijhoff, Dordrecht, 1985.
22. R.L. Tate III, 'Soil Organic Matter. Biological and Ecological Effects', John Wiley, New York, 1987.

23. C. Arden-Clarke and R.D. Hodges, Biol Agric. Hortic., 1988, 5, 223.
24. R.G. Burns and J.A. Davies, 'The Role of Microorganisms in a Sustainable Agriculture', J.M. Lopez-Real and R.D. Hodges, eds., AB Academic Publishers, Berkhamsted, 1986, p. 9.
25. A. Pera, G. Vallini, I. Sireno, M.L. Bianchin and M. de Bertholdi, Plant and Soil, 1983, 74, 3.
26. A.J. Low, J. Soil Sci., 1972, 23, 363.
27. E.W. Russell, Phil. Trans. Roy, Soc., B, 1977, 281, 209.
28. C. Arden-Clarke and R.D. Hodges, Biol. Agric. Hortic., 1987, 4, 309.
29. D.J. Linehan, 'Soil Organic Matter and Biological Activity', D. Vaughan and R.E. Malcolm, eds., Martinus Nijhoff, Dordrecht, 1985, p. 403.
30. U. Mueller-Wegener, 'Humic Substances and Their Role in the Environment', F.H. Frimmel and R.F. Christman, eds., John Wiley, Chichester, 1988, p. 179.
31. D. Vaughan, R.E. Malcolm and B.G. Ord, 'Soil Organic Matter and Biological Activity', D. Vaughan and R.E. Malcolm, eds., Martinus Nijhoff, Dordrecht, 1985, p. 77.
32. J.M. Lynch, 'Soil Organic Matter and Biological Activity', D. Vaughan and R.E. Malcolm, eds., Martinus Nijhoff, Dordrecht, 1985, p. 151.
33. J.M. Lynch, 'The Role of Microorganisms in a Sustainable Agriculture', J.M. Lopez-Real and R.D. Hodges, eds., AB Academic Publishers, Berkhamsted, 1986, p. 57.
34. J.N. Ladd, 'Soil Organic Matter and Biological Activity', D. Vaughan and R.E. Malcolm, eds., Martinus Nijhoff, Dordrecht, p. 175.
35. N.A. Krasil'nikov, 'Soil Microorganisms and Higher Plants' The Israel Program for Scientific Translations, Jerusalem, 1962.
36. M.M. Kononova, 'Soil Organic Matter' Pergamon Press, Oxford, 1966.
37. R.J. Cook, 'Plant Disease', J.G. Horsfall and E.B. Cowling, eds., Academic Press, New York, Vol. 1, p. 23.
38. C. Schueler, J. Biala and H. Vogtmann, Agric. Ecosyst. Envir., 1989, 27, 477.
39. Vogtmann, H., 'The Quality of Agricultural Produce Originating from Different Systems of Cultivation', The Soil Association, Bristol, 1979.
40. A. Meier-Ploeger, R. Duden and H. Vogtmann, Agric. Ecosyst. Envir., 1989, 27, 483.
41. J.P. Reganold, L. F. Elliott and Y.L. Unger, Nature, 1987, 330, 370.
42. H. Bolton, L.F. Elliott, R.I. Papendick and D.F. Bezdicek, Soil Biol. Biochem., 1985, 17, 297.
43. L.F. Elliott and R.I. Papendick, 'The Role of Microorganisms in a Sustainable Agriculture', J.M. Lopez-Real and R.D. Hodges, eds., AB Academic Publishers, Berkhamsted, 1986, p. 45.
44. A. Wild, ed., 'Russell's Soil Conditions and Plant Growth', Longman, Harlow, 11th. edn., 1988.

Refuse-derived Humus: a Plant Growth Medium

A.A. Keeling[1], J.A.J. Mullett[1], I.K. Paton[1], N. Bragg[2], B.J. Chambers[3], P.J. Harvey[4], and R.S. Manasse[4]

[1] SECONDARY RESOURCES PLC. TAMESIDE DRIVE, BIRMINGHAM B35 7AG, UK
[2] MAFF/ADAS, WOLVERHAMPTON, UK
[3] MAFF/ADAS, SOIL SCIENCE DEPARTMENT, GOVERNMENT BUIULDINGS, BROOKLANDS AVENUE, CAMBRIDGE CB2 2DR, UK
[4] DEPARTMENT OF BIOLOGY, IMPERIAL COLLEGE, LONDON SW7 2AY, UK

1 INTRODUCTION

The disposal of domestic waste is now becoming a problem as landfill capacity decreases. An alternative to landfill is the recycling of waste into useful products. A domestic waste recycling plant at Castle Bromwich (operated by Secondary Resources Plc) converts putrescible matter (30% of the waste stream) into humus (RDH; HUMOS 90) by bacterial (mesophyllic and thermophilic) and fungal aerobic digestion (1). Putrescible matter (all varieties of food and garden waste) is extracted from the waste stream by passage through a rotating screen containing 5 cm apertures. In addition, heavy materials (glass and metal) enter the composting process line but are substantially removed by passage through an extraction unit. The moisture content is then adjusted to 60% to facilitate the composting process, and the material is added to 18 m high silos with a capacity of 1000 m^3. The material is kept aerobic by blowing air through the silo. Temperatures rise rapidly to 70°C or more and under conditions of good aeration, moisture content and insulation these can be maintained for a number of weeks.

The final products of the process are humus, carbon dioxide and water. A relatively odour-free humus can be achieved after 10-20 days composting, though quality may be improved by drying and pelletising.

The material is a rich source of organic matter and nutrients and has the potential to replace peat as a topsoil ameliorant and in potting composts. Previous attempts to use composted domestic waste have been hampered by the poor quality of the product, due to the presence of glass (25% wet weight), phytotoxins (unspecified) and high concentrations of heavy metals[2]. The Secondary Resources process has overcome many of these problems to produce a material containing less than 4% glass and substantially reduced heavy metal concentrations.

Typical analyses of refuse-derived humus are given in Table 1.

Table 1 : Typical analyses of Secondary Resources refuse-derived humus (Humos 90).

METALS (mg/kg)		GROWING MEDIUM ANALYSIS (SR RDH)*	
Zn	712	pH	7.7
Cu	265	Conductivity	530 µS (Index 4)
Ni	40	Phosphorus	55 mg/l (Index 6)
Pb	244	Potassium	228 mg/l (Index 4)
Cr	29	Magnesium	28 mg/l (Index 4)
Cd	3	NO3-N	< 6 mg/l (Index 0)
Hg	5	NH4-N	120 mg/l (Index 3)

*MAFF/ADAS method.

A number of biochemical analyses and growth trials have been performed to assess the performance of Humos 90 as a plant growth medium. A well-matured humus (SR 11) and a less mature humus (SR 20) were compared with unfertilized peat and a commercial fertilised peat-based growing media, also a less mature humus (SR 7) was used in the identification of phytotoxic materials present in refuse-derived humus.

2 MATERIALS AND METHODS

Ryegrass Growth Trials
Amenity grass was grown in Humos 90 (SR 11 and SR 20), unfertilized peat and a commercial peat-based growing medium for comparison. Troubadour ryegrass seed (Lolium perenne) was sown on the surface of the medium at a rate of 1 g per 127 mm round plastic pot.

Separate trials were performed by MAFF/ADAS (Wolverhampton) and by Secondary Resources plc. Four replicates were used in the MAFF trial, growth in Humos 90 (SR 11) being compared with unfertilised peat and fertilised peat-based medium. Grass was harvested three times over six months. Three replicates were used in the Secondary Resources trial. Yields in immature humus (SR 20), mature humus (SR 11) and a commercial peat-based medium were compared over three months. Grass was harvested fortnightly. Performance in both experiments was determined from fresh weights and nitrogen, phosphorus and potassium cocentrations (determined using standard MAFF/ADAS procedures; 3). Zinc concentration of the grass was also determined in the MAFF/ ADAS trial.

Annual Salvia Growth Trial

Young annual Salvia plants (Salvia splendens) were obtained commercially and transplanted (2 per 100 mm pot) into either Humos 90 (SR 11; 16 plants) or a commercial peat-based growing medium (14 plants). Plant heights and flower numbers were recorded regularly for two months post transplanting.

Tomato Growth Trial

Seedling tomatoes (Lycopersicon esculentum) at the 2 to 4 leaf stage were transplanted into three replicate trays containing either a commercial peat-based growing medium, Secondary Resources Humos 90 (SR 11) and humus derived from aerated static pile composting (12 plants per tray). They were harvested 45 days later and analysed using standard MAFF/ADAS procedures (3).

Phytotoxicity of refuse-derived humus

The cress germination assay is widely used in the determination of compost phytotoxicity. It measures both the percent germination and root length in germinating cress seeds (Lepidum sativum) in water extracts of a compost. The nature of the phytotoxins can be further clarified by paper chromatography and gas chromatography of the extracts.

Humos 90 (SR 7; 100 g wet weight) was stirred in distilled water (100 ml) for 1 h, then filtered and loaded onto 50 cm x 25 cm Whatman 3MM filter papers. The chromatograms were run and allowed to dry prior to division into 6 equally sized strips with Rf values 0-0.167, 0.167-0.333, 0.333-0.499, 0.5-0.667, 0.667-0.833 and 0.833-1.0. Cress seeds pre-incubated at 27°C in the dark were grown (20 per dish in plastic petri dishes) on these strips re-wetted with water (6 ml). Root length and numbers of seeds germinating were used to estimate phytotoxicity within each fraction. Tentative identification of phytotoxins was carried out by gas chromatography of aqueous washes of the paper chromatograms. Extracts were added to methanol (2 ml), n-valeric acid (5 µl; 40% in methanol) and sulphuric acid (0.75 ml; 50% v/v in water). After incubation for 30 min at 50°C and cooling, distilled water (1 ml) and chloroform (0.5 ml) were added and the samples shaken for 5 min prior to storage at -20°C. Acetate, formate, lactate, pyruvate, succinate and propionate standards were prepared in a similar way using 0.2% solutions. Analyses of the samples were performed using a Phillips PU 4500 gas chromatograph.

Paper chromatograph strips 1 to 6 were allowed to soak in water which was then squeezed out for use in spectrophotometric studies at pH 6 and pH 12. Difference spectra (I∆S spectra) were generated, and the location of relative optical density peaks determined.

3 RESULTS

Growth Trials

In the early stages of ryegrass growth post germination and up to the first harvest, higher yields of perennial ryegrass were obtained from the fertilised peat-based

Table 2 : Analysis of ryegrass yields in different growing media (experiment conducted by ADAS)

HARVEST (Date)	GROWTH MEDIUM	FRESH WT(g/pot)	N(%)	P(%)	K(%)	Zn (mg/kg)
1 (7-3-90)	SR11	35.8	5.36	0.35	4.72	219
	PEAT (a)	17.7	3.34	0.22	3.20	28
	PBGM (b)	63.4	-	-	-	-
2 (9-4-90)	SR 11	42.7	3.80	-	-	323
	PEAT	4.4	2.14	-	-	67
	PBGM	12.7	2.68	-	-	58
3 (22-6-90)	SR 11	31.8	2.00	0.30	3.00	225
	PEAT	2.9	-	-	-	69
	PBGM	17.9	2.0	0.23	2.3	207

(a): Unfertilized (b): Peat-based growing medium; fertilized

Table 3 : Analysis of ryegrass yield in different growing media (experiments conducted by Secondary Resources plc)

DAY OF HARVEST	GROWTH MEDIUM	N(%)	P(%)	K(%)
28	SR 11	3.34	0.22	3.92
	SR 20	5.25	0.22	2.55
	PBGM(a)	4.22	0.60	5.55
42	SR 11	3.19	0.31	3.59
	SR 20	6.06	0.27	4.62
	PBGM	2.21	0.62	3.33
56	SR 11	3.43	0.32	3.36
	SR 20	4.36	0.36	4.54
	PBGM	*	*	*

(a): Peat-based growing medium   * : Insufficient material

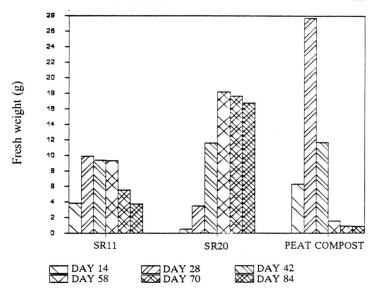

Figure 1 Fresh weight of ryegrass harvested from various growth media (experiment conducted by Secondary Resources plc).

growth media (MAFF/ADAS data). At 2nd and 3rd harvests higher yields were obtained from the Humos 90 growth medium. (see Table 2). The zinc concentrations in the grass grown in Humos 90 did not approach toxic thresholds (4). Similar results were obtained in the Secondary Resources trial (see Table 3), but there were substantial differences between plants grown in mature humus (SR 11) and immature humus (SR 20) (see Figure 1). At all harvests from SR 20 had a higher nitrogen concentration than those from both SR 11 and the peat-based medium.

In the Salvia growth trial, the mean heights and flower numbers were less in the plants grown in Humos 90 compared with those grown in peat-based growth medium, but the results were not statistically different (determined by a t-test of significance; (Figures 2(A) and 2(B)).

Seedling tomatoes grown in Humos 90 (SR 11), aerated static pile RDH and a peat based growing medium showed significant differences in growth and nutrient/heavy metal concentration. High Mn and Ca and low P concentrations were identified in plants raised in SR 11 (Table 4). The phosphorus concentration was below the minimum satisfactory level for tomatoes (0.2%), while Zn, Cu and Ni were below toxic concentrations (5).

Table 4 : Analysis of tomatoes raised in a commercial peat-based growth medium (PBGM), Humos 90 (SR 11), and aerated static pile refuse-derived humus (SPH).

ANALYSIS (Unit)	TREATMENT		
	PBGM (a)	HUMOS 90 (SR 11)	SPH (b)
Dry matter (%)	12.8	15.5	10.8
Total Zn (mg/kg)	26.0	74.0	84.0
Total Cu (mg/kg)	14.0	29.0	28.0
Total Mn (mg/kg)	97.0	525.0	99.0
Total P (%)	0.33	0.11	0.24
Total K (%)	1.83	2.68	2.12
Total Mg (%)	0.90	0.43	0.39
Total B (mg/kg)	40.0	49.6	59.0
Total S (%)	0.38	0.57	0.57
Total Ca (%)	1.70	8.50	4.14
Total Ni (mg/kg)	0.30	1.50	-
Total Fe (mg/kg)	99.6	116.0	-
Kjeldahl N (%)	3.31	2.52	-

(a): Peat-based growing medium; fertilized
(b): Static-pile refuse-derived humus

Identification of phytotoxins in Humos 90

Cress (Lepidum sativum) germination assays demonstrated the presence of phytotoxins in Secondary Resources Humos 90 (SR 7). Paper chromatography identified one component of the phytotoxic fractions to be highly water soluble (fraction 6) and one to be less soluble (fractions 3 - 4; Figure 3). Gas chromatography of the aqueous chromatographic extracts suggested that the toxic components of fraction 6 were volatile fatty acids, notably, formate. UvS peaks were pronounced at 250 and 300 nm in fractions 3 to 5, and maximally in fraction 4 (Figure 4), indicating the presence of phenolic compounds. It was therefore considered likely that phenol derivatives contributed to the phytotoxicity of fraction 4.

Refuse-derived Humus: A Plant Growth Medium

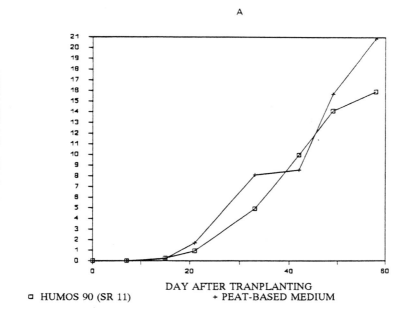

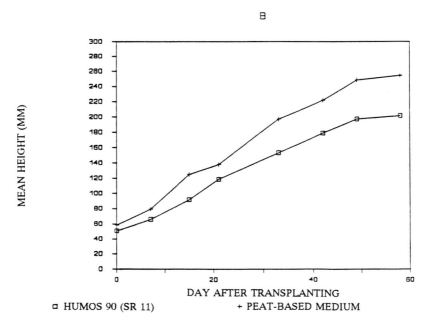

Figure 2 (A) flower numbers and (B) heights of annual Salvias grown in Humos 90 and a peat-based growth medium.

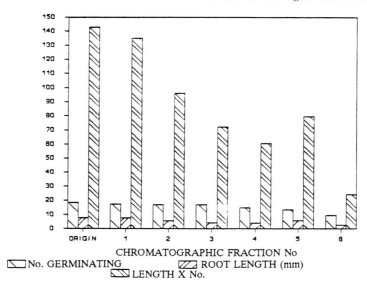

Figure 3 Phytotoxicity in paper chromatographic extracts of Humos 90 (SR 7).

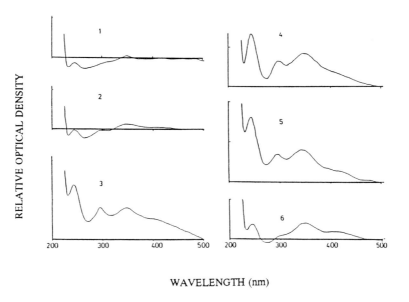

Figure 4 IAS spectra of paper chromatographic fractions 1-6 of refuse-derived humus (SR7).

4 CONCLUSIONS

The growth of perennial ryegrass and Salvias in RDH (Humos 90) compared favourably with growth in commercial peat-based growing media, indicating its potential value as a growth medium, either alone or as part of mixtures.

The long term ryegrass growth trials also demonstrated the slow nitrogen-release properties of RDH and its potential as a topsoil ameliorant for impoverished land, supplying organic matter, nitrogen and other nutrients over prolonged periods. Further work is required to determine application rates for improving soil physical properties and to characterise its nitrogen releasing properties.

Refuse-derived humus can contain small amounts of phytotoxic low molecular weight fatty acids and phenolic compounds which inhibited early root growth in cress (Lepidum sativum). Tomatoes (Lycopersicon esculentum) showed inhibited growth and phosphorus deficiency in early stages of growth, also likely to be due to phytotoxins inhibiting phosphorus uptake. Less sensitive species such as Salvia splendens and ryegrass are tolerant of RDH, as are others (e.g. wallflower and pansies, not reported here). Phytotoxicity therefore seems to be species and age specific.

ACKNOWLEDGEMENTS

We thank Marilyn Hughes for preparation of the manuscript.

REFERENCES

1. R.P. Bardos and J.M. Lopez-Real, 'The composting process: susceptible feedstocks, temperature, microbiology, sanitisation and decomposition'. In: Compost Processes and Waste Management. Workshop Proceedings, Commission of the European Communities pp. 179 - 190, 1989.

2. F. Zucconi, M. Forte, A. Monaco and M. de Bertoldi, Biocycle, 1981, 22, 27.

3. MAFF. The analysis of agricultural materials (3rd. Edn.). MAFF/ADAS Reference Book 427 (1980). HMSO London.

4. P. Bould, E.J. Hewitt and P. Needham. Diagnosis of mineral disorders in plants, vol. 1. Principles. (1983) MAFF/AFRC. Ed. J.B.D. Robinson. HMSO London.

5. G. Winsor and P. Adams. Diagnosis of mineral disorders in plants, vol. 3. Glasshouse crops. (1987) MAFF/AFRC. Ed. J.B.D. Robinson. HMSO London.

Land Disposal of Meat Processing Plant Effluent

M.R. Balks and R.F. Allbrook

EARTH SCIENCES, UNIVERSITY OF WAIKATO, HAMILTON, NEW ZEALAND

1 INTRODUCTION

In New Zealand the meat industry is a major part of the economy contributing 19.3% of export earnings (NZ$2000 million) and processing 45 million animals annually. Each meat processing plant, of which there were 43 in 1986, contributes a pollution load equivalent to a town of 60-100 000 people (Cooper et al. 1979)[1]. Since effluent discharge directly into waterways is becoming both socially and environmentally unacceptable, nine of these meat processing plants are now using land disposal systems. The benefits of these systems are that not only can they handle large volumes of water, up to 20 000 m^3/day, thus overcoming water shortage in drought prone areas but there is an apparent build up of soil fertility resulting from the added nutrients giving an increase in pasture production.

The objectives of this project were to determine the long term effects of disposal of effluent from meat processing plants on soil properties, with particular emphasis on soil physical properties.

2 METHODS

Six meat processing plants were visited. At each plant the land management history was noted and general observations were recorded on the methods and effectiveness of the operations and any problems that had arisen. At each locality a site receiving effluent and a comparable site that had not been irrigated with effluent were identified. These localities represent a wide range of soil types, climate and effluent treatments. Pits were excavated and described following the method of Taylor and Pohlen (1979)[2]. The saturated and unsaturated hydraulic conductivities at a tension of 0.4 m were measured using a disc permeameter (Perroux and White, 1988)[3]. Soil samples were collected for laboratory analysis: carbon using the Walkley - Black method, nitrogen following the kjeldahl method, pH in water and 1M KCl at a soil:solution ratio of 2.5:1 and electrical conductivity following the methods in Blakemore et al, (1987)[4]. Bulk density was measured on undisturbed cores.

3 RESULTS

Table 1 gives the results of the physical analysis. No general trends due to the application of effluent were shown. At Fairton increase in bulk density with distance along the border and a decrease in the saturated hydraulic conductivity suggests that either too much effluent is being added or the return period is too short. At Wakanui both hydraulic conductivity and bulk density suggest an improvement due to effluent addittion. The three other sites show no real effects

Table 1   Soil Physical Data

Plant	Sat.Hyd. Cond.×10⁻⁴ m s⁻¹	S.D.	Blk.Dens. Kg m³	S.D.	Unsat. Hyd. Cond×10⁻⁷ m s⁻¹	S.D.	Horizon Border
Fairton							
Site 1	0.5	0.3	1100	50	4.4	4	A
Site 2	0.26	0.14	690	60	36	12	6 m
			1030	80			155 m
			1160	100			280 m
Lorneville							
Site 1	1.05	0.5	810	30			A
Site 2	1.16	0.4	760	50			A
Makarewa							
Site 1	0.39	0.14	870	10	75	40	A
			1450	20			B
Site 2	0.37	0.17	1090	30	11	5	A
			1370	20			B
			1410	20			C
Taumaranui							
Site 1	0.19	0.06	680	20	17	1.7	A
			570	20			B
			650	20			C
Site 2	0.11	0.06	680	30			A
			560	30			B
			560	30			C
Wakanui							
Site 1	0.17	0.05	1340	50	25	6	A
			1200	60			B
Site 2	very rapid		1100	40	23	6	A
			1250	–			B

* S.D.- standard deviation

of adding effluent.
   Table 2 gives the results of chemical analysis. Although possibly not significant there appears to be an increase in pH due to the effluent. Fellmongery is calcium rich and accounts for the pH rise at sites except for Taumaranui and Wakanui. Only at Fairton is there a significant rise in electrical conductivity, carbon and nitrates where border dike irrigation is practised and hence the sites at the top end will get a higher rate of effluent.
   4 DISCUSSION AND CONCLUSIONS

Soil Physical Properties
   In general no serious long term changes were detected at the sites visited but it is evident that, if too much effluent is applied too often, severe problems of ponding leading to surface crusting and sealing causing a decrease in soil hydraulic conductivity can develop. At most sites ponding is confined to small areas as a result of run off. This was illustrated at the Taumaranui site and at Makarewa although permeabilities at this site were faster than those at non-ponded sites due to the cracking of the surface. Ponding

Table 2   Soil Chemical Data

Plant	pH H_2O	pH KCl	Elec.Cond. µs	Carbon %	Nitrogen %	Horizon Border
Fairton						
Site 1	5.8	4.9	55	3.7	0.24	A
Site 2	5.8	4.6	750	9.6	0.96	6 m
	5.7	4.8	320	5.9	0.42	155 m
	5.6	4.6	75	4.4	0.27	280 m
Lornville						
Site 1	5.2	4.2	80	12.0	0.45	A
Site 2	5.6	4.6	80	12.6	0.46	A
Makarewa						
Site 1	5.8	4.8	50	5.1	0.3	A
	6.0	4.5	15	1.2	0.1	B
Site 2	6.2	4.9	40	4.6	0.3	A
	5.9	4.3	70	1.6	0.1	B
Taumaranui						
Site 1	4.9	4.4	360	10.4	0.6	A
	5.4	4.4	10	2.7	0.2	B
Site 2	5.1	4.1	110	9.1	0.6	A
	5.1	4.3	60	3.4	0.2	B
Wakanui						
Site 1	6.3	5.3	55	3.1	0.2	A
	5.6	4.4	20	0.9	0.1	B
Site 2	6.5	5.7	130	3.3	0.2	A
	5.4	4.4	110	0.8	0.1	B

may be avoided by applying effluent at a rate slow enough to prevent surface saturation and hence overland flow,or by grading of fields in order to remove any hollow areas.Ripping,ploughing and mole drainage have also been applied as ameliorative measures.Further research is needed to define how much effluent constitutes "too much",how often effluent can be applied and the effect the treatment of effluent has received prior to irrigation.

Many of the soils studied displayed higher surface than subsurface hydraulic conductivities as a result of biopores being absent in the subsurface horizons. If irrigation is carried out at rates faster than the subsoil can sustain, it may be better to irrigate in short but repeated bursts to take advantage of the high surface hydraulic conductivity but allowing time for the effluent to move down the profile. This would prevent water perching on subsurface horizons, saturation of the surface and hence leading to run-off,ponding and anaerobic conditions.

Only at the Wakanui plant was there any marked increase in hydraulic conductivity with a corresponding decrease in bulk density. This appeared to be due to a large increase in earthworm population on the effluent treated site.At Lornville,however,large earthworms were present on the control but not on the treated sites.Since they are potentially very beneficial,as they contribute to the breakdown of the effluent,maintain permeability,aeration and soil structure their presence or absence is crucial.The lack of an observed increase at other sites may be because of suitable species adapted to the type of nutrients present in effluent from meat works.The high pH of fellmongery effluent,10.2, may also have a detrimental effect on earthworm numbers since the range of pH tolerance of Allobophora caliginosa lies between pH 5.8 and 8.3 (Kuhnelt,1976)[5].

The effect of increasing volumes of effluent is clearly illustrated at Fairton where at the top end of the border bulk densities were significantly lower than at the bottomend.A tendency for the suspended load to settle out at the top end could be a contributing factor.

No damage to the soil structure was observed at any of the sites visited.Where fellmongery was the source of effluent the high levels of calcium relative to sodium would prevent dispersion. Churchman and Tate (1986)[6] found that fellmongery led to a weakening of the bulk soil structure but they found no change in microaggregate stability.Some indication of water repellency was observed and the implications of this require further investigation.

Soil Chemical Properties

Only at Fairton were there marked changes in chemical properties measured.The pH,electrical conductivity,carbon and nitrogen contents all increased down the border corresponding to the changes in soil physical properties.The electrical conductivity at the top end is likely to inhibit plant growth and may account for the poor pasture growth.The increase in carbon and nitrogen recorded here confirms work reported by Wells and Whitton (1970)[7] and Ross et al.(1982)[8]. None of these increases appears to be detrimental to the long term viability of the scheme.They are likely to be more marked in a border dike system with limited effluent treatment rather than a spray system with major tertiary treatment.

There were different restrictions imposed by local authorities for land disposal of effluent at the different sites. At Makarewa the water right set out the maximum volumes and depths of application and a minimum return period between irrigations. Water rights may reflect the differing local conditions or only the limited information available.This highlights the need for further research on what limits and standards shoud be set. The risk of nitrate contamination of ground water should always be taken into account when designing a scheme.

Conclusions

1. Current land disposal of meat processing plant effluent in New Zealand appears to be a successful means of disposing of the large volumes of effluent without any long-term detrimental effects on the soil. It avoids direct discharge to water ways and can provide benefits through increase in both water and nutrients to plants.

2. Losses in soil permeability were observed at every site visited at areas of high effluent impact.

3. High electrical conductivity and reduced soil permeability could be remedied by leaching or ploughing respectively.

4. Earthworms are clearly beneficial in any scheme. High pH of effluent containing fellmongery may inhibit earthworms.

5. Further research is needed to gain a better understanding of the factors that contribute to surface crusting,permeability loss and development of water repellancy.

Acknowledgements

The New Zealand Ministry for the Environment provided the funding for this study which forms part of the Ph.D. studies of the senior author.

REFERENCES

1. R.N. Cooper, J.F. Heddle, J.M. Russell, Characteristics and treatment of slaughterhouse effluents in New Zealand, Prog. Water Techn., 1979, 11, 15.
2. N.H. Taylor and I.J. Pohlen, Soil Survey Method. Pub. DSIR, New Zealand 1970.
3. K.M. Perroux and I. White. Designs for disc permeameter, Soil Sci. Soc. Amer. J., 1988, 52, 1205.
4. L.C. Blakemore, P.L. Searle and B.K. Daly, Methods of Chemical Analysis of soils. N.Z. Soil Bur. Sci Rpt. 80 1987.
5. W. Kuhnelt, Soil Biology, 2nd edit. Faber and Faber, 1976.
6. C.J. Churchman and K.R. Tate, Effect of Slaughterhouse effluent and water irrigation upon aggregation in seasonally dry New Zealand soil under pasture. Austr. J. Soil Res., 1986, 24, 505.
7. N. Wells and J.S. Whitton, The influence of meatworks effluent on soil and plant composition, N.Z. Agr. Res., 1970, 13, 494.
8. D.J. Ross, K.R. Tate, A.Cairns, K.F. Meyrick and E.A. Pansier, Effects of slaughterhouse effluent and water on biochemical properties of two seasonally dry soils under pasture, N.Z. J. Sci., 1982, 25, 341.

The Influence of Organic Matter and Clay on Adsorption of Atrazine by Top Soils

L.V. Vaidyanathan and D.J. Eagle

MAFF/ADAS, SOIL SCIENCE DEPARTMENT, GOVERNMENT BUILDINGS, BROOKLANDS AVENUE, CAMBRIDGE CB2 2DR, UK

SUMMARY

Recommendation for doses of many soil acting herbicides are based on soil texture, implying clay as the predominant component controlling activity and mobility of the chemicals. It is also recognised that organic matter in soils affects the partitioning of such chemicals between soil solution and the solid phase.

In fields under long-term arable cropping with no recent grass or organic manuring history, there is a good positive relationship between clay and organic matter, attributable to the retention of humified material. Texture in these soils, identifying clay level as a surrogate for organic matter can thus help in judging the appropriate dose of chemicals. There are cases where clay:organic matter relationship is not reliable and guidance by soil texture may be misleading.

Results from ADAS studies are presented comparing the roles of clay and organic matter in a range of arable soils in the adsorption of residual herbicides, using atrazine as a typical representative.

The impact of the impending ban on straw burning and the consequence of retaining crop residues in soil on herbicidal efficiency are considered and contrasted with the effect of the ash after burning straw.

INTRODUCTION

The effectiveness of soil acting herbicides such as the triazines and substituted ureas is influenced by moisture content and the extent to which they are adsorbed by soil components. The mechanism of adsorption is likely to be the interaction between negatively charged sites on soil solids and the N-H groups of the chemicals[1]   pH dependent charge development and protonation of the chemicals may be involved. London-van der Waals forces, entropy change, hydrogen bonding, ligand exchange and charge transfer bonds are all considered possible means of adsorption [2,3]. Ionisation of carbonyl groups of soil organic matter will thus offer potential sites for interaction between chemicals and soil solids.

There is an assumption that soil texture (the feel of the moist soils between thumb and finger caused by the combined effect of their clay, silt, sand and organic matter composition, with a dominant role for the clay component) and visual judgement of colour are good indicators of organic matter level. Fields in arable cropping without much addition of bulky organic manures are expected to contain a steady state level of indigenous organic matter that remains in association with the clay in the soil. Soils identified as belonging to defined texture classes by hand texturing are, therefore, assigned attributes (non quantitative) of interaction with soil acting chemicals. This has been the basis for the recommended dose to be varied according to texture, greater doses being prescribed for "heavy" soils with large clay concentration than for "light" soils with less clay. This helps decisions on the dose of chemical chosen for maintaining enough in solution to be effective as herbicide and, *pari passu*, guard against possible dispersion by leaching to lower depths in the profile whereby selectivity may be jeopardised and also minimise the risk of the harmful chemicals escaping via excess water draining into public water resources.

Doses appropriate to "light" soils are inadequate for weed control on more adsorptive "heavy" soils; conversely, the "heavy" soil dose may cause crop damage on light soils besides the risk of the chemical being displaced downwards by leaching.

The relative importance of the clay and organic matter levels of soils in determining their adsorption of the triazine, atrazine, was investigated for four groups of soils of known organic matter and clay. As the soon to be introduced straw burning ban is likely to cause some increase in soil organic matter when straw is continuously incorporated over successive seasons, this study is of current interest.

MATERIALS AND METHODS

Determinations of atrazine adsorption were made on four sets of soils available for other purposes.

(i) ADAS standard texture group - These were eleven standard soils kept as typical representatives of the

(ii) Paired soils identical in texture except that one of each pair was classed as 'organic' because it was collected from areas with a much larger than normal level of organic matter. Five pairs of soils were studied.

(iii) Ten 'organic' texture i.e. soils with larger than usual organic matter derived from the remnants of a peat cover which has degraded.

(iv) General texture group: A random selection of 31 soils received in the laboratory for routine major nutrient analysis for fertiliser recommendations.

Soil organic matter was determined by oxidation with dichromate as in method 56 of MAFF Reference Book 427, pp 172-174[4].

Clay was determined by particle size distribution analysis after oxidation of organic matter with H_2O_2 before dispersion in 0.2% sodium hexametaphosphate solution (method 57, MAFF Reference Book 427, pp 175-180[5].

Atrazine adsorption was measured by shaking 20 g of air dry soil passing 2 mm sieve with 40 ml of 0.01 M $CaCl_2$ solution containing 3 mg l^{-1} of atrazine. After overnight end over end shaking (16 rpm), suspensions were centrifuged at 3000 rpm for 15 minutes. Atrazine in clear supernatant was measured by the method of Williams 1968[6] after extraction into chloroform and the residue remaining after evaporating all solvent dissolved in acetone. Atrazine was determined by gas liquid chromatography using a thermionic nitrogen sensitive detector and suitable standards. Triplicate measurements were made on each soil of groups i and iii, quadruplicates on each soil of group ii and single only on soils of group iv. Average values of the calculated amounts of adsorbed atrazine were used to assess the influence of organic matter and clay.

RESULTS AND DISCUSSION

Table 1 gives the data. Precision of atrazine measurement was variable and reproducibility was less than desirable in some soils, the reasons for which could not be fully investigated in this study.

Table 1. SOIL PROPERTIES AND AMOUNTS OF ATRAZINE ADSORBED

SOIL GROUP	HAND TEXTURE CLASS		% ON AIR DRY SOIL		ATRAZINE ADSORBED mg kg^{-1} AIR DRY SOIL		
	OLD	CURRENT	OM	CLAY	ADSORBED	SD	CV%
i	LS	LS	2.1	4.4	2.16	0.938	43.4
STANDARD	LCS	LCS	0.8	3.0	3.08	0.901	29.2
TEXTURES	LFS	LFS	1.9	3.7	3.59	0.574	16.0
AVERAGE	SL	SL	3.3	10.5	3.56	2.210	62.1
OF 3	FSL	FSL	2.5	12.0	2.56	0.393	15.3
REPLICATES	L	SZL	2.6	18.6	3.88	1.414	36.5
	ZYL	SZL	3.3	16.4	3.35	1.514	45.1
	CL	CL	3.3	34.2	3.58	0.421	11.8
	ZYCL	ZCL	3.2	22.6	2.98	0.789	26.5
SUBSOILS)	SC	SC	1.5	22.0	2.12	1.185	55.9
)	C	C	1.3	50.6	2.90	1.401	48.3
ii	VFSL	SZL	3.9	13.5	3.12	0.435	13.9
	VFSL	SZL	1.2	12.2	2.10	0.593	28.2
PAIRED	ZYL	SZL	4.1	16.6	3.14	0.128	5.1
TEXTURES	ZYL	SZL	2.7	18.9	2.48	0.420	18.0
AVERAGE	ZL	ZL	4.6	27.6	3.38	0.558	16.5
OF 4	ZL	ZL	2.7	27.9	2.51	0.821	32.8
REPLICATES	ZYCL	ZCL	5.2	42.0	4.00	0.225	5.7
	ZYCL	ZCL	3.1	39.4	2.43	1.071	44.2
	SCL	SCL	4.0	33.2	2.53	0.440	14.0
	SCL	SCL	2.7	31.9	2.34	0.475	19.1
iii	LS	LS	4.9	12.8	1.29	0.418	44.0
"ORGANIC"	SL	SL	3.7	11.7	0.95	0.446	60.4
TEXTURES	SL	SL	3.7	18.8	0.74	0.355	14.4
AVERAGE	VFSL	SZL	10.3	23.9	2.83	0.402	55.5
OF 3	ZYL	SZL	7.8	41.9	2.46	0.149	11.6
REPLICATES	SCL	SCL	4.3	30.0	0.72	0.355	12.5
	SCL	SCL	6.6	21.9	1.28	0.527	33.1
	SCL	SCL	15.6	41.9	2.77	0.312	11.2
	ZL	ZL	6.4	26.1	1.59	0.662	51.4
	ZYC	ZC	13.5	37.6	3.00	0.780	26.0

Table 1. CONTINUED

SOIL GROUP	HAND TEXTURE CLASS		% ON AIR DRY SOIL		mgkg^{-1} AIR DRY SOIL ATRAZINE ADSORBED
	OLD	CURRENT	OM	CLAY	
iv	LS	LS	1.6	10.5	1.97
GENERAL	LCS	LCS	4.4	6.0	3.63
TEXTURES	LCS	LCS	1.6	8.5	1.86
	LCS	LCS	2.1	8.5	2.45
	LCS	LCS	1.6	5.5	1.72
SINGLE	LCS	LCS	1.9	4.5	2.17
DETERMINATIONS	LCS	LCS	1.7	4.0	1.93
	FSL	FSL	2.0	16.0	2.46
	FSL	FSL	1.9	14.0	2.35
	FSL	FSL	1.7	17.0	2.27
	FSL	FSL	2.5	14.5	2.84
	SL	SL	1.6	11.5	2.11
	SL	SL	1.9	11.0	2.00
	SL	SL	2.1	12.5	2.66
	SL	SL	4.0	16.5	3.13
	VFSL	SZL	2.3	13.0	2.80
	VFSL	SZL	2.3	15.0	2.66
	ZYL	SZL	2.8	17.5	2.81
	ZYL	SZL	2.6	20.0	3.05
	ZYL	SZL	2.4	23.5	2.78
	ZYL	SZL	2.8	18.0	2.87
	ZYL	SZL	2.6	31.5	2.70
	ZYL	SZL	2.4	12.0	2.83
	ZYL	SZL	3.2	8.5	3.63
	ZL	ZL	2.8	17.0	3.18
	ZL	ZL	3.7	-	3.95
	ZL	ZL	2.5	22.0	2.85
	ZYCL	ZCL	2.9	35.5	3.53
	ZYCL	ZCL	2.9	28.0	3.56
	SCL	SCL	4.2	16.0	3.34
	CL	CL	3.4	18.0	3.64
	CL	CL	4.5	26.5	4.13

Simple regression of adsorbed atrazine as a linear function of % organic matter and % clay was used to assess which of the two soil properties accounted for most of the variance in such experimental relationships.
Table 2 includes the regression constants and slopes, correlation coefficients and their statistical significance. There is little doubt that organic matter and not clay is the soil component associated with adsorbed atrazine. Four of the group i standard texture soils were clear outliers for unexplained reason. Regression omitting these produced a very significant correlation. The soil groupings were not based on any specific criteria and were assembled at different times for other purposes. No critical importance should, therefore, be attributed to the slopes (atrazine adsorbed per unit of OM). The organic matter in soils of group iii, "organic" texture seems to be considerably less adsorptive than that in soils of groups i, ii and iv. Most soils of this "organic" iii group were collected from fields in the fringes of the "black fen" area in East Anglia; the organic matter in these will contain much of the ancient recalcitrant and humified remnants of the lowland eutrophic peat. Whether this type of organic matter interacts in a characteristically different manner with the soil-acting herbicide remains speculative.

A multiple linear regression of the adsorbed atrazine on both organic matter and clay emphasises the unequivocal role of organic matter on its own and the trivial significance of clay in this interaction.

Hand texture grouping of soils is still used, if only as a first approximation, in specifying commercial label recommendations of herbicide dosages; in the majority of such specifications, a threshold level of %OM is included in the label to exclude soils with a larger organic matter content from the label recommendations. Routine analysis of soils for pH, lime requirement and availability of major (P, K, Mg) nutrients by ADAS usually includes hand texturing. However, the formal reporting of the data includes the explicit caveat: "These hand-textures should NOT be used for decisions on use of soil-acting herbicides". Results of this study have been examined in an attempt to reconcile any apparent incongruity between label recommendations and the ADAS's reluctance to approve hand texturing as the basis for herbicide dosage.

Averages of organic matter concentrations in soils grouped by hand texture are reasonably well related to corresponding averages of clay concentrations. Groups that had only single representatives and textures collected from subsoils (because such textures are rare in topsoils) are the outliers; regressing average %OM as a linear function of average % clay, omitting these exceptions, produces a very significant ($P<0.0001$) correlation (%OM = 1.283 + 125 x % clay for average values

Table 2: RELATIONSHIPS BETWEEN SOIL PROPERTIES AND ATRAZINE ADSORPTION

SOILS Group	No	REGRESSION INTERCEPT mg kg^{-1} soil			REGRESSION SLOPE mg kg^{-1} soil (%OM) or (% clay)				CORRELATION % variance accounted
		Value	SD	F Probability	Value	SD	F Probability	F Probability	

(%OM)$^{-1}$

Group	No	Value	SD	F Prob	Value	SD	F Prob	F Prob	% var
i(a) Standard	11	2.42	0.502	<0.0001	0.278	0.201	0.199	0.199	8.4
i(b) Standard	7	0.676	0.389	0.143	0.811	0.138	0.002	0.002	84.9

omit LCS, LFS, L and C - for unexplained reason.

ii Paired	10	1.26	0.286	0.002	0.451	0.080	<0.0001	<0.0001	77.6
iii "Organic"	10	0.256	0.283	0.393	0.197	0.033	<0.0001	<0.0001	79.5
iv General	31	1.060	0.171	<0.0001	0.675	0.063	<0.0001	<0.0001	78.7
i(b) and ii	17	1.424	0.290	<0.0001	0.452	0.088	<0.0001	<0.0001	63.7
i(b), ii and iv	48	1.328	0.159	<0.0001	0.537	0.054	<0.0001	<0.0001	66.8

(% Clay)$^{-1}$

| i(a) Standard | 11 | 3.046 | 0.311 | <0.0001 | 0.0013 | 0.0138 | 0.927 | 0.927 | 0.0 |
| i(b) Standard | 7 | 2.457 | 0.519 | 0.005 | 0.0255 | 0.0264 | 0.378 | 0.378 | 0.0 |

omit LCS, LFS, L and C - outliers in %OM regression above.

ii Paired	10	2.421	0.531	<0.0001	0.145	0.0189	0.464	0.464	0.0
iii "Organic"	10	0.1105	0.498	0.830	0.064	0.0179	0.007	0.007	56.9
iv General	31	2.105	0.218	<0.0001	0.043	0.0126	0.002	0.002	26.1
i(b) and ii	17	2.538	0.336	<0.0001	0.0135	0.0134	0.331	0.331	0.0
i(b), ii and iv	48	2.340	0.174	<0.0001	0.0253	0.0085	0.005	0.005	14.2

(%OM)$^{-1}$ and (% Clay)$^{-1}$

| i(b), ii and iv | 48 | 1.346 | 0.160 | <0.0001 | 0.516 | 0.061 | <0.0001 | (%OM)$^{-1}$ |
| | | | | | 0.0014 | 0.0061 | 0.815 | (% Clay)$^{-1}$ |

F Probability <0.0001
Variance accounted for 66.1%

representing texture classes) accounting for 77% of the variance. Based on this, there is a good linear relationship between averages of atrazine adsorbed and corresponding averages of % clay for the soils grouped into hand texture classes - the basis for label recommendations related to texture. Averages of % organic matter in soils grouped by texture classes accounts for corresponding average amounts of adsorbed atrazine less convincingly than their surrogate % clay counterparts. The disclaimer in the ADAS soil analysis reports excluding the use of hand texture for decisions on herbicide use is to ensure that full account is taken of the need for field specific properties to avoid harmful consequences of depending on broad generalisations derived from aggregate averages of wide ranging data.

Changes in organic matter resulting from major alterations in soil management and crop husbandry practices assume much importance for decisions on soil acting herbicide usage. The ploughing out of pastures for arable cropping that occurred during the nineteen fifties and again in the mid to late eighties - consequence of dairy quotas leading to large reductions in herd size and numbers - is an example. In the few years after ploughing up of leys, soil organic matter declines rapidly[7] and then gradually towards a steady state. Now that burning of crop residues is to be proscribed, albeit with few exceptions, consequences of continuously incorporating straw on a wide and large scale need attention. Recent ADAS experiments comparing the effects of burning straw (vigorously encouraged between early seventies to mid eighties to facilitate minimum cultivations and direct drilling) show that, within a span of four to seven seasons, a measurable accumulation of soil organic matter is evident[8]. Given that adsorption and immobilisation of soil acting herbicides increase proportionally with increasing organic matter level, such accumulation of this component in soils is cause for concern. However, experience during the long years of burning of straw for early establishment of autumn sown crops after minimal or zero tillage clearly showed that the ash retained much partially combusted and charred residues that were far more adsorbant than soil organic matter as reflected in the "Kd values"[9] shown below:

> "Kd values" in 0-2.5 cm surface layer from continuous wheat growing areas where straw was burnt every year. Treatments compared were preparing a seedbed after ploughing in of ash or direct drilling without any prior cultivation at all. Soils were sampled in autumn 1981 after 7 years of experiment.

Treatment	% OM	"Kd value"
Ploughed	3.6	2.5
Direct drilled	5.4	10.8

Soils with "Kd values" exceeding 6 are usually strong adsorbers of soil acting chemicals. Values in the 4.5-6.0 range is marginal for effectiveness.

There is more recent evidence that partially decomposed remnants of both incorporated straw, those left on the soil surface in the autumn seedbed and those remaining apparently undegraded in the following spring do not interact with soil acting herbicides and do not impair their effectiveness.

"Kd values" for three straw disposal methods at Boxworth EHF. Measurements were made on top soils sampled in autumn 1989 from treatments continuing from autumn of 1983. Cultivation was with tines throughout the period. Straw was chopped on the combine harvester and spread.

Treatment	"Kd value"
Burnt	3.9
Baled	3.3
Incorporated	3.2

In an adjacent experimental area, methods of incorporation by tine cultivation or ploughing were being compared during the same period.

Tined cultivation	3.9
Ploughing in	3.0

It seems that small increases in soil organic matter in the short to medium term may not necessitate larger doses of residual herbicides. Progressive accumulation of humified organic matter in the long term may alter this perception and will need to be regularly reviewed.

Triazine herbicides are used widely by the nonagricultural sectors for weed control - keeping railway tracks, highways and their immediate verges and open yards surrounding industrial premises free of vegetation. These are situations with small inherent organic matter levels compared with land cropped continuously. Successive applications of the herbicides are made mostly by contractors who, for commercially justifiable reasons, may err on the side of generous doses to ensure effective curbing of vegetation rather than having to revisit and repeat the application in the event of unsuccessful treatment. The maximum dose for agricultural use is 4.5 kg ha^{-1} whereas that for nonagricultural applications is 45 kg ha^{-1}.

It does not require a genius to see that leakage of such chemicals applied in these nonfarming premises with little organic matter to adsorb them may far exceed that from the much smaller usage on cropped land. Atrazine is a synthetic chemical found in public water sources at concentrations inconsistent with amounts used on farms[10].

Acknowledgement: Mr C West on his sandwich course year from the Bristol Polytechnic assisted with atrazine measurements. We thank colleagues in the Analytical Chemistry Department of Eastern Region ADAS for the organic matter and clay determinations. Mr R Cross and A D Rochford in Soil Science Department helped with statistical analysis.

REFERENCES:

1. Chassin, P., Calvet, R. and Terce, M. (1981). Proc. E.W.R.S. Symposium on the theory and practice of the use of soil applied herbicides. Adsorption de atrazine et du chlortoluron par les acides humiques pp 10-17.

2. Calvet, R. (1980). Adsorption - desorption phenomena. In: Interactions between herbicides and the soil (Ed. by R.J. Hance) pp 1-30. Academic Press, New York.

3. Embling, S.J., Cotterill, E.G. and Hance, R.J. (1983). Effect of heat-treating soil and straw on the subsequent adsorption of chlortoluron and atrazine. Weed Research, 23, 357-363.

4. M.A.F.F. Reference Book 427 (1986). HMSO. Method 56, 172-174.

5. M.A.F.F. Reference Book 427 (1986). HMSO. Method 57, 175-180.

6. Williams, J.D.H. (1968). Adsorption and desorption of simazine by some Rothamsted soils, Weed Research, 8, 327-335.

7. Jenkinson, D.J. (1990). The turnover of organic carbon and nitrogen in soil, Phil. Trans. Roy. Soc., Lond. B, 329, 361-368.

8. Rule, J.S., Turley, D. and Vaidyanathan, L.V. (1990). Straw incorporation into soils compared with burning during successive seasons - Impact on soil nitrogen supply and crop husbandry. This symposium.

9. Talbert, R.E. and Fletchall, O.H. (1965). The adsorption of some s-triazines in soils, Weeds, 13, 46-52.

10. Croll, B.T. (1981). The effect of the agricultural use of herbicides on fresh waters. Paper 13 - a Water Research Centre Conference in conjunction with the World Health Organisation.

Subject Index

Acetal carbon, 55
Acid rain, 232
Actinomycetes, 63, 275-279, 282
  enzymes, 222, 275-277, 279, 281, 282
Afforestation, 103, 104
Aggregate:
  composition, 154
  hierarchy, 174
  microporosity, 197
  soil, 175, 185
  stability, 150, 153, 154, 156, 160, 175, 182, 207, 291
  water stable, 163
Aggregated system, 198
Aldrin, 94, 133-136
Alfisol, see Soil order
Algae, 231
Aliphatic acids, 341
Aliphatic biopolymers, 61, 62
Aliphatic carbon, 77
Alkyl carbon, 50, 61
Alkyl groups, 72
Amino acid analysis, 33
Ammonia, 89, 112
Ammonification, 113
Analytical pyrolysis, see Pyrolysis
Annual addition of fresh organic matter, 143
Anoxic condition, 107
Anoxic metabolism, 213
Arable forest, 267
Arable soils, 108, 113
Aromatic carbon, 51, 61, 64, 66
Aromatic content of humic substances, see Humic substances
Atrazine, 381-387, 390
Autochthonous organisms, 253
Autotrophs, see Microorganisms

Azospirillum, 236

Bacterial denitrification, see Denitrification
Barley, 233
Basal respiration, see Respiration
Basal respiration rate, see Respiration
Beech, see Fagus
Benzophenone, 48
Binding:
  agents, 145, 173, 174, 231
  mechanisms, 173
Biogeochemical process, 239
Biomass, see Microbial biomass
Biosynthesis, 97
Biota, 164
Black Fen soil, 316, 318, 386
Blacktoft series, see Soil series
Bomb-^{14}C (nuclear weapons derived), 239
Bulk density, see Soil
Brazilian soils, see Soil family
Broadbalk continuous wheat experiment, 303, 330, 331
Bruker CXP 100, 47

^{14}C, 24
^{14}C bomb, 239, 240, 250
^{14}C labelling, 239
^{14}C mineralisation, 83
$C_{mic}:C_{org}$ ratio, 255, 256, 262, 265
C-H stretching, 72
C:N ratio, 49, 71, 109, 112, 234, 300, 348
^{13}C-NMR, 25, 31, 32

^{13}C-NMR spectroscopy, 24, 25
^{14}CO$_2$, 233
C-O stretching, 72
C=O stretching, 72
CPMAS, 36, 86
CPMAS, ^{13}C-NMR, 25, 29, 31, 49, 63, 71, 116, 118
Carbendazim, 136
Carbon:
  budget, 272, 273
  cellulose, 74
  dynamics, 240
  hemicellulose, 74
  isotopes (^{12}C, ^{13}C and ^{14}C), 239
  isotopic measurement, 23
  mineralisation rate, 84
  quantity and quality changes, 269
  root-translocated, 269
  transfer, 240
  turnover, 207
Carbonyl carbon, 50, 72, 74, 77
Carbonyl groups, 382
Carboxyl carbon, 55, 72
  content, 57
  groups, 56
Cation exchange capacity, 23, 185, 187, 189, 192
Cellulomonas, 235
Cellulolysis, 234
Cellulose, 71, 218, 234, 235
  measurement, 218, 226
CFI method (chloroform fumigation incubation), 208
Chemical ionisation (CI), 148
Chemical shift limits, 47
Chemisorption, 94
Chernozem, 262, 263
Chitin, 52, 53
CIMS/MS, 48, 54
Clay:
  microstructure, 171-173
  organic complex, 185
Clostridium butyricum, 235
Colluvium, 131-135, 137
Comminution of litter, 77
Comminution of plant material, 217, 229
Composts, 126, 232, 310
  industrial, 150
CO$_2$ elevated atmosphere, 268, 269
Conversion rate (m), 143
Cotton strip assay, 217

Crop:
  optimum yield, 305, 306, 349
  rotations, 356, 357
  yield, 304, 305
Crops:
  C$_3$, 268
  C$_4$, 268
Cross polarisation, 73
Cross polarisation magic angle - CPMAS, 25
CuO oxidation, 64, 65
Cutin, 61, 62

Dead organic matter, see Organic matter
Decomposition, see Organic matter
Decomposition rate constant, see Organic matter
Dehérain experiment, 330, 334
Denitrification, 99, 109, 207
Deuteromycetes, see Fungi Imperfecti
Dieldrin, 94, 133-136
Diffuse reflectants infrared FOURIER transform (DRIFT), 72
Dipolar dephasing (NMR), 64
Disc permeameter, 375
Dissolved organic carbon (DOC), 25, 55, 115-120
Dissolved organic matter (DOM), see Organic matter
Downholland series, see Soil series
Drainage water:
  pH, 104
  sulphur content, 104
Dynamometer, 340

Earthworm population, 377, 378
Ecosystem, 185
Effective humus, see Humus
Effluent from meat processing plant, 375
Electron microscopy, 174
Electron spin echoes, 36
Electron spin resonance (ESR), 35
Electrophoretic analysis, 25
Emulsion:
  asphalt, 149
  latex, 150
Enchytraeid worms, 93
ENDOR, 36
Enterobacter cloacae, 235
Episodic flow, 102

Subject Index

EPR, 36
EPS (extracellular polysaccharide), 175, 176
Erodibility, 146, 150
  indicator, 150
  K (annual decomposition rate), 147, 148
  nomograph, 146, 147
  soil, 146
Erosion, 148
  gully, 148
  rill, 148
Erosivity (R) index, 147
ESPAS (experimental soil plant atmosphere system) growth chambers, 270
ESR, 41
Eutrophication, 93, 109
Exogenous organic chemicals, 121
Exudates, 233

Fagus, 262-265
Fallow, 80, 258, 261
Farm woodland, 285, 291
Farmyard manure, 197, 201-205
Fatty acids, 52-54
  monounsaturated, 48
Fellmongery, 376-378
'Flow country', 104
Forest soils, 61, 62, 64, 67, 111, 240, 246
  aromatic content of humic acids, 64
Fractionation of humic substances, 7-9, 18, 19
Electrophoresis, 7
  gel chromatography, 8
  isoelectric focusing, 7
  ultrafiltration, 8
Free radicals, 35
Free sugar molecules, 25
Fulvic acids, 23-26, 98, 100-104, 186, 192-194, 232
  amino acid content, 26
  definition, 5
  fraction, 7
  groundwater, 29
  marine, 26
  monosaccharide content, 26
  soil, 26
  stream, 26
  structural aspects, 18
Fumigation, $CHCl_3$ method, 81
Fungi Imperfecti, 63

Gas chromatography, 67

Genesis of humic substances, 3-5
  browning reaction, 4
  degradative processes, 4
  synthetic processes, 4
Gisburn Forest, Lancashire, 219, 220, 221, 225
Gley soil, 81
Glucose:
  ^{13}C, 57, 58
  carbon, 58
  natural, 55
Grass, 293-295, 301
Grassland (^{14}C), 246
Green manure, see Organic fertilisers
Groundwater, 136
Growth chambers (see ESPAS growth chambers)
Growth regulators, 234

^1H-NMR, 23-25, 31-33
Hamble series, see Soil series
Harps, Minnesota, see Soil family
Hemicellulose, 71, 235
Herbicide, 235
Herbicide sorption, 121, 122
Heterocyclic compounds, 97
Histosol, see Soil order
Hoos field continuous barley experiment, 306, 307, 330, 333
Humic acids, 23-25, 98, 100-102, 192, 193, 232
  amino acid content, 26
  CEC, 8
  definition, 5
  forest soil, 64
  frictional ratio, 9, 10
  FTIR, 8
  interactions, 18, 19
  isolation from soil, see Isolation of humic acids from soil
  molecular weight, 8, 9
  structures, 18-20
  random coil concept, 18
Humic components, 23
Humic materials, 231
Humic structures, 18-20
Humic substances, 23, 153, 155, 157, 161, 194
  amino acids, 19
  Archard's extraction 1786, 97
  aromatic content, 64
  'backbone', 9, 11, 13, 19, 20

browning reaction, 4
CEC-pH dependent, 8
fatty acids, 19
fractionation, see Fractionation of humic substances
genesis, see Genesis of humic substances
interactions with chemicals, 18, 19
isolation from soil, 5-7
mixtures, 5
physicochemical investigations, 9
purification, see Purification of humic substances
pyrolysis, 15
structure, 87
structural concepts, 18-20
structural studies, see Structural studies of humic substances
sugars, 19
titration, 20
Humification, 62, 66
Humification coefficient, 142
Humin, 23, 24, 190, 192-194
definition, 5
Humus, 62, 84, 299-302, 308, 312, 313
active, 89
carbon content, 85
moder, 62, 65
mor, 62, 67, 68, 71, 74, 76, 77
mull, 62, 71, 72, 75-77
pH, 77
refuse-derived, 365, 366, 373
steady state equilibrium, 79
structure, 86
young, 84, 85, 88, 89
Hydraulic conductivity, 97, 375
Hydrophilic acid, 98-102, 105

Incubation, 57, 58, 81, 84, 110, 112, 157
anaerobic, 341
Inorganic nitrogen fertiliser, 329
International Biological Programme, 240
International Humic Substances Society, 97
Ionisation, 382
Ionophores, 234
IR, 59

Isolation of humic acids from soil, 5-7
DMSO, 6, 7
sodium hydroxide, 6
sodium pyrophosphate, 6
Isotherms:
effective, 126, 127
sorption, 94, 126

K_D, see Mobility classes
K_{OC}, see Mobility classes

Leaching, 109
Legumes, 233
Lethbridge rotation experiments, 331, 335, 336
Ley arable experiments, 301, 305
Light intensity, 270
Lignaceous material, 77
Lignin, 61, 64, 71, 84, 97, 234, 235
biodegradation, 63, 64
chemical degradation, 64
fungi, 84, 253
mineralisation, 84
mineralisation rates, 83
Lignocellulose, 275, 276, 279, 282
degradation, 275, 276, 282
Lincolnshire, 293, 295
Linear regression model, 119
Liquid scintillation counting, 13, 208, 209, 243, 271
Liquid-state ^{13}C-NMR, 29
Litter, 253, 262
Locations overseas:
Clearwater, Victoria, Australia, 115-119
Fairton, New Zealand, 375-378
Lawless, South Australia, 115-117, 119
Lethbridge, Alberta, Canada, 331, 335
Lorneville, New Zealand, 376, 377
Makarewa, New Zealand, 375-378
Mt. Lofty Ranges, South Australia, 115-117, 119
Neuhof, Germany, 79-90
Otway Ranges, Victoria, Australia, 115-119
Puch, Germany, 79-84, 86, 87, 89
Redwater, Victoria, Australia, 115-119
Retreat, South Australia, 115-117, 119

Subject Index

Taumaranui, New Zealand, 376, 377
Wakanui, New Zealand, 375-377
London - van der Waals forces, 186, 382

Macroaggregates, 166, 167, 171, 173, 189, 190, 192, 194
Macropores, 198
Maize, 233
Management, 260, 261
Manuring, 197
Marine fulvic acid, see Fulvic acid
Mass spectrometry, 67
  pyrolysis CI tandem, 48
Meathop Wood, Cumbria, 240, 242, 243
Melanins, 4
Melanoidin, 4, 11, 232
Metabolic activity, 109, 207
Metabolic quotient, 81, 208-211, 214
Microaggregates, 168-171, 173, 189, 190, 192, 194
Microbes, see Microorganisms
Microorganisms, 55
  actinomycetes, see Actinomycetes enzymes
  algae, 23
  anaerobic, 234
  autotrophic, 231, 232
  soil, 215
Microbial biomass, 49, 62, 79, 109, 207, 211-214, 253, 256, 261, 265
  activity, 79
  carbon, 256, 257, 260
  death rate, see qD
Microbial carbon 254
  successions, 253
Microbial metabolism, 333
Micropore, 198
Microporosity, 199, 201, 205
Mobility classes, 129
  K_{OC}, 129, 130
  K_D, 129, 347, 388, 389
Moder humus, see Humus
Mollisol, see Soil order
Monoculture, 80, 83
Mor humus, see Humus
MS, 59
Mucopeptides, 100
Mulches:
  organic, 149
  straw, 149
Mull humus, see Humus

Mycorrhiza, 166
m/z, 48, 88

Nitrate leaching, 107, 108
  arable soils, 108, 113
  crop residues, 108
  straw disposal, 109
  tillage 108
  winter losses, 109
Nitrification, 107, 111-113
  forest humus, 112
  liming, 110
  peat, 109
  soil pH, 110
Nitrogen, 97, 118, 234, 263, 265, 300, 324, 363, 365
  amino acid, 102, 103
  cycle, 97
  fixing bacteria, 235
  fulvic acid, 104
  humic acid, 104
  mineralisation rates, 246
  nitrate, 104, 298, 312, 313
  soil mineral (SMN), 319, 320, 322
  soil supply (SNS), 319
  total soil organic, 348
  "unknown" forms, 102
Nitrogenase, 236
Nitrogen immobilisation, 109, 110
  arable soils, 109
  crop residues, 110
  net, 352
  reseeded peatland, 110
Nitrogen mineralisation, 108, 111, 332, 334
  forest humus, 112
  forest soils, 111
  mixed litters, 111, 113
  net potential, 315, 317
  plant available, 97
  rate, 317, 329
  reseeded peatland, 110
NMR, 36, 37, 59
Norway Spruce, 101
N_+, 264
Nutrient cycling, 240, 244

O-alkyl carbon, 50
O-alkyl (carbohydrate) carbon, 77
O-alkyl resonance, 5
Organic acids, 235
Organic carbon, 190, 191, 285
  accumulation, 285
  distribution, 285

Walkley-Black determination, 199
Organic farming, 80, 149, 230, 258
  systems, 355, 356
  maintenance of organic matter level, 357, 358
  organic matter function, 358-361
Organic fertilisers:
  animal manures, 150
  farmyard manure, 197, 200, 204, 303, 305-307, 309-311, 341
  green manure, 149, 256
  sewage sludge, 126, 310, 311
Organic matter:
  colloidal, 105, 229
  dead, 98
  decomposition, 234
  decomposition rate constants, 143, 144, 235
  dissolved (DOM), 93, 121-123
  mineralisation rate, 85
  raw, 143
  turnover, 107, 113
Organic nitrogen fertilisers, 329
  annual additions, 143
  temperate forest litter, 143
  temperate prairie litter, 143
  tropical forest litter, 143
  tropical savannas litter, 143
Oxisol, see Soil order

Palaesol, 239
Paramagnetic species, 46
Particle size, 71, 75, 77, 165, 199
Peat, 232, 236, 383
  eutrophic, 386
Peatland, 104
Pedogenic factors, 24
Pesticide:
  mobile, 130
  movement, 133
  residue, 133, 134
  total, 129
pH, 7, 112, 123, 200, 232, 240, 258, 262, 264, 366, 386
Phenolic carbon, 73
Phosphorus, 118, 258, 262-265, 300, 366, 386
  mineralisation rate, 246
Phototroph inocula, 176
Phytotoxins, 232, 365-367, 370, 372

Pinus, 254, 262
Pine, 263, 264
Plant growth media, 367-369
  growth trials, 366, 367
  Salvia splendens, 367
  Lepidum sativum, 367
  Lolium perenne, 366
  Lycopersicon esculentum, 367
Plant materials, 49
Plant residue, 234
Polysaccharides, 61-63, 145, 153-157, 160, 182, 231, 234
Ponding, 376
Pore functions, 164
Porosity, 199, 201
  total, 199, 201
Potash, 340
Potassium, 264, 366, 386
Prairies, 141
Primary producers, 231
Primary productivity, 232
Primary screen, 177
Proteins, 97
Pseudomonads, 235
Purification of humic substances, 7-9
  ^{13}C-NMR spectroscopy, 16, 17
  ESR spectroscopy, 16
  FTIR spectroscopy, 8
  gel chromatography, 8
  sephacryl S-200 gel, 8
  ultrafiltration, 8
Pyrolysis, 48
  analytical, 61
  CIMS/MS, 59
  curie point, 67
  curie point Py/MS analysis, 32
  mass spectrometry, 23, 87
  Py/MS, 31-33
  tandem mass spectrometry (PTMS), 45, 47, 48, 52
Pythium, 235

qCO_2, 230, 255-258
qD, 230, 255-258, 260

Refuse-derived humus, 365, 366, 373
Rendzina, 262
Residual glucose ^{13}C, 58
Residual glucose carbon, 57, 58
Residue incorporation, 207
Respiration, 233, 258, 259, 262
  basal, 208
  basal rate, 255
  rate, 261-263
Rhizodeposition, 233, 267

Subject Index

Rhizosphere, 234, 267, 268
Romney series, see Soil series
Root, 231, 329
  decomposition, 232
  exudates, 244, 267, 329
  exudation, 232
  maize, 268
  translocated carbon, 269

Saturated fatty acids, 53
Savannas, 143
Scanning electron microscopy, 50, 174
Seedbed consolidation, 346
Sediment:
  deposit, 141
  entrainment, 141
Sedimentation, 141
Sewage sludge, see Organic manures
Silt lands, 293
Simazine, 136
SIR (substrate induced respiration) method, 208
Sitka Spruce, 113, 246
Sodium dithonite, 46
Soil, 121, 123, 185
  aggregation, 145, 153, 175
  arable, 110, 267, 301
  biomass, 232
  biota, 269
  Brazilian, 185
  bulk density, 73, 119, 208, 375, 377
  chalky boulder clay, 349
  clay content, 119
  compaction, 200, 207
  composition, 122, 200
  decomposition, 217, 218, 220-226
  ecosystem, 267
  electrical conductivity, 375, 378
  erodibility, 146
  forest, see Forest soils
  fulvic acid, see Fulvic acid
  high organic content, 258
  humus, see Humus
  hydraulic conductivity, 375, 377
  inoculants, 176, 181
  microaggregate stability, 378
  microbiomass, see Microbial biomass
  microflora, 207
  mineral nitrogen (SMN), 315-321, 324, 326

  nitrogen, 258
  nitrogen supply (SNS), 318-321
  net mineralisation potential, see Nitrogen mineralisation net potential
  organic carbon, 85, 254, 257
  pH, 378
  porosity, 207
  structural stability, 129, 131
  structure, 164, 185, 197, 201, 202
  surface crusting, 150, 378
  tropical, 186
  water, 200, 204
  water repellancy, 378
Soil class:
  clay loam, 130, 286
  loamy sand, 130, 270
  sandy clay loam, 154, 200, 286, 293
  sandy loam, 148, 200, 293, 300, 303, 304, 312
  silty clay loam, 154, 200, 297, 300, 302, 312
Soil family:
  Brazilian, 185
  Harps (Minnesota), 25
  Meadow (Australia), 45, 46, 54, 56-59
  Millicent (Australia), 45-47, 49-53
  Urrbrae (Australia), 45-47, 49-53
  Webster (Iowa), 25
Soil nitrogen, 285, 291
  distribution, 286, 288
  mineralisation, 286, 291
Soil order:
  Alfisol, 148, 163, 166-172
  Histosol, 9
  Latosol, 185, 186
  Luvisol, 80, 258
  Mollisol, 163, 171, 172
  Oxisol, 143, 163, 168, 170, 171, 173, 186, 192
  Vertisol, 143
Soil organic carbon, 254
  annual decomposition rates (K), 143
  conversion rate, 143
Soil organic matter, 111, 112, 175, 231, 258, 292, 294, 295, 383
  annual mineralisation rate, 85
  forest, 62, 111
  microbial turnover, 269
  mineralisation coefficient, 142

Soil series:
  Blacktoft, 293
  Downholland, 293, 294
  Hamble, 177
  Hanslope, 252,
  Romney, 293
  Stirling, 153
  Tickenham, 286
Solid-state ^{13}C-NMR, 45-47, 50, 51, 55, 56, 59, 64
Solid-state CPMAS ^{13}C-NMR, 29
Sorption, 121, 123
Specific activity:
  CO_2, 272
  plant material, 272
  rate constant, 25
Specific surface area, 116
Spectral quality (signal:noise ratio), 46
Spinning side bands, 54
Stannous chloride, 46
Straw, 231, 232, 235, 275-278, 289, 308, 309
  addition, 297
  burning, 339, 340, 343, 344, 381, 382
  decomposition, 212
  incorporated, 109, 340, 343-348, 351, 389
  solubilisation, 275-277, 279, 282
Stream fulvic acid, see Fulvic acid
Streptomycetes, 277-282
Structural studies of humic substances, 9, 10
  oxidative degradation, 11
  phenolic, 13, 19
  reductive, 13, 14
  sulphide, 13
Suberin, 61, 62
Sugar analysis, 33
Sugar beet, 133, 210, 258, 261, 317
Sugar content, 25

Sugar fungi, 253
Sulphur content, see Drainage water
Swedish 'M' series, soil fertility experiments, 330, 335

Tannins, 67
Temperature, 260, 262, 264
Thermomonospora mesophila, 276
Tomato, 233
Total amino acid, 26
Total signal intensity, 47
Tree growth, 248
Triazines, 381, 382
Trichoderma harzianum, 235
Trichoderma species, 234
Turnover, see Organic matter

Unhumified free sugar, 25
Upland soil, 24

Waste recycling, 365
Water, 262
  courses, 133
  erosion, 141
  interstial, 31
  potential, 254
  repellancy, 378
Webster soil, see Soil family
Wheat, 210, 233, 235, 270, 273, 316, 317, 320, 321, 343, 344, 347
White-rot fungi, 63
Wood, 236

XAD-8 resin, 23-25, 32
X-ray fluorescence, 46

Yield:
  coefficients, 256
  economic, 349
  response, 322, 323

Zymogenous organisms, 253